Neurochemistry and Clinical Disorders: Circuitry of Some Psychiatric and Psychosomatic Syndromes

Editors

Fuad Lechin, M.D.

Titular Professor
Department of Physiology
and
Chief
Department of Psychopharmacology and Psychosomatic Medicine
Central University of Venezuela School of Medicine
Caracas, Venezuela

Bertha van der Dijs, M.D.

Chief of Chromatography
Department of Physiology
Central University of Venezuela School of Medicine
Caracas, Venezuela

CRC Press, Inc.
Boca Raton, Florida

Library of Congress Cataloging-in-Publication Data

Neurochemistry and clinical disorders: circuitry of some psychiatric
    and psychosomatic syndromes / editors. Fuad Lechin, Bertha van der
    Dijs
            p. cm.
        Includes bibliographies and index.
        ISBN 0-8493-6595-3
        1. Psychopharmacology. 2. Neuropsychopharmacology.
3. Neurochemistry. 4. Psychoneuroendocrinology. I. Lechin, Fuad.
1928-   . II. Dijs, Bertha van der, 1937-   .
        [DNLM: 1. Neural Pathways—physiopathology. 2. Neurochemistry.
3. Mental Disorders—physiopathology. 4. Psychophysiologic
disorders. WM 100 N4924]
RC483.N37 1989
616.89′071—dc19
DNLM/DLC
for Library of Congress   88-19263
                                                                              CIP

Direct all inquiries to CRC Press, Inc., 2000 Corporate Blvd., N.W., Boca Raton, Florida, 33431.

©1989 by CRC Press, Inc.

International Standard Book Number 0-8493-6595-3

Library of Congress Number 88-19263
Printed in the United States

DEDICATION

A human being is much more than the sum of blood, bones, and viscera. In the same way, each fragment of truth in itself is a lie; therefore, the accumulation of unintegrated scientific facts does not protect us against ignorance.

In the measure that we interrelate a greater number of fragments, the closer we can come to truth, although truth as an absolute is unattainable.

We dedicate this book to those doctors and scientists who believe that the work of a good specialist or researcher must draw on the widest possible interrelated knowledge in many disciplines, in order to avoid their objective findings — clinical or experimental — being left like loose pieces of a puzzle, without universal context and lacking meaning.

Fuad Lechin
Bertha van der Dijs
Editors

PREFACE

Differences must be drawn at the outset between the craft of medicine and medical science. In the former, rudimentary guidelines of diagnosis and therapy, laid down by legally invested bodies, are applied by practitioners. In the latter, basic mechanisms are explored in an attempt to explain the symptoms of disease and the effect on them of therapeutic drugs. It is possible to practice medicine with only rough knowledge of the basic sciences but, as more diagnostic tools and powerful drugs are introduced, the inability to make global diagnoses increases and with it the danger of iatrogeny.

The practitioner of today grows more dependent on sophisticated diagnostic aids, he becomes tempted to forego the use of his intellect and his ability to integrate knowledge. Further, his use of pharmacotherapy in the blind fashion dictated by rigid medical school precepts, although limiting his medical efficiency, allows him to rest safely within a framework of legality, so that even when he cannot help his patient, he is protected against professional criticism, moral reproaches, and legal suits.

Ostensibly to protect the patient, such professional limitations arise from the circumstance that medical schools are aware of the poverty of the education they impart and their low level of requirements. The reduction in effort expected of the practitioner allows him more free time for personal interests.

One attempt to make up for the practitioner's partial training is team practice, compensating for the lack of individual knowledge through group cooperation. The rise of specialization and super-specialization reduces 'per capita' labor and dilutes responsibilities.

In our opinion, superspecialization has led to the mistaken concept that the practitioner need not know what lies outside his specialty because supposedly it is not his legal concern. His personal contribution grows ever narrower.

Such superspecialization has led to the mystification of new diagnostic tools. Developed by experts in physics, electronics, optics, biochemistry, pharmacology, physiology, immunology, genetics, etc., these tools bestow on the practitioner in whose hands they are used an aura of wisdom, almost of infallibility and omnipotence. The dazzling light of the new tool may obstruct a view of the global process of the disease affecting the patient. For example, when a gastroduodenal ulcer is revealed through endoscopy, the gastroenterologist will concentrate on curing the ulcer rather than the condition which caused it, like a mason plastering holes in a wall weakened by leaky plumbing. If an ulcer is produced by oversecretion of peptoacid, the specialist administers high doses of drugs to halt the secretion and allow cicatrization; there ends his job. He designs numerous double-blind tests to determine whether the H2 blockaders are more effective than the anticholinergic or prostaglandinic agonists; and once the ulcer is healed the gastroenterologist is satisfied. His fragmentary and partial training prevents him from understanding that the hypersecretion of peptic acid is the final manifestation of a chain of physiopathological events beginning in the central nervous system and that, although the ulcer may be healed, the patient is still ill. Furthermore, the gastroenterologist's scanty knowledge of the central nervous system (CNS) stands in the way of his fully understanding all the neuroendocrinal and neurochemical effects provoked by the administration of the drugs.

Anxiolytics are administered by practitioners in such prolonged and exaggerated doses that they now account for the majority of all drugs consumed. Yet how many practitioners know in what way and where the benzodiazepines act? What do they know about the GABA system, the main target of these drugs? Why do anxiolytics lose effectiveness after prolonged usage? What do their paradoxical effects mean? How does one correct the symptoms of exaggerated anxiety present in a high percentage of patients who have taken benzodiazepines for some time?

Should a practitioner be allowed to administer to his patients drugs acting on the CNS, when he has no adequate knowledge of the CNS morphology, physiology, biochemistry, and pharmacology? For example, in the case of clonidine, an α_2-agonist whose action is fundamentally central, should this drug be handled by a practitioner ignorant of the central noradrenergic (NE) neuronal organization and the possible pre- or postsynaptic mechanisms of clonidine? How can he interpret the paradoxical effects of clonidine, which in certain patients does not provoke hypotension but rather hypertension? Similarly, how can he interpret the neuroendocrinal effects exercised by clonidine on growth hormone, cortisol, etc., or on glycemia?

We feel that the mystification of diagnosis leads the practitioner to put modern diagnostic aids foremost, relegating the patient to a subordinate or secondary plane. Thus, the lowering of blood pressure levels becomes more important than treating the patient who suffers from high blood pressure. Is the practitioner aware that drugs which lower blood pressure by inducing a depletion of central and peripheral monoamines also provoke neurochemical imbalances similar to those found in stress and depression? Is he aware of the relationship between stress, immunology, and cancer? How can the practitioner be certain that such a treatment will not predispose the patient to tumoral growth of an undiagnosed neoplasy or to an immunological deficiency which would make the patient susceptible to a latent viral infection?

We should ask, then, what is the minimum basic knowledge that a practitioner should have? Where does one specialty end and another begin? Does gatroenterology stop at the serosa of the gastrointestinal tract? Should the cardiologist and endocrinologist, who handle drugs which act on the hypothalamus, mesencephalus, mesolimbus, and cerebral cortex, also be trained in neurophysiology, neuroanatomy, and neuropharmacology? Can a team of superspecialists take the place of a practitioner who possesses adequate basic information in areas outside the field in which he was trained?

We have written this book because we believe it to be useful and necessary. Over the course of 35 years of medical practice, we have attended some 80,000 patients suffering from almost every kind of upset. In the last 2 decades we have witnessed the eruption in therapy of a vast array of psychoactive drugs, not counting those previously believed to have little effect on the CNS but which are now known to be centrally active.

Antibiotics such as penicillin, for example, have been proved to possess a GABA-antagonist effect; other antibiotics which can inhibit protein synthesis and cross the blood brain barrier are able to inhibit synthesis of the REM sleep factor. Likewise, differences have been shown between the action mechanisms of synthetic and natural steroids, because the former do not bind beyond the hypothalamus while the latter interfere powerfully in the synaptic transmission of serotonin at mesolimbus and cerebral cortex levels. (Moreover, natural steroids inhibit the hypophysis-suprarenal axis, acting preferentially at the hypothalamus level by inhibiting CRF secretion, while synthetic steroids act preferentially on the hypophysis by inhibiting secretion of ACTH).

Drugs used to halt gastric secretion (H2-antagonists, prostaglandinics, and anticholinergics), widely prescribed for treating gastroduodenal ulcers, have marked effects at the CNS level and give rise to psychoneuroendocrine alterations. Antidepressive drugs such as Doxepin and chlorimipramine have been shown to be very effective in the treatment of gastroduodenal ulcers, even when administered in small doses. Likewise, antipsychotic drugs such as thioproperazine and centrally acting, antihypertensive drugs such as clonidine, are succesfully used in treating idiopathic ulcerative rectocolitis, psychosis, and Gilles de la Tourette syndrome.

In our own experience we have found the employment of psychoactive drugs highly effective in the treatment of bronchial asthma, irritable colon, Chron's disease, rheumatoid arthritis, multiple sclerosis, female infertility, primary amenorrhea, dysthyroidism, skin allergies, and various kinds of neoplasias. With respect to neoplasias, we have been able to induce, through administration of psychoactive drugs, varying degrees of improvement in some 200 patients suffering from different kinds of cancer. Paralleling improvement in the cancerous condition of these patients was an increase in the cytotoxic activity of NK cells plus a reduction of OKT4/OKT8 ratio. This same result was obtained in rats inoculated with Walker's carcinoma.

We could extend the list of examples but those mentioned suffice to give an idea of the importance that psychoactive drugs have now and will have in the future in the treatment of so-called somatic diseases. Therefore, it becomes imperative that doctors acquire working knowledge of the anatomy, biochemistry, physiology, physiopathology, and pharmacology of the CNS.

In support of the above, the growing use of biological markers of depression and psychiatric cases in general should be noted. Such markers are of different kinds: gastrointestinal motility, hormonal, metabolic, haematic, cutaneous, pupillary, immunological, etc. Medical literature regarding immunology now abounds with experimental and clinical studies demonstrating the close relationship between the CNS and immunological activity.

Today a therapeutic arsenal of potent psychoactive drugs is available to practitioners. In view of this and the fact that many of these drugs, believed to excercise their main effect peripherally, actually cross the blood brain barrier to generate powerful central effects, the practitioner's poverty of knowledge concerning the CNS becomes not only incongruous but dangerous. Without adequate information about the drugs' numerous effects, practitioners freely administer benzodiazepines, synaptic reuptake inhibitors, MAO inhibitors, receptor agonists and antagonists (pre- and postsynaptic), inhibitors of neurotransmitter synthesis, amine depletors, antagonists of calcium channels, etc. Yet all drugs in current medical usage act on so many levels, by means of such diverse mechanisms, that it is very difficult to determine which action produces benefits and which generates iatrogeny.

No drug exists which acts on only one site through one mechanism. Like a chord played by ten fingers together, or a ray of light which diffracts on penetrating a prism, drugs act in different places in the body. Moreover, when the practitioner administers a β-receptor antagonist, a GABA mimetic, or a calcium antagonist, he cannot aim their direction to a certain central or peripheral zone. Such drugs will act on all the β-receptors, all the GABA systems, and all the cell walls in the organism. Truly specific and selective drugs have not as yet been found which would allow us to be certain they will produce a single effect and act in a single region, e.g., a β-receptor blocker which would act only in the posterior hypothalamus but not in the mesencephalus, the cardiopulmonary sphere, or the gastrointestinal tract.

One approach to minimize drug nonspecificity of action and nonselectivity of place is the use of minimum effective doses. Although it is true that drugs can have many effects at various sites, this solution has worked well for us over many years, apparently because the degree of receptivity varies in different places. As a result, the more the dosage is reduced, the greater is the drug specificity. Further, the drugs act preferentially on those receptors and mechanisms which are most activated at the moment of the drug administration. For example, when clonidine (an α_2-agonist which inhibits the NE neurons) is administered, its effect will be registered on those NE neurons which at the time are most active. If the locus coeruleus (group A6 neurons) are most active at that moment, these are the neurons which will be inhibited by clonidine, resulting in lowered arterial pressure, an effect which is registered even when the dose of clonidine is very low. On the other hand, if large doses of clonidine are needed to provoke a hypotensor effect, this may indicate that the A6 group was not very active at the time and that the effect obtained was possibly postsynaptic (at the level of the sympathetic preganglionic neuron which is cholinergic in nature, located in the intermediolateral horn of the spinal cord and in the reticular nucleus of the medulla, in whose neurons there are α_2-receptors.

Low doses have the advantage, besides reducing iatrogeny and side effects to a minimum, that the mechanisms of tolerance to the drug are also kept to a minimum. It is our experience that low dosage allows prolonged use of the drugs without their loss of effect with time. This phenomenon could have the following explanation: if an α_2-agonist, for example, is administered in minimum doses, the firing of the hyperactive NE neurons could be inhibited without inducing hyperpolarization of the membrane and hyposensibilization of these receptors. Likewise, if propalanol is administered to block the β-receptors of a certain central or peripheral zone, a partial and incomplete blockade capable of reducing beta mechanisms in the target area, without suppressing them totally, might avoid the subsequent proliferation of receptors (supersensibilization) which obliges the practitioner to escalate the dosage of propalanol, at the risk of provoking its well-known and undesirable side effects.

In our opinion, drugs are employed in unnecessarily heavy doses, leading to counterproductive results in the medium and long term. The widespread use of large doses derives from the mistaken assumption that human dosage can be extrapolated from experiments on rats, and that double-blind studies on large groups of patients can establish accurately the minimum dose effective for a single patient. We arrive at the minimum effective dose through adjustments after frequent communication with the patient. In our experience, when large doses of a drug are needed to suppress a symptom, then the chosen drug is not the right one and it is better to use another.

Benzodiazepine is commonly administered as an anxiolytic. Yet, benzodiazepine drugs are GABA-mimetics which act on all neurological circuits, among which the GABA system is just one. Although these drugs stimulate the GABA system responsible for blocking monoaminergic neurotransmissions generating anxiety, their prolonged administration can stimulate other GABA systems and block other monoaminergic circuits. As a result, benzodiazepine drugs can throw a patient from a state of anxiety into a depressive or even psychotic state. Sadly, this phenomenon is a daily occurrence.

In this book we have gathered enough information to propose a model for the anxiety circuit. A considerable body of clinical and experimental research supports the NE, dopaminergic (DA), and serotonergic (5HT) mechanisms believed to be involved in anxiety. Whether or not our proposed model is simplistic, incomplete, or topographically inexact, it has the virtue of allowing a therapeutic approach broad enough to cover, besides the benzodiazepines, a whole range of other anxiolytic drugs (5HT antagonists, DA blockers, α-antagonists, β-antagonists, dopamine liberators, etc.). Our therapeutic approach, and the small doses we employ, have led us to surprising findings, widening our practical knowledge to the point where we can formulate hypotheses and propose model circuits, some of which are put forward in this book.

Perhaps the most important conclusion we have derived from our work with drugs is that, in the face of neuropharmacological advances, it is no longer possible to continue authorizing the practitioner to administer powerful psychoactive drugs if he is unprepared in the anatomy, biochemistry, physiology, physiopathology, and pharmacology of the central autonomous nervous system (ANS). No matter what his specialty, the practitioner should be required to know with some degree of accuracy what is the effect on the organism of the powerful drugs he administers. He should not handle such drugs according to the rough rules set out by the legal medical authorities which make the physician into a simple medical artisan. A deeper level of knowledge in the scientific areas composing modern medicine should be required of the practitioners of today.

In his book, *Structure of Scientific Revolutions*, Khun puts forward the eternal confrontation between what is accepted as the offical truth and new findings which must struggle for a place within the old structures of "paradigms". He also speaks of the phenomenon of scientific revolution which, once accepted officially, becomes paradigmatic, resisting changes brought by new scientific knowledge. The only logical conclusion to be drawn from this cycle is an acceptance of continuous revolution, in accordance with the one constant of our universe: change.

Poets dream, philosophers reflect and spin hypotheses, scientists test these hypotheses, and artisans apply the new knowledge. The practitioner belongs in the last category and therefore mocks the poet, denies the philosopher, and attempts to ignore the scientist while glorifying his own craft although, without the preceding links, his craft would not exist. When a medical specialist applies diagnostic or therapeutic innovations, in his skill he often overlooks the debt he owes to those who first dreamed of the technique or tool, others who conceived it, later tested it, and finally produced it.

In the field of medicine it is difficult if not impossible to determine where science ends and craft begins. Over the centuries, the interdisciplinary nature of medicine has led to close cooperation by chemists, physicists, mathematicians, biologists, and other scientists with the physicians directly concerned with people's health. This has given rise to areas of convergence such as physiology, physiopathology, pharmacology, genetics, immunology, etc., and, as if this were not complex enough, psychology and psychiatry call in all the above scientists and philosophers, sociologists, and writers as well.

However, the interdisciplines are so numerous in medicine today they are no longer able to intercommunicate. Like the biblical Tower of Babel, the higher the specialization, the deeper the cracks in communication. With few exceptions, the superspecialist has become the purest craftsman of the entire medical community. In the end he can only communicate with the few who are perched on his peak and so affords a typical example of what conceptual philosopher Ernesto Lechín calls "tunnel vision" as opposed to "peripheral vision". Only through the latter is it possible to maintain an overall view of knowledge, avoiding the risk of its psychotic fragmentation. By way of illustration, the author and thinker contrast the inhabitant of an island or a valley who regards his limited surroundings as his universe with the passenger on a space ship whose view of the planet makes him see the need to integrate knowledge. Naturally, neither of the two positions can stand alone since knowledge is an infinite succession of analyses and syntheses.

Empirical knowledge leads to lineal thinking or, in other words, A leads to B. In order to form hypotheses and then build models, systemic thinking is required in which each point is related to all others. Instead of a straight line, the association of more than two variables leads to polyhedra of three, four, five, or six sides and more, ad infinitum. Scientific knowledge must have recourse to systemic thinking in order to approach the absolute truth, even though the absolute is unobtainable, of course. All truths known and accepted as such are conventions and therefore are constants, each forming one of the infinite tangents of a circle, the only form which can adequately represent systemic thinking.

In current biomedical research, fragmentary and partial experimental knowledge predominates in chains of lineal thinking. Some researchers have reached the extreme of saying, "I do not think, I investigate." Such an attitude arose as a defense against the speculative tendency derived from Cartesian rationalism. Scientific journals oblige researchers who sumbit articles for publication to limit discussion of their results and rule out inferences, or have their work rejected. Therefore, when a researcher has a great deal of interdisciplinary information to which he could relate his study and he is cut short in his capacity to report it, his readers are denied this wide and potentially useful context.

In this way, scientific literature accumulates an enormous body of new, fragmentary knowledge which in the long run is underused. Further, the superspecialist all too often lacks information from areas other than his own and so remains unaware that his personal ocean is, in fact, a pond. Countless specialized "oceans", deprived of intercommunication, become sterile in the end. Those of us who read medical journals widely are constantly astonished by the great number of closely related studies whose authors are apparently ignorant of similar research, despite journal indexing and the development of information technology. Fragmentation, when combined with poor communication and the superspecialist's restricted knowledge, leads to costly repetition of research and a continual "rediscovery of the wheel".

Review articles are published with the aim of fitting together loose pieces in the jigsaw puzzle of experimental research. But it is our observation that most such reviews are timid attempts at integration which do not risk making structures of any complexity or proposing true models.

The inability of modern medicine to solve human health problems, despite vast resources and efforts invested in cures, appears to justify our Tower of Babel simile, our fears of the failure of the ambitious project of medicine to reach heaven. Karl Popper, Ph.Sc., states that no scientist can extend the bounds of knowledge unless he has complete knowledge of the work of previous scientists. Although empirical science struggles to make knowledge as objective as possible, uncontaminated by subjective speculation, thinkers and philosophers teach us that this is not possible. Albert Einstein in his *Theory of Relativity* and Werner Heisenberg in his *Uncertainty Principle* assert the impossibility of avoiding interaction between the observer (subject) and the observed (object), which are the two main structural components of the scientific method.

In this book, when we launch a hypothesis and propose models of greater or lesser complexity (which are certainly simplistic in the description of real structures or behaviors), we are not concerned that our models may describe the mechanisms, say of mamalian brain function, less than perfectly. Our view is pragmatic; what counts is the soundness and broad base of information on which our models are based, and the direct usefulness of the model in therapy for our patients. We know that we can never own the truth and that we must conform our goals to approximate or even probable truths. Neither are we worried that future models may substitute ours; we believe we must work on the basis of theoretical models which can be tested by experimentation. Karl Popper states that a theory is scientific if its set of propositions has a logical/rational structure permitting it to be compared with reality, and that a scientific theory does not attempt to explain all the facts deriving from an experience. Popper accepts that in order to refute a scientific theory experimental findings must contradict the hypothetical postulations. Finally, as the criterion of validity, Popper replaces "empirical verification of scientific hypothesis" with his thesis of "empirical refutation of scientific hypothesis".

Rudolph Carnap and Werner Heisenberg believe that contemporary science should substitute "probable predictive hypothesis" in place of the conventional "exact predictive hypothesis" which presupposes control of all variables intervening in the empirical phenomenon of observation or experimentation. If we take into account the thinking of Einstein, Popper, and Heisenberg that empirical data arising from observation may be subjective and that, vice versa, rational hypotheses may be objective, it becomes impossible to attain control of all variables.

Mario Bunge, Ph.Sc., says that today scientific progress is measured more by theoretical progress than by the accumulation of data. For this author, contemporary science should offer something beyond experience: theory, to which is added experience — planned, directed, and understood in the light of hypotheses. He believes that theoretical models can and should be represented in the form of mathematical models, and that reality is better represented through theoretical models than empirical data. Bunge concludes that a scientific prediction is one in which experimental data, obtained by the scientific method, are integrated to formulate a set of hypotheses or a theory. This is what we have attempted to do in the present book.

According to Bunge, no theory which has survived examination of its rational construction can be entirely false, while no theory is totally true even when it has triumphed in the test of experimentation. Bunge believes that science does not require absolute certainty but rather the possibility of correction and that, further, although one or several of the hypotheses making up a theoretical model may be refuted, the general theory may still stand.

Popper, like Bunge, is of the opinion that the process of scientific research begins by posing problems within the body of theoretical scientific knowlege already acquired. To carry out research, therefore, one must read widely and gather full and up-to-date information.

Ernesto Lechín postulates that it is impossible to be aware without conceiving; the two processes of perception and conception are inseparable. Yet the current of empiricism in biomedical research of today to a certain extent restricts and even prohibits conception. The introduction of ideas in the discussion of a scientific paper is usually criticized and rejected by the referees appointed by editors of the scientific journals on the grounds of "unscientific speculation". In fact, present biomedical science is almost totally immersed in the positivist theory of Ernst Mach that "sensations" (the observable) are the only means of perceiving reality. Marshall McLuhan says that the tools of empirical research are just extensions of our five senses. Yet, since the discovery of invisible phenomena such as atoms and the unconscious, the positivist theory must be considered obsolete.

This book may not meet great acceptance among the positivists because our models are based on ideas as well as observation.

The book is organized into six chapters covering (1) neuroanatomical basis, (2) anxiety-like syndrome, (3) depressive syndromes, (4) psychotic syndromes, (5) blood pressure regulation, and (6) biological markers.

Chapter 1 reviews information on morphology, electrophysiology, histochemistry, and neurochemistry in general which contribute direct and indirect evidence of the existence of neuronal nuclei, their areas of projection, and their interconnections. Although all this information is found in hundreds of specialized papers, as far as I know it has not been previously gathered, ordered, or assembled into book form. Until now, this information has been the exclusive property of superspecialists, beyond the reach of most practitioners who therefore were totally cut off from a possible understanding of psychoneuropharmacology and psychoneuroendocrinology. This chapter, in essence "positivist", is in no way controversial since it draws only on data from proven experimental findings and we have abstained from conceptual or hypothetical considerations.

In Chapter 2 a model is proposed for the monoaminergic circuits involved in this syndrome. The model is based on hundreds of experimental and clinical findings demonstrating that there are NE, 5HT, and DA nuclei which are activated during the appearance of the syndrome, and which, at the same time, are believed to inhibit other monoaminergic nuclei. This model has led the authors to a succesful therapeutic approach which goes beyond the simple use of drugs known as anxiolytics. The chapter also proposes connections between anxiety manifestations coexisting with certain types of depressive syndromes.

In Chapter 3 we have tried to avoid being trapped by the numerous classifications used in the area of depressive syndromes. We have attempted to simplify the definitions and limits of depressive syndromes which are in truth rather imprecise. The considerable confusion and disagreement surrounding these definitions are aggravated by similar symptoms shown by subjects during certain stages of stress.

Some models of animal depression which have been employed in neurochemical investigations and later extrapolated for human study are reviewed in this chapter, as are some of the biological markers used in diagnosis or classification of depressive syndromes. Largely because our therapeutic methodology is based on the use of psychoactive drugs in much smaller doses than those conventionally employed, we have been able to establish subtle differences between antidepressive drugs which potentiate the NE and 5HT systems. We have become convinced that the size of the dose employed not only allows a different therapeutic approach but also a more accurate diagnosis.

Our definition of depressive syndrome is based on two types: (1) anxious depression and (2) anergic depression. The former is accompanied by peripheral sympathoexcitation (sympathoexcitatory side effects), while the latter coexists with peripheral sympathoinhibition. We accept that, rather than two types of depression, the conditions are really alternating states of the same syndrome. Naturally, there are subjects in whom one of the two states predominates in intensity and frequency. Our theoretical standpoint is reinforced by the practical results emanating from treatment of hundreds of patients. We have drawn useful conclusions from our therapeutic successes but more so from our failures, for these have led to successful reformulation of therapies. Although we have published our findings in some specialized journals, these papers far from reflect the store of information we have accumulated through years of day-to-day individual treatment of patients. By comparison, we feel that double-blind studies are clumsy tools of clinical research which exclude the other, more profound side of medical practice — that direct knowledge which every physician takes to his grave, too often without transmitting it to his colleagues.

Psychotic syndromes, perhaps the area in which we have greatest clinical and therapeutical experience, is discussed in Chapter 4. Based on this experience and on an exhaustive review of published information, we classify psychotic syndromes according to two large groups: schizophrenics and schizoaffectives. The first are psychotic patients characterized by a lack of libido and affectiveness (weak affections and libido), dangerous aggressive conduct, and a history in which truly normal periods are absent. In contrast, schizoaffectives include psychotic patients showing an excess of affectivity and libido, pseudo-aggressive behavior (more apparent than real), and a history of long periods of normal family life. Schizoaffectives treat their partners and children with affection and enjoy positive, aggreable social activity. While the schizophrenic, even in his "normal" periods, appears cold, unaffectionate, different from others, it is hard to distinguish a schizoaffective patient from other people during periods of normalcy. We believe we have achieved an effective therapy for these patients, superior to any we have seen reported in medical literature. Our therapy differs from the conventional approaches and for many years we have lamented that our papers published on the subject in different scientific journals have not drawn adequate attention to the benefits of effective pharmacotherapy. We therefore continue our work without attempting to convince others of the usefulness of our physiopathological and therapeutical approach.

The section discussing aggressive behavior analyzes results of numerous experimental studies provoking aggressive behavior in animals, in particular the muricidal behavior experimental model. Also discussed are findings of studies on humans showing homicidal and suicidal behavior, with special reference to biochemical analyses of the cerebrospinal fluid *in vivo*, and postmortem brain studies of suicide victims. We also present the indirect evidence arising from therapeutic trials.

The section on manic syndrome is defined by the symptoms which, according to the DSM III, are the indispensable requisites for labeling a syndrome as manic. This is a section which sparks controversy among the various schools of psychiatry. However, it is not our intention to endorse any particular current or to enter discussions on diagnostic criteria; for this reason we have adhered to the DSM III convention. We simply offer possible anatomical and physiological bases for the symptoms accepted as making up the manic syndrome. As in earlier and later chapters, our theoretical work is reinforced by considerable therapeutic experience in the use of psychoactive drugs on patients, in this case manic subjects.

In Chapter 5 the possible mechanisms involved in blood pressure regulation puts forward an excellent model of what in our opinion should be considered a psychosomatic disorder, high blood pressure. We attempt to show that the so-called psychosomatic illnesses may operate through neuronal circuits similar to those involved in psychiatric syndromes. The fact that these neurochemical disorders produce in certain subjects only somatic manifestations, with little or no psychic alteration, obliges us to seek additional physiopathological factors at work, possibly genetic ones. Nevertheless, we believe that the imbalance in neuronal circuits could be the point where somatic and psychosomatic illnesses converge.

In this chapter on blood pressure regulation we wish to draw attention to the fact that a majority of scientists involved in neurochemical research fall into the error of treating the autonomous CNS as a homogeneous system whose only differentiation is in the type of neurotransmitter or modulator which is synthetized or released. In fact, each of the neuroautonomic systems is composed of antagonic pairs and we place great emphasis on presenting the exhaustive information now demonstrating this. For example, there are sympathoexcitatory and sympathoinhibitory NE systems. There are 5HT systems which are active during wakefulness and others which are active during sleep. Further, there are DA systems favoring motor activity and others paralyzing motor activity. Although this antagonic pairing is familiar to all who seriously read scientific literature dealing with this matter, it is not yet taken into account by most neurochemical researchers and much less by practitioners.

In Chapter 6 we focus on the two biological markers we employ in our search for a diagnostic approach to psychosomatic syndromes: (1) intestinal pharmacomanometry and (2) intramuscular clonidine test.

The introduction to the chapter is a summary of the anatomical and functional interactions among components of the ANS: NE, DA, 5HT, and acetylcholinergic (ACh) systems. How the drugs (agonists and/or antagonists to these systems) modify distal colon motility (DCM) and how we draw inferences from these drug-induced changes are discussed.

In the section discussing procedure, we give details of intestinal pharmacomanometric methodology, the significance of the two components of DCM — intestinal tone (IT) and phasic activity (PA) or waves, the influence of emotional factors on DCM, and the effects of different psychoactive drugs.

We put forward experimental and clinical evidence supporting the hypothesis that drug-induced DCM changes are central and not peripheral effects.

We also discuss the difference between sigmoidal and rectal responses to drugs.

In other sections we summarize the results obtained in treatment of psychotic, affective, and psychosomatic disturbances.

Pharmacomanometric, metabolic, hormonal, and neurochemical evidence strongly suggesting the existence of two antagonistic DAs is also provided. This first line of research supporting a DA-antagonistic receptor hypothesis (1981), has been reinforced by other evidence. However, the new findings are less impressive than those drawn from intestinal pharmacomanometry.

Lastly, we describe how intestinal pharmacomanometry is used to guide psychoactive drug therapy of psychotic and depressive syndromes. These two examples are used to illustrate the procedures we follow for psychosomatic syndromes.

In the section discussing intramuscular clonidine test, we present the results obtained with the test in three groups of subjects: normal, depressed, and severely ill. Although response of plasmatic growth hormone (GH) levels to intravenous clonidine injection is widely used as a biological marker of depressive syndromes, intramuscular clonidine testing (introduced by us) offers some advantages over the former. Intramuscular administration of the drug is a weaker stimulus to α_2-anterior hypothalamic receptors, hence this test gains in sensitivity. On the other hand, differentiation of depressive patients into two groups, (1) low-IT + high NE plasma levels and (2) high-IT + low NE plasma levels, which show different responses to clonidine challenge, constitutes a valuable biological marker. In addition, intramuscular clonidine test performed according to our methodology introduced three other parameters of evaluation: (1) plasma cortisol response, (2) plasma NE response, and (3) diastolic blood pressure (DBP) response. This trio, in addition to GH responses, allows us to evaluate not only depressive syndromes but exacerbation of chronic illnesses. The fact that during exacerbation periods clonidine-induced responses are similar to those obtained during experimental stress situations leads us to postulate that stress plays some role in triggering such exacerbation periods in severely diseased patients.

Finally, the authors of this book ask those who do not agree with our point of view to allow us the recognition due to professionals who have worked unceasingly for many years. We know it is improbable, even impossible, to "sell" our viewpoint to the majority. However, this does not detract from the value of the vast bibliographic review gathered here for use by the reader; we believe, like Popper and Bunge, that in order to begin research, the investigator must obtain as much information as available. Perhaps our most important message is that those who intend to begin diagnosis or therapy in the areas of psychoneuropharmacology and psychoneuroendo-crinology should have a prior theoretical model which can be rectified or ratified with use.

Science, according to Bunge, does not consist of accumulating experimental data but in the interpretation of the findings. Men have always observed the sun rising in the east and setting in the west, yet it took a scientist to interpret the fact as due to the rotation of the earth around the sun.

ACKNOWLEDGMENTS

The authors wish to express their sincere gratitude to Mrs. Hilary Branch and Mrs. Judith Cristina Bermúdez for their revision of the text, their valuable secretarial assistance, and careful typing and revision of the manuscript.

This work was supported by a FUNDAIME grant.

THE EDITORS

Fuad Lechin, M.D., is Titular Professor of General Pathology and Physiopathology and Post-Graduate Professor of Psychopharmacology at the Central University of Venezuela School of Medicine. He is also Chief of Psychopharmacology and Psychosomatic Sections of the Institute of Experimental Medicine, Caracas, Venezuela. Dr. Lechin received his M.D. degree from the Central University of Venezuela, graduating summa cum laude. After internship and residency in internal medicine and gastroenterology at the Vargas and Bello Hospitals, Caracas, Dr. Lechin furthered his clinical training with postgraduate studies in several disciplines (physiology, pharmacology, psychoanalysis, mathematics, statistics, neurochemistry, and neuroendocrinology). He has contributed more than 150 publications to scientific literature. His papers in the fields of clinical psychopharmacology and biological psychiatry have appeared in many specialized scientific journals (*Biological Psychiatry, American Journal of Psychiatry, British Journal of Psychiatry, Lancet, Experientia, British Medical Journal, Journal of Clinical Pharmacology, Journal of Affective Disorders, Psychology, Psychiatry, and Behavior, Digest of Disorders in Science, Journal of Clinical Gastroenterology, American Journal of Gastroenterology, Neuroendocrinology, Psychoneuroendocrinology, Diabetologia,* etc.). Dr. Lechin is the author of *The Autonomic Nervous System; Physiological Basis of Psychosomatic Therapy.* A member of many scientific organizations, he is past president of the Venezuelan Clinical Pharmacology and Therapeutical Society and a member of the Argentinian and World Federation of Biological Psychiatry Societies.

Bertha van der Dijs, M.D., is Assistant Professor of Biochemistry at the Central University of Venezuela School of Medicine, Bioanalysis Department. She received her medical degree from the Central University of Venezuela School of Medicine. After internship and residency in internal medicine at the University Hospital, she has carried out biochemical research in catecholamines and indoleamines, in both clinical and experimental animal studies. An expert in high pressure liquid chromatography (HPLC), she is currently devoted to this research area. She is chief of the Section of Chromatography at the "Hans Selye" Institute of Psychosomatic Medicine, Caracas. Dr. van der Dijs has authored and coauthored over 60 articles in many scientific journals on psychopharmacology, biological psychiatry, and internal medicine published in many international scientific journals. She is a member of numerous scientific societies and coauthor of related books.

CONTRIBUTORS

José Amat, M.D., Ph.D.
Assistant Professor
Department of Physiopathology
Central University of Venezuela School of
 Medicine
and
Chief
Neurophysiology Section of the Institute of
 Experimental Medicine
Caracas, Venezuela

Luis Arocha, M.D.
Chief
Medical Education Office
and
Head
Department of Scientific Methodology
Central University of Venezuela School of
 Medicine
Caracas, Venezuela

Francisco Gomez, M.D.
Staff Member
Department of Psychopharmacology and
 Psychosomatic Medicine
Experimental Medicine Institute
Caracas, Venezuela

Alex E. Lechin, M.D.
Staff Member
Department of Internal Medicine
Hospital Vargas
Caracas, Venezuela

Fuad Lechin, M.D.
Titular Professor
Department of Physiology
and
Chief
Department of Psychopharmacology and
 Psychosomatic Medicine
Central University of Venezuela School of
 Medicine
Caracas, Venezuela

Marcel Lechin, M.D.
Staff Member
Department of Internal Medicine
Hospital Perez Carreño
Caracas, Venezuela

Scarlet Lechin, M.D.
Assistant Professor
Division of Immunogenetics
Dana Farber Cancer Institute
Harvard Medical School
Boston, Massachusetts

Bertha van der Dijs, M.D.
Chief of Chromatography
Department of Physiopathology
Central University of Venezuela School
 of Medicine
and
Staff Member
Section of Psychopharmacology
Experimental Medicine Institute
Caracas, Venezuela

Simon Villa, M.D.
Staff Member
Department of Psychopharmacology and
 Psychosomatic Medicine
Experimental Medicine Institute
Caracas, Venezuela

TABLE OF CONTENTS

Chapter 1

NEUROANATOMICAL BASIS

Fuad Lechin, Bertha van der Dijs, José Amat, and Marcel Lechin

TABLE OF CONTENTS

I. THE NORADRENERGIC SYSTEM

Noradrenergic (NE) neurons are confined to three groups in the pons and medulla oblongata (see Figure 1): (1) the well-defined locus coeruleus (LC) or A6 cell group, (2) a more diffuse but continuous lateral and ventral group that arches through the pons and medulla, and (3) a third dorsal medullary group known as A2 cell group, centered in the dorsal motor nucleus of the vagus. NE-LC or A6 nucleus and NE-A2 nucleus are dorsally located in the tegmental area, whereas the lateral groups are ventrally located with respect to the former and include A7, A5, and A1 cell groups. NE-A1 cells are the most caudally located in ventrolateral medulla oblongata. The LC (A6) with approximately 1600 neurons (in the rat) and the A5 with approximately 340 neurons are the largest in size of all NE cell groups.

There is general agreement that the basic organization of the central catecholaminergic (CA) system consists of cell bodies located in caudal brain stem that give rise to ascending and descending fiber systems terminating in widespread areas of the brain. However, substantial areas of disagreement still exist in results obtained with different methods and by different investigators as to precise pathways and functions.

The NE brain stem cell groups (except LC) do not form compact nuclei but are dispersed among non-NE cells. However, they send axons which terminate in a larger number of apparent pericellular arrays, particularly in certain cranial motor nuclei where they are capable of exerting direct, potent control on neuronal postsynaptic activity. LC innervation, affecting only sensory and association nuclei (in the brain stem), also sends profuse innervation to telencephalic and diencepalic structures. In these projection areas, LC axons terminate in a uniform, sparse to moderately dense plexus which may contribute modulatory control over other neuronal input here.

Three NE pathways are known to project rostrally: (1) dorsal NE bundle (DNB) or dorsal tegmental tract, originating in LC; (2) ventral NE bundle (VNB) or ventral tegmental tract which collects fibers from A1, A2, A5, and A7 cell groups, plus some fibers arising from LC and NE cells lying ventral to LC (subcoeruleus group); and (3) dorsal periventricular tract originating in A2 cell group which collects fibers from NE cells of LC and subcoeruleus areas.

The DNB projects mainly to brain cortex, dorsal hippocampus, striatum, and some mesolimbic and hypothalamic structures. The VNB projects mainly to hypothalamus and some mesolimbic structures; its axons never reach hippocampus or brain cortex. The dorsal periventricular tract projects mainly to central gray or periventricular area (see Figures 1 to 3).

NE-LC (A6) efferent projections are found in brain cortex, putamen-caudate, globus pallidus, amygdala, hippocampus, septum, olfactory tubercles, nucleus accumbens, posterior hypothalamus, mediobasal hypothalamus (nucleus arcuate, nucleus ventromedial and nucleus paraventricular), and median eminence of hypothalamus (outside the blood brain barrier, bbb). Profuse projections to the brain stem structures (sensory and association but not motor nuclei) have also been demonstrated, as well as a clearly defined projection to the ventral spinal horn (see Figure 4).

The following areas are known to receive NE-A5 axons, based on unilateral decreases in NE levels following A5 nucleus lesion: caudate nucleus, piriform cortex, interstitialis nucleus stria terminalis, medial forebrain bundle, medial preoptic nucleus in anterior hypothalamus, and median eminence. NE-A5 neurons send also axons to pons, medulla oblongata, and spinal cord. With respect to this, NE-A5 axons constitute the main innervation of sympathetic preganglionic cells located in the intermediolateral (IML) spinal horn, in the rat (see Figure 5).

NE-A1 efferent projections send axons to the hypothalamus (n. paraventricularis, n. ventromedial, n. arcuate, anterior preoptic area, etc.), septum, and other mesolimbic structures. The axons do not reach the hippocampus and brain cortex. Pons, medulla oblongata, and spinal cord receive NE-A1 efferents. Important NE-A1 fibers also reach NE-LC and NE-A2 cell groups (see Figure 6).

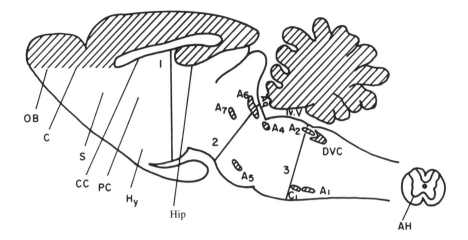

FIGURE 1. Sagittal section of the rat brain showing the dorsoventral and rostrocaudal location of the noradrenergic cell groups. (1) Divisory line between diencephalon and mesencephalon, (2) divisory line between mesencephalon and pontis, and (3) divisory line between pontis and medulla oblongata. OB = olfactory bulb, C = cortex, S = septum, CC = corpus callosum, Hy = hypothalamus, Aq = aqueduct, IV V = fourth ventricle, DVC = dorsal vagal complex, AH = anterior spinal horn, and Hip = hippocampus.

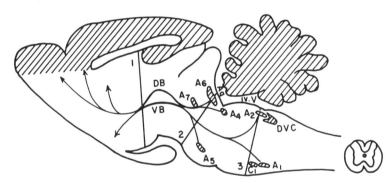

FIGURE 2. Sagittal section of the rat brain sources and projections of the dorsal and ventral noradrenergic bundles (DB and VB).

The NE-A2 efferents cell group sends its axons to the same central nervous system (CNS) areas as the NE-A1 cell group. A2 efferents also reach parasympathetic and sympathetic preganglionic cells. Parasympathetic preganglionic cells are located in nucleus ambiguus and n. dorsal motor vagii, and are also dispersed in the reticular formation. Sympathetic preganglionic cells are found in the lateral reticular formation, nucleus reticularis lateralis, IML spinal horn, etc. (see Figure 7).

A. Neuroanatomical Connections between Noradrenergic Cell Groups

The different methodologies employed in numerous investigations of the neuroanatomical connections between NE cell groups have given rise to some disagreements. Not surprisingly, some discrepancies occur because of the different animal species studied; others are due to the fact that the NE cell groups can interact not only through direct monosynaptic but also polysynaptic mechanisms.

The LC complex (A6 + subcoeruleus cell area + the caudally located A4 cell group)

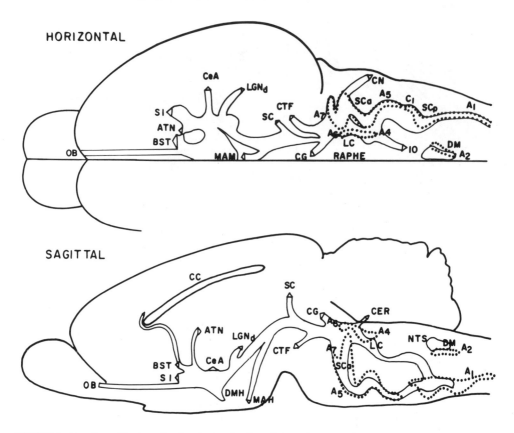

FIGURE 3. Drawings of horizontal and sagittal projections of the rat brain showing the location of adrenergic cell groups (dotted lines), the principal adrenergic fiber bundles, and major terminal field.

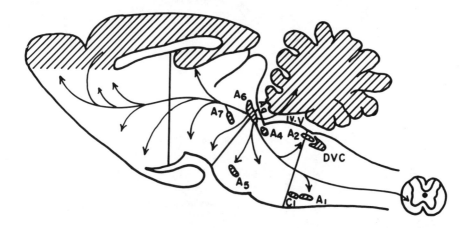

FIGURE 4. Sagittal section of the rat brain showing the locus coeruleus (A6 cell group) efferents.

innervates primary sensory and association brain stem nuclei located in the pontine gray and medullary reticular formation. It does not contribute innervation to somatic motor nuclei or discrete visceral motor nuclei of the brain stem. However, many physiological studies provide evidence that the LC has a central role in peripheral pressor, depressor, micturition, respiration, and other mechanisms which involve brain stem reticular formation (see Figure 8).

The lateral tegmental NE cell groups (A7, A5, A1) and the dorsal paramedian NE-A2 cell

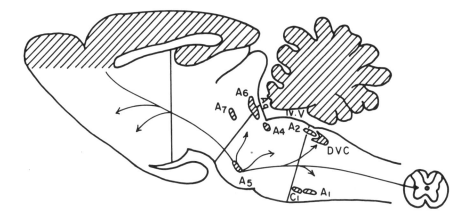

FIGURE 5. Sagittal section of the rat brain showing A5 cell group efferents.

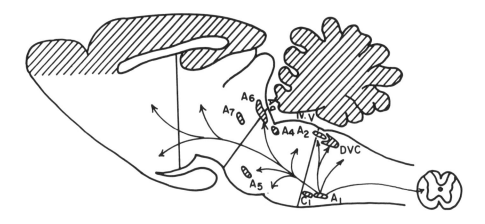

FIGURE 6. Sagittal section of the rat brain showing A1 cell group efferents.

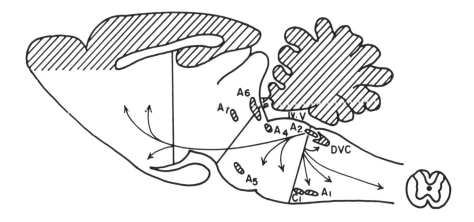

FIGURE 7. Sagittal section of the rat brain showing A2 cell group efferents.

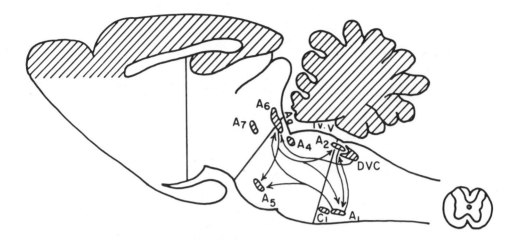

FIGURE 8. Sagittal section of the rat brain showing interconnections among noradrenergic cell groups.

group densely innervate the somatic motor and visceral motor nuclei. It is of note that innervation of the hypothalamus also arises mainly from these groups, suggesting that they may have a general function in regulating autonomic, visceral, and neuroendocrine activity.

The NE-A6 (LC) cell group sends axons to A5, A2, and A1 cell groups. In turn, NE-A1 group densely innervates A6, A5, and A2 groups. NE-A2 neurons send a well-defined innervation to NE-A1 group which, however, is not NE in nature.

A study of inputs to antidromically identified neurons of the LC has shown that stimulation of the vagus nerve produces inhibition of LC neurons, an inhibition followed by excitation, while stimulation of the splacnic nerve and sciatic nerve produces excitation of LC neurons. These findings suggest that afferent projections to LC arise from the dorsal vagal complex in which are included NE-A2 neurons. This connection, however, does not emanate from the NE-A2 neurons but from the nucleus of the solitary tract (NTS) which is the principal recipient of first order vagal afferent input. The NTS projects to preganglionic cell groups of both divisions of the autonomic nervous system — sympathetic and parasympathetic, a series of relay nuclei in the brain stem, and a number of cell groups in the hypothalamus and limbic region of the telencephalon which control autonomic, neuroendocrine, and regulatory behavioral responses. Because cell groups receiving direct NTS inputs project back to this region and/or the vagal motor nuclei (dorsal motor nucleus of the vagus and nucleus ambiguus), they are therefore in a position to influence vagal motor outflow. Such vagal motor outflow is also under the influence of the NE-A2 and NE-A5 neurons which send important projections to vagal motor nuclei (see Figure 8).

B. Neuroanatomical Connections between Noradrenergic Nuclei and Other Monoaminergic Nuclei

Although a great bulk of evidence demonstrates direct monosynaptic pathways between NE and serotonergic (5HT) or dopaminergic (DA) neurons, some of these pathways are inferred from pharmacological, biochemical, and behavioral evidence. Frequently the various monoaminergic brain stem nuclei are so closely joined that electrophysiological and microinjection studies are difficult to perform and produce conflicting and even contradictory results. However, as the new methodology develops, results obtained will be more precise. Below, we summarize the most convincing evidence supporting our postulations. Our propositions, however, undergo continuous revision and rectification.

1. Neuroanatomical Connections between Noradrenergic and Serotonergic Systems

Fuxe demonstrated fine CA terminals within the dorsal raphe (DR) 5HT nucleus (see Figure

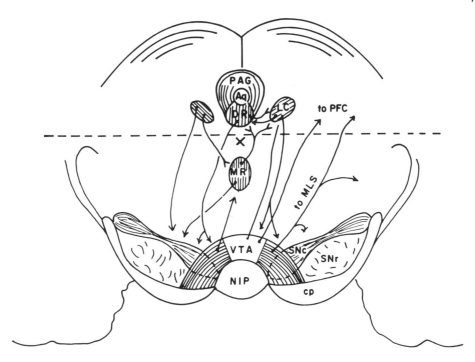

FIGURE 9. Two transverse sections of mesencephalic (bottom) and pontis (top) levels. Interconnections between NE locus coeruleus (LC) nucleus, 5HT dorsal raphe (DR) nucleus, 5HT median raphe (MR) nucleus, DA ventral tegmental area (VTA) and DA substantia nigra (SN) nucleus. PFC = prefrontal cortex, NIP = interpeduncular nucleus, MLS = mesolimbic structures, CP = cerebral peduncle, PAG = periaqueductal gray, and Aq = aqueductus.

9); Loizou reported that some of these terminals disappear following lesions of the LC; Baraban et al. demonstrated recently NE-LC innervation of 5HT neurons in the DR. These anatomical findings are supported by others showing that NE innervation of the DR is not exclusively provided by LC neurons since lesioning of LC nucleus does not eliminate the NE content of DR, which is among the richest brain stem nuclei. Furthermore, Roizen and Jacobowitz reported that VNB contributes most NE input to the raphe area. Finally, Gallager and Aghajanian demonstrated that complete transection at the pontine reticular formation level, separating the lower brain stem from the raphe area, abolishes the depressant effects on raphe firing of antipsychotic agents, piperoxane, and other α_2-antagonists. This finding may be interpreted as the well-known inhibition of 5HT neurons exerted by NE innervation through α_2- adrenoceptors; this inhibition arises at least partially from caudal brain stem NE nuclei. According to the above, NE innervation arising from LC (A6) nucleus would excite 5HT-DR neurons through α_1-adrenoceptors, while non-A6 NE innervation would inhibit 5HT-DR neurons through α_2-adrenoceptors.

It is well known that the largest 5HT content of LC is significantly reduced after DR lesioning (86.7%). Experimental findings also show that LC receives dense innervation from DR. These DR axons are proven to be 5HT and to exert an inhibitory role on NE neurons of the LC, since lesioning the DR results in an increase in tyrosine hydroxylase activity of LC neurons. However, it should be noted that 5HT axons arising from the median raphe (MR) nucleus also innervate the LC with inhibitory fibers; destruction of MR produces a long-lasting stimulation of NE-LC neurons.

We know that MR receives important NE innervation from LC and lateral tegmental NE nuclei (non-A6 nuclei) because LC lesions produce only partial reduction of MR norepinephrine content. In contrast with well-established excitatory effects of NE on 5HT-DR neurons, NE innervation exerts an inhibitory effect on 5HT-MR neurons. This effect is mediated through α_1-adrenoceptors. In effect, both microinjection of α_1-agonists and lesioning of the MR decrease

5-HIAA in those forebrain structures innervated by 5HT-MR axons. This NE-5HT antagonism between NA and MR-5HT systems would explain the increased 5HT turnover observed following administration of the CA-specific neurotoxin 6-OHDA, of the tyrosine hydroxylase inhibitor α-methyl-*p*-tyrosine, and of dopamine-β-hydroxylase inhibitors. Furthermore, marked increases in serotonin synthesis in the telencephalon are produced by lesioning the rostral third of LC, a fact consistent with the existence of an LC-MR inhibitory fiber system.

CA = NE + DA are present in varying amounts in all raphe nuclei. Their activity in these areas is found to be of the same order of magnitude as that of tryptophane hydroxylase, an enzyme responsible for the formation of serotonin. Whereas administration of dopamine-β-hydroxylase inhibitor results in over 90% depletion of NE according to recent experiments, 5HT levels are not significantly changed except in raphe magnus (RMg) and MR nuclei, where 5HT concentrations significantly increased. No change is seen in 5HT levels in DR nucleus. This finding would ratify the fact that NE input exerts an inhibitory effect on MR and RMg nuclei.

Other studies demonstrate that 5HT-MR neurons send prominent innervation to LC and that this 5HT input inhibits NE neurons. In effect, destruction of MR system produces a long-lasting stimulation of NE neurons and a significant rise in the LC concentration of tyrosine hydroxylase (see Figure 9).

2. Neuroanatomical Connections between Noradrenergic and Dopaminergic Systems

Neurons of LC and non-LC NE systems send and receive axons to and from the two brain stem DA systems: the substantia nigra (SN) = A8 + A9 cell groups, and the ventral tegmental area (VTA) = A10 cell group. Moreover, NE and DA axons frequently converge on the same CNS areas populated by various types of postsynaptic NE and DA receptors. Furthermore, NE terminals have been shown to possess inhibitory DA receptors, while DA terminals possess excitatory α_2 adrenoceptors. All these findings show that NE-DA interactions are complex and difficult to determine precisely. This complexity is accentuated by the existence of polysynaptic pathways which permit indirectly exerted influences between both systems.

There are NE terminals in the SN and VTA nuclei (see Figure 9). The NE in the SN seems to exert an excitatory effect, since lesions of NE cell bodies in LC produce increased sensitivity of striatal postsynaptic DA receptors on the operated side (in turning experiments), perhaps due to interruption of NE neurons from LC which normally facilitate the nerve impulse flow in nigro-striatal DA neurons. However, other NE influences exert inhibitory effects on the nigrostriatal system. In effect, lesion of VNB increases nigrostriatal activity. In our opinion, this finding might be due to a direct inhibitory effect by NE axons arising from NE-LC neurons which are facilitatory on the nigrostriatal DA system. With respect to this, it has been shown that lesioning the VNB provokes increase of DA activity in the nucleus accumbens along with reduction of DA activity in the prefrontal cortex. Other experiments show that NE innervation of VTA exerts an inhibitory influence on DA mesolimbic system, as reflected by a decreased DA turnover in the olfactory tubercles after microinjection of NE in the VTA DA area. This apparent inhibitory NE-DA interaction has been suggested in a variety of experimental studies. The NE innervation of VTA region, which exerts an inhibitory influence on DA mesolimbic system, originates in the LC. However, this same LC-VTA NE fiber system has been proven to exert an excitatory influence on DA mesocortical system. This DA system consists of DA neurons located in the median and anterior parts of VTA region whose axons terminate in the deepest layers of the prefrontal cortex. The DA neurons innervating subcortical mesolimbic structures are located in the lateral and posterior parts of VTA (see Figures 9 to 11).

Further experimental studies demonstrate that lesions of VNB produce opposite effects to the above mentioned lesions of NE-LC neurons. In effect, VNB lesions do not decrease prefrontal cortical DA activity but, paradoxically, increase dopamine at this cortical level by as much as 40%. This opposite effect gives rise to the often-suggested hypothesis that LC and non-LC NE systems behave as two opposing NE systems. The fact that cortical prefrontal and mesolimbic subcortical DA systems are able to inhibit each other, through direct and indirect pathways, fits

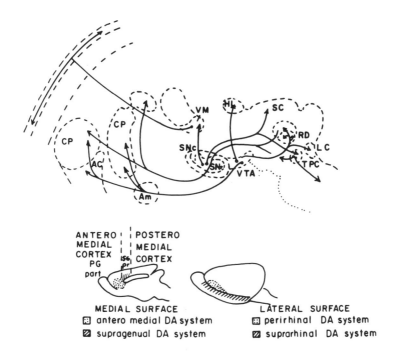

FIGURE 10. Top: diagrammatic representation of the efferent relationships of the pars compacta (SNc), pars reticulata (SNr), and ventral tegmental area (VTA). Bottom: schematic representation of the distribution of the mesocortical DA systems. Not included in the diagram are the DA VTA projections to the prefrontal cortex and entorhinal cortex. AC = nucleus accumbens, LC = locus coeruleus, RD = dorsal raphe, Am = amygdala, CP = caudate putamen, HL = lateral habenular nucleus, SC = superior colliculis, SNc = substantia nigra compacta, and SNr = Substantia nigra reticulata.

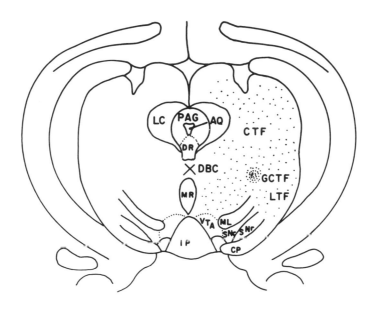

FIGURE 11. Transverse section of the rat brain at the mesencephalic level. LC = locus coeruleus, PAG = periaqueductal gray, AQ = aqueductus, DR = dorsal raphe, DBC = decussatio brachium conjunctive, MR = median raphe, VTA = ventral tegmental area, IP = interpeduncular nucleus, SNc = substantia nigra compacta, SNr = substantia nigra reticulata, CTF = central tegmental field, LTF = lateral tegmental field, GCTF = nucleus giganto cellularis tegmental field, ML = medial lemniscus, and CP = caudate putamen.

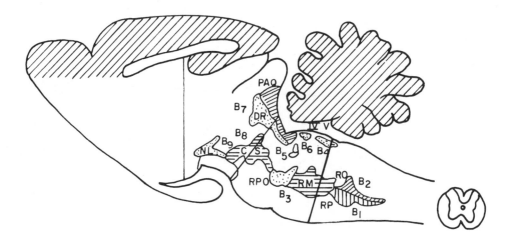

FIGURE 12. Sagittal section of the rat brain showing the dorsoventral and rostrocaudal location of the serotonergic cell groups.

well with the postulation of two opposing DA systems. This matter will be discussed in other chapters.

Many findings show that SN and VTA DA neurons send efferent projections to LC and other NE nuclei located in the brain stem reticular formation, thus giving anatomical support to the above hypothesis postulating the existence of two opposite NE-DA systems. One of them would be constituted by the NE-LC and DA-SN + DA mesocortical, whereas the other, operating antagonistically to the former, would be constituted by the NE-non-LC system and the DA-mesolimbic system (see Figure 9).

II. THE SEROTONERGIC SYSTEM

The 5HT neurons are grouped in nuclei located along the midline (raphe) of the brain stem. There are two chains of 5HT raphe nuclei, one dorsal, the other ventral. The most rostral nucleus of the former chain is the DR nucleus or B7 cell group. DR limits with the aqueduct and possesses rostral (mesencephalic) and caudal (pontine) parts. The raphe pontine B6 and B5 cell groups and the raphe pontine-medullary B4 cell group constitute the three other 5HT components of the dorsal chain.

The most rostral nucleus of the ventrally located raphe chain is the nucleus linearis (B9 cell group) which is located in the mesencephalon. The nucleus centralis superioris or MR nucleus (B8 cell group) is caudal to the former and possesses a mesencephalic and a pontine part. MR nucleus is ventral to DR and is separated by the decussation of brachii conjunctivi. Caudally to MR are located the pontine nucleus raphe pontis oralis (RPO) and the pontine-medullary RMg nucleus. RPO + RMg constitute the B3 cell group. The medullary raphe obscurus (RO) = B2 and raphe pallidus (RP) = B1 cell groups are the most caudally located 5HT nuclei integrating the ventral 5HT chain (see Figures 11 and 12).

The lateral and dorsal parts of the aqueduct are occupied by 5HT neurons composing the periaqueductal gray system (PAG). Yet another 5HT system has been postulated with neurons located at hypothalamic level.

The mesencephalic and pontine 5HT neurons innervate the anterior regions of the CNS: the telencephalon, diencephalon, and mesolimbic structures, whereas pontine and medullary-5HT neurons innervate pontine, medullary, and spinal structures. PAG-5HT system innervates anterior brain stem and some telencephalic-diencephalic structures. Rostral 5HT nuclei are interconnected with caudal 5HT nuclei (see Figure 13).

DR projects to brain cortex, mainly in the temporoparietal area, as well as to the striatum,

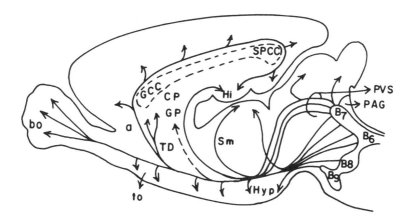

FIGURE 13. Schematic representation of the major organizational features of the ascending 5HT systems of adult rat brain, as revealed by light microscope radioautography after intraventricular administraton of [³H] 5HT. bo = bundle olfactorius, to = tractus opticus, a = nucleus accumbens, GCC = genu corporis callosi, SPCC = splenium corporis callosi, CP = caudate putamen, TD = tractus diagonalis (broca), GP = globus pallidus, Sm = stria medullaris thalami, Hyp = hypothalamus, Hi = hippocampus, PVS = periventricular system, and PAG = periaqueductal gray.

amygdala, nucleus accumbens, lateral septum and hypothalamus (ventromedial mainly). The median eminence, outside the blood brain barrier, is also innervated by DR axons. Some DR axons project to dorsal hippocampus and brain stem reticular formation.

MR projects to brain cortex, mainly the prefrontal, hippocampus, septum, and other mesolimbic structures. The anterior preoptic hypothalamic area is selectively innervated by MR axons. Further, MR innervates brain stem reticular formation and DR nucleus.

The RPO nucleus sends axons to anterior spinal motor neurons and to preganglionic sympathetic neurons located in the IML spinal horn. These 5HT projections exert an excitatory influence in both spinal regions.

A 5HT nucleus which is closely involved in nociceptive functions, the nucleus RMg sends its axons to the spinal cord dorsal horn. This sensory area in turn sends projections to the RMg. All medullary 5HT nuclei send axons to all medullary non-5HT nuclei and to brain stem reticular formation. RMg nucleus sends projections to hypothalamus.

Although PAG-5HT system is preferentially interconnected with brain stem structures, it also interconnects with some mesolimbic and hypothalamic structures (see Figures 12 and 13).

A. Neuroanatomical Connections between the Different Serotonergic Systems

Although there is no direct projection from DR to MR, the latter sends axons to the former. Since serotonin iontophoretically injected on 5HT DR-perikarya exerts an inhibitory effect on the firing rate of these neurons, the MR to DR fiber system would represent an inhibitory control of MR over DR. There is also evidence of anatomical connections between PAG and DR, PAG and MR, and MR with raphe pontine nuclei (RPO and RMg). Indeed, caudal 5HT nuclei are considered to constitute true relay stations to rostral 5HT nuclei.

One of the best established connections linking 5HT nuclei is the excitatory input of PAG on RMg. Supposedly mediated through glutamate or aspartate fibers, this pathway would be involved in analgesic mechanisms depending on opiates to release PAG activity from a GABA inhibitory neuron (see Figures 9 and 13).

B. Neuroanatomical Connections between Serotonergic and Noradrenergic Nuclei

It is well established that DR sends axons to LC and receives afferents from LC and non-LC NE nuclei. MR and PAG nuclei also receive afferent projections from NE systems. In turn, MR sends axons to both NE systems. However, 5HT fibers innervating NE-A2 nucleus arise from medullar 5HT nuclei, preferentially (B4 cell group) (see Figures 9 to 11).

Intermingled 5HT and NE axons frequently innervate the same central areas. Both types of terminals possess α_2-presynaptic receptors capable of exerting a modulatory effect at those central projection areas (see Figure 9).

C. Neuroanatomical Connections between Serotonergic and Dopaminergic Nuclei

The DR is proven to send axons to both SN and VTA. The resulting inhibitory effect would be exerted by serotonin through 5HT receptors located on DA neurons. However, evidence exists that DR bridles only DA neurons of the lateral regions of VTA, sparing DA neurons of the anterior VTA nucleus. Lateral VTA neurons send axons to subcortical mesolimbic structures, while anterior VTA-DA neurons provide axons composing the mesocortical DA fiber system which innervates the deepest layers of the prefrontal cortex.

Both DR and MR 5HT nuclei send axons to VTA DA region. These 5HT fiber systems exert an inhibitory effect on DA mesocortical system. In turn, there are clearly established DA afferents to MR arising from VTA, which give an excitatory input (see Figures 9 to 11).

Abundant evidence demonstrates that MR innervates the SN. Although this 5HT input to SN is inhibitory, it would have a different functional role to that displayed by 5HT input arising from DR. In effect, while the latter 5HT input is associated with punishment behavior, suppression of MR input to SN fails to modify this behavior (see Figure 9).

D. Neuroanatomical Connections between Serotonergic Nuclei and Brain Stem Reticular Formation

The nuclei of brain stem reticular formation (BSRF) all contain about the same amount of 5HT. Excitatory and oscillatory as well as inhibitory responses to raphe stimulation have been found in BSRF neurons. However, although BSRF is densely charged with serotonin, there is only a scattered 5HT input to BSRF neurons. Despite the fact that both the sensory and motor cranial nerve nuclei receive 5HT innervation, motor nerve nuclei seem to contain more of the amine than do sensory nuclei.

Evidence suggests that DR receives an important cholinergic input. Since microinjection of acetylcholine in DR induces synchronization, it is logical to assume that this ACh input to DR exerts an inhibitory inflluence (see Figures 9 and 13).

III. THE DOPAMINERGIC SYSTEM

Two well-defined DA cell groups are found in the mesencephalon: the SN = A8 + A9 cell groups, and the adjacent VTA = A10 cell group. VTA nucleus is medially located and posterior to SN. Pars compacta of the SN = SNc sends axons to the striatum whereas pars reticulata (SNr), besides its striatal axons, sends projections to the thalamus, tectum, and pontomesencephalic tegmentum. Pars compacta (SNc) also sends axons to DR, MR, and hypothalamus, both inside and outside the blood brain barrier (see Figures 10 and 11).

The VTA nucleus sends axons to ventromedial striatum, nucleus accumbens, olfactory tubercles, thalamus, habenula, amygdala, posterior hypothalamus, lateral hypothalamus, anterior hypothalamus, lateral septum, prefrontal cortex, enthorhinal cortex, DR, MR, and LC. It has been found that VTA is composed of two different groups of DA cells, those located in its lateral and posterior regions and those located in the anterior and medial regions. The former group known as DA mesolimbic system, provides the axons which innervate subcortical mesolimbic structures, whereas the second group of cells, the DA mesocortical system, gives rise to axons innervating the prefrontal cortex (see Figure 10).

DA cells are also differentiated according to their electrophysiological characteristics. Thus, mesocortical DA cells possess a much faster firing rate (9.3 ± 0.6 per second) than mesolimbic DA cells (5.9 ± 0.5 per second) and mesostriatal DA cells (3.1 ± 0.5 per second).

Most DA cells of the SN-striatum system receive an inhibitory GABA-afferent loop from the striatum, emanating from DA-innervated area. Such a loop has not been demonstrated for DA cells of VTA region although GABA is present within this VTA area.

DA neurons contain their own transmitter in their dendrites. When dopamine is applied iontophoretically in the vicinity of DA cell bodies, it provokes an inhibition of DA neuronal activity. Evidence suggests that this somatodendritic inhibitory mechanism is more important for DA-VTA neurons than DA-SN neurons. For instance, whereas systemic picrotoxin, a GABA antagonist, reverses the depressant effects of systemic D-amphetamine on DA-A9 impulse flow, it does not reverse this effect on DA-A10 impulse flow. This would indicate that the inhibition of A10 activity may be more strongly influenced by dendritic release of DA, which would stimulate autoreceptors located on the somatodendritic areas. However, if we accept recent evidence that mesocortical DA neurons lack autoreceptors, then inhibitory mechanisms for these neurons would depend on other mechanisms. With respect to this, an effective inhibitory mechanism of further synthesis and release has been found in the reuptake of dopamine by DA terminals in the transmitter released at synaptic level (see Figure 10).

A. Neuroanatomical Connections between Substantia Nigra and Ventral Tegmental Area Dopaminergic Nuclei

The term SN refers to a complex structure lying immediately dorsal to the cerebral peduncle, composed largely of medium-sized cells of a fairly uniform type. Such cells appear in Nissl material to be more darkly stained and more closely spaced in the dorsal area called SNc, than in the larger subjacent SNr. As demonstrated through histofluorescence, SNc is composed largely, if not entirely, of DA neurons while most but not all SNr cells are not dopaminergic. The same technique reveals that, in the rat at least, SNc = A9 cell group of Dahlström and Fuxe does not have the flat dorsal border traditionally ascribed to it, but instead emits a pair of large dorsal excrescences lacking the dense cell packing of SNc. The larger medial one of these protruding cell masses, dopamine cell group A10, extends dorsomedially into the VTA, while the smaller, more caudal and lateral extrusion, dopamine cell group A8, invades a ventrolateral region near the SN caudal pole. At caudal levels of the SN, cell groups A10 and A8 are interconnected by an irregular array of cells that extends transversely over the dorsal border of the medial lemniscus; many of these cells synthetize dopamine (supralemniscal cell group, retrorubral nucleus) (see Figure 10).

DA-VTA cells send axons to the ipsilateral SN through which they are distributed over the entire rostrocaudal extent of SNc, and in lesser number to the most dorsal zone of SNr. This rather homogenous distribution suggests a termination of VTA efferents in contact with either somata of compacta neurons, or compacta dendrites oriented parallel to the dorsal border of SN (see Figures 9 and 10).

B. Neuroanatomical Connections between Dopaminergic and Noradrenergic Brain Stem Nuclei

VTA but not SN sends axons to LC; however, LC sends axons to both SN and VTA-DA nuclei. There is evidence showing that the VNB which collects axons from non-LC NE neurons, preferentially, innervates VTA. On the other hand, the DA-A10 cells (VTA) send projections to both types of NE nuclei (coeruleus and noncoeruleus).

Interactions between NE and DA systems are complex and both cooperation and antagonism can be observed. This phenomenon is found not only at cell body level but also at terminal level (see Figure 9).

C. Neuroanatomical Connections between Dopaminergic and Serotonergic Brain Stem Nuclei

SNc sends efferents to both DR and MR-5HT nuclei. The SNc projections exert an inhibitory influence on 5HT neurons. In turn, both 5HT nuclei inhibit SNc-DA neurons through DR and

MR efferents. Similarly, DA-A10 cell group interchanges axons with DR and MR-5HT neurons and, again in these cases, DA and 5HT influences are inhibitory. There is also strong evidence showing that DR projects to and inhibits mesolimbic DA cells, while MR projects to and inhibits mesocortical DA cells (see Figure 9).

IV. THE CHOLINERGIC SYSTEM

Cholinergic neurons are so widely diffused throughout the CNS that it is difficult, if not impossible, to group these dispersed neurons into one or several systems functioning as units. An attempt to transfer to the CNS the peripheral autonomic model of two sides, one sympathetic and the other parasympathetic, according to the corresponding norepinephrine or acetylcholine neurotransmitters, is fruitless because the activity of certain central cholinergic pathways may lead to peripheral sympathetic hyperactivity, while the activity of some central catecholaminergic pathways may lead to reduced peripheral sympathetic activity and consequently predominance of the peripheral parasympathetic system. Despite such objections, there are sufficient experimental findings to document concrete proposals concerning the existence of central cholinergic pathways. We mention here the most significant results of such research.

Injection of physostigmine, a cholinesterase inhibitor, either i.v. or into the cerebral ventricles (icv) and specific brain regions, evokes a centrally mediated rise in arterial blood pressure. This response requires functional brain acetylcholinesterase and brain acetylcholine (ACh). The blood pressure rise is mediated peripherally through an increase of sympathetic activity. The icv injection of various ACh agonists in dogs and rats, or into specific brain areas of rats, also raises blood pressure. Recent evidence from several laboratories suggests that brain ACh is involved in the elevated blood pressure observed in spontaneous hypertensive rats. Moreover, cholinergic substances also appear to influence spinal autonomic mechanisms and sympathetic reflexes.

Electrical stimulation of pontine NE cell nucleus, LC, increases the turnover of peripheral norepinephrine via the sympathetic system. For this reason it has been suggested that LC acts as a sympathetic nucleus situated in the brain with extensive parts of the CNS as its target regions. The LC receives rich ACh input and the activity of NE-LC cells is enhanced by microinjection of ACh agonists.

The posterior hypothalamus appears to be an intermediate link in the neural pathway mediating sympathetic responses to LC stimulation since destruction of posterior hypothalamus significantly inhibits such sympathetic responses. Further, LC has been shown to be a major source of NE innervation of the posterior hypothalamus. Superfusion of this area with ACh and carbachol (an ACh agonist) induces increased peripheral sympathetic activity.

Stimulation of the so-called medullary reticular pressor area with nicotinic but not with muscarinic agonists provokes increased peripheral sympathetic activity. The lateral reticular nucleus is the area of medullary reticular formation most related to sympathetic activation.

Besides the above central areas in which cholinergic mechanisms are associated with vasopressor responses and thus with peripheral sympathetic hyperactivity, there are other central regions in which cholinergic mechanisms elicit vasodepressor responses supposedly mediated through reduction in peripheral sympathetic activity. For example, while ACh agonists evoke hypertensive response when injected in posterior hypothalamus, injections of these agents into dorsomedial and anterior hypothalamus of rats result in hypotension. Similarly, areas of medullary reticular formation such as the nucleus tractus solitarius (NTS) and paramedian reticular formation behave as vasodepressor regions when stimulated with ACh agonists.

The NTS is the primary termination site of afferent fibers of cranial nerves IX and X including those arising from arterial and cardiopulmonary mechanoreceptors (see Figures 14 and 15). As such, NTS plays a critical role in integrating the cardiovascular reflexes arising from those

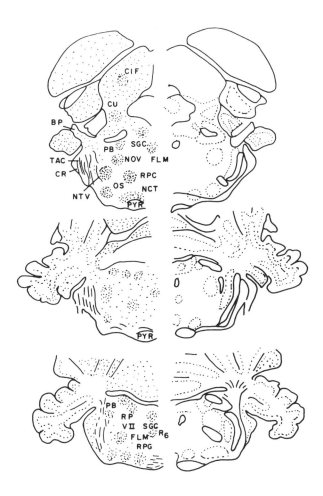

FIGURE 14. Coronal sections (transverse sections) of adult rat brain at 3 different levels (pontis and medullary). CIF = collicular inferior, CU = area cuneiformes, BP = brachium pontis, SGC = striatum griseum centrale, NOV = nucleus originis nervi trigemini, FLM = fasciculus longitudinalis medialis, RPC = nucleus reticularis pontis caudalis, OS = oliva superior, NTV = nucleus terminationis nervi trigemini, NCT = nucleus corpus trapezoides, Pyr = pyramidis, RG = nucleus reticularis gigantocellularis, RPG = nucleus reticularis giganto paracellularis, and V II = nervus fascialis.

receptors. Injections of ACh lower blood pressure and heart rate, depending where the injection is applied. This cardiovascular response is elicited from the intermediate one third of NTS, termination site of baroreceptor afferents. A pressor response has been observed after injections outside the NTS. Such findings suggest that activation of different populations of neurons produces differing cardiovascular responses.

Hypotensive response to ACh is blocked by administration of ACh antagonists such as atropine, but not hexamethonium. This observation itself implies that the action of ACh agonists released in NTS results directly from activation of cholinergic muscarinic receptors, not cholinergic nicotinic receptors.

As implied by the conflicting responses to centrally administered ACh mentioned earlier, cholinergic systems elsewhere in the CNS may have opposite effects on the peripheral autonomic nervous system. Sympathetic activity arises following activation of cholinergic mechanisms in posterior hypothalamus, lateral reticular formation, and poorly defined areas

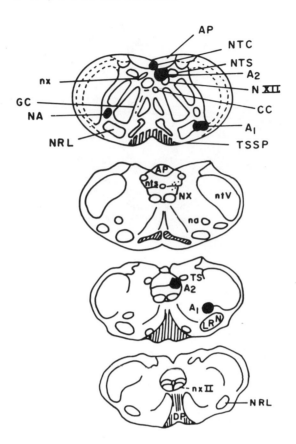

FIGURE 15.   Sections at different levels of the rat brain stem
from 0.6 mm of the spinal cord-medulla border (bottom) until 2
mm of the spinal cord-medulla border (top). AP = area postrema,
NTC = nucleus tractus commisuralis, NTS = nucleus tractus
solitarii, NX = nucleus originis dorsalis vagi, nx II = nucleus
originis nervi hypoglossi, CC = central canal, NA = nucleus
ambiguus, NRL = nucleus reticularis lateralis, DP = decussatio
pyramidis, ntV = nucleus of the spinal tract of the trigeminal
nerve, TS = tractus solitarius, and GC = nucleus reticularis
giganto cellularis.

around the fourth ventricle. On the other hand, sympathetic activity declines upon activation of
cholinergic mechanisms in some other hypothalamic areas and some paramedian reticular
medullary areas.

In order to explain the existence of opposing central ACh mechanisms, we suggest that these
paradoxical effects would depend on the postsynaptic neuron being activated by the cholinergic
neuron. Hence, activation of sympathetic preganglionic neurons, which are cholinergic in
nature, results in peripheral sympathetic hyperactivity, while activation of parasympathetic
preganglionic neurons, also cholinergic like the parasympathetic postganglionic neurons,
results in a diminished peripheral sympathetic activity and in peripheral parasympathetic
predominance.

Cholinergic neurons are widely disseminated in most CNS structures: the cerebral cortex,
hippocampus, striatum, septum, hypothalamus, and other brain areas (see Figures 14 and 15).
All these cholinergic interneurons receive heavy input from ACh system disseminated along the
BSRF. The BSRF projects to cholinergic preganglionic sympathetic and parasympathetic
neurons. In turn, preganglionic neurons make contact with and stimulate postganglionic
sympathetic and parasympathetic neurons, respectively.

One of the most thoroughly studied cholinergic projections in the CNS emanates from the septum to the hippocampal formation. This septal-hippocampal pathway has been related to analepsis. In effect, activation of this pathway increases EEG synchrony and theta activity in the hippocampus.

There is also evidence that BSRF activates cerebral cortex cholinergic interactions. It is probable that ACh is indeed the final mediator at the cerebral cortex level, responsible for behavioral and EEG arousal subsequent to BSRF activation.

Although it is now widely accepted that ACh may be a synaptic transmitter in the cerebral cortex, its precise role has not been clearly defined and pathways which mediate impulses via cholinergic synapses have as yet not been identified. A number of studies suggest that cholinergic cortical endings transmit input from mesencephalic reticular formation to cortex. Systemic application of ACh produces behavioral and EEG arousal similar to that elicited by electrical stimulation of BSRF. In contrast, systemically applied atropine induces slow waves in EEG and increases the threshold for EEG stimulation, probably by acting not only on BSRF but also on the cortex. The hypothesis that ACh participates in cortical arousal is supported by numerous investigations showing that the ACh release rate from the cortical surface increases during spontaneous or induced periods of alertness and EEG desynchronization. Furthermore, histochemical studies have demonstrated that ACh-esterase-containing fiber systems to the cortex probably arise in various subcortical structures which may serve as relayers for the BSRF activating system.

The effect of ACh on single neurons in the cerebral cortex has been studied using multibarreled micropipettes. These studies show two kinds of responses depending on the cortical area investigated. In the visual cortex, for instance, an excitatory response is observed which mimics the excitatory effect of stimulating BSRF. On the other hand, prominent depressant effects of ACh are observed on neurons in the pericruciate cortex and other cortical areas. In theses studies strychnine and muscarinic and nicotinic cholinolytics were found to block depressant effects produced both by ACh and by synaptic stimulation of cortical surface, pyramidal tract, lateral hypothalamus, and BSRF. Hence, ACh is postulated as a transmitter for those inhibitory cortical interneurons involved in responses. Effects of reticular stimulation on single neurons of the pericruciate cortex have been reported variously to be predominantly inhibitory, excitatory or mixed.

Although ACh neurons are spread widely along BSRF, they are also found in some groups of nuclei, i.e., nucleus reticularis lateralis (NRL), nucleus reticularis pontis caudalis (RPC), and nucleus gigantocellularis (NGC). BSRF cells receive diffuse endings from the various NE, 5HT, and DA brain stem nuclei (see Figures 14 and 15). In turn, BSRF cells send axons to those monoaminergic nuclei. The two pontine reticular nuclei, RPC and GC, receive afferent projections only within or caudally to the pons medulla; they receive no afferents from structures rostral to the pons. Such afferent cholinergic projections are compatible with the proposal that both RPC and GC nuclei may serve as connections within the brain stem for an ascending system. In addition, two cholinergic nuclei send efferent projections to brain stem 5HT raphe nuclei, from which they receive afferent projections.

BSRF cells have extensive ascending and descending projections. Golgi studies in rodents indicate that a considerable number of reticular neurons give off an axon which dichotomizes and has a long ascending branch and a long descending branch. The number of reticular cells with ascending projections diminishes towards the caudal area. The relative sparsity of reticular projetions ascending from medulla oblongata to midbrain or beyond in rats is in keeping with data on mammals (see Figures 11, 14, and 15)

REFERENCES

1.   **Dahlström A, Fuxe K.** Evidence for the existence of monoamine containing neurons in the central nervous system. I. Demonstration of monoamines in the cell bodies of brain stem neurons. *Acta Physiol Scand Suppl* 232: 1-55, 1964.

2.   **Konig JF, Klippel RA.** *The Rat Brain. A Stereotaxic Atlas of the Forebrain and Lower Parts of the Brain Stem.* Williams & Wilkins, Baltimore, 1963.

3.   **Levitt P, Moore RY.** Origin and organization of brainstem catecholamine innervation in the rat. *J Comp Neurol* 186: 505-528, 1979.

4.   **Anden NE, Dahlström A, Fuxe K, Larsson K, Olson L, Ungerstedt U.** Ascending monoamine neurons to the telencephalon and diencephalon. *Acta Physiol Scand* 67: 313-326, 1966.

5.   **Berman AL.** *The Brainstem of the cat. Cytoarchitectonic Atlas with Stereotaxic Coordinates.* University of Wisconsin Press, Madison, 1968.

6.   **Lindvall O, Björklund A.** The organization of the ascending catecholamine neuron systems in the rat brain as revealed by the glyoxylic acid fluorescence method. *Acta Physiol Scand Suppl* 412: 1-48, 1974.

7.   **Moore RY, Bloom FE.** Central catecholamine neuron systems: anatomy and physiology of the norepinephrine and epinephrine systems. *Annu Rev Neurosci* 2: 113-168, 1979.

8.   **Olson L, Fuxe K.** Further mapping our of the central noradrenaline neurons systems: projections of the "subcoeruleus" area. *Brain Res* 43: 289-295, 1972.

9.   **Palkovits M, Jacobowitz DM.** Topographical atlas of catecholamine and acetyl-cholinesterase-containing neurons in the rat brain. II. Hindbrain (mesencephalon, rombencephalon). *J Comp Neurol* 157: 29-42, 1974.

10. **Swanson LW, Hartman BK.** The central adrenergic system. An immunofluorescence study of the location of cell bodies and their efferent connections in the rat utilizing dopamine-beta-hydroxylase as a marker. *J Comp Neurol* 163: 467-506, 1975.

11. **Jacobowitz DM, Palkovits M.** Topographic atlas of catecholamine and acetylcholinesterase-containing neurons in the rat brain. I. Forebrain (telencephalon, diencephalon). *J Comp Neurol* 157: 13-28, 1974.

12. **Touret M, Valatx JL, Jouvet M.** The locus coeruleus: a quantitative and genetic study in mice. *Brain Res* 250: 353-357, 1982.

13. **Amaral D.G, Sinnamon HM.** The locus coeruleus: neurobiology of a central noradrenergic nucleus. *Progr Neurobiol* 9: 147-166, 1977.

14. **Swanson LW.** The locus coeruleus: a cytoarchitectonic, Golgi and immunohistochemical study in the albino rat. *Brain Res* 110: 39-56, 1976.

15. **Chu NS, Bloom FE.** The catecholamine-containing neurons in the cat dorsolateral pontine tegmentum: distribution of the cell bodies and some axonal projections. *Brain Res* 66: 1-21, 1974.

16. **Jones BE.** *Catecholamine-Containing Neurons in the Brain Stem of the Cat and Their Role in Waking.* Imprimerie Des Beaux-Arts, Lyon, 1969, 1.

17. **Blessing WW, Frost P, Furness JB.** Catecholamine cell groups of the cat medulla oblongata. *Brain Res* 192: 69-75, 1980.

18. **Blessing WW, Goodchild AK, Dampney RAL, Chalmers JP.** Cell groups in the lower brainstem of the rabbit projecting to the spinal cord, with special reference to catecholamine-containing neurons. *Brain Res* (cited in *Brain Res Rev* 4: 275-325, 1982).

19. **Kobayashi RM, Palkovits M, Kopin IJ, Jacobowitz DM.** Biochemical mapping of noradrenergic nerves arising from the rat locus coeruleus. *Brain Res* 77: 269-279, 1974.

20. **Arbuthnott G, Christie J, Crow T, Eccleston D, Walter D.** The effect of unilateral and bilateral lesions in the locus coeruleus on the levels of 3-methoxy-4-hydroxyphenylglycol in neocortex. *Experientia* 29: 52-53, 1973.

21. **Korf J, Aghajanian G, Roth R.** Stimulation and destruction of the locus coeruleus: opposite effects on 3-methoxy-4-hydroxyphenylglycol sulfate levels in the rat cerebral cortex. *Eur J Pharmacol* 21: 305-310, 1973.

22. **Loizou L.** Projections of the nucleus locus coeruleus in the albino rat. *Brain Res* 15: 563-566, 1969.

23. **Maeda T, Shimizu N.** Projections ascendantes du locus coeruleus et d'autres neurones aminergiques pontiques au niveau du prosencéphale du rat. *Brain Res* 36: 19-35, 1972.

24. **Olson L, Fuxe K.** On the projections from the locus coeruleus noradrenaline neurons: the cerebellar innervation. *Brain Res* 28: 165-171, 1971.

25. **Segal M, Pickel V, Bloom F.** The projections of the nucleus locus coeruleus: an autoradiographic study. *Life Sci* 13: 817-821, 1973.

26. **Ungerstedt U.** Stereotaxic mapping of the monoamine pathways in the rat brain. *Acta Physiol Scand* 367(Suppl):1-48, 1971.

27. **Brownstein M, Saavedra JM, Palkovits M.** Norepinephrine and dopamine in the limbic system of the rat. *Brain Res* 79: 431-436, 1974.

28. **Descarries L, Watkins KC, Lapierre Y.** Noradrenergic axon terminals in cerebral cortex of the rat. III. Topometric ultrastructural analysis. *Brain Res* 133: 197-222, 1977.

29. **Fallon JH, Koziell DA, Moore RY.** Catecholamine innervation of the basal forebrain. II. Amygdala, suprarhinal cortex and entorhinal cortex. *J. Comp. Neurol* 180: 509-532, 1978.

30. **Blackstead TW, Fuxe K, Hökfelt T.** Noradrenaline nerve terminals in the hippocampal region of the rat and guinea pig. *Z Zellforsch* 78: 463-472, 1967.

31. **McBride RL, Sutin J.** Noradrenergic hyperinnervation of the trigeminal sensory nuclei. *Brain Res* 324: 211-221, 1985.

32. **Feuerstein TJ, Hertting G, Jackish R.** Endogenous noradrenaline as modulator of hypocampal serotonin (5-HT)-release. Dual effects of yohimbine, rauwolscine and corynanthine as alpha-adrenoceptor antagonists and 5-HT-receptor agonists. *Naunyn-Schmiedeberg's Arch Pharmacol* 329: 216-221, 1985.

33. **Jonsson G, Fuxe K, Hökfelt T.** On the catecholamine innervation of the hypothalamus, with special reference to the median eminence. *Brain Res* 40: 271-281, 1972.

34. **Moore R.Y, Björklund A, Stenevi U.** Plastic changes in the adrenergic innervation of the rat septal area in response to denervation. *Brain Res* 33: 13-35, 1971.

35. **Pickel VM, Segal M, Bloom FE.** A radioautographic study of the efferent pathways of the nucleus locus coeruleus. *J Comp Neurol* 155: 15-41, 1974.

36. **Sachs C, Jonsson G, Fuxe K.** Mapping of central noradrenaline pathways with 6-hydroxydopamine. *Brain Res* 63: 249-261, 1973.

37. **Tohyama M, Maeda T, Shimizu N.** Detailed noradrenaline pathways of locus coeruleus neuron to the cerebral cortex with use of 6-hydroxydopamine. *Brain Res* 79: 139-144, 1974.

38. **Moore RY.** Catecholamine innervation of the rat forebrain. I. The septal area. *J Comp Neurol* 177: 665-684, 1978.

39. **Palkovits M, Záborszky L, Feminger A, Mezey E, Fekete MIK, Herman JP, Kanyicska B, Szabo D.** Noradrenergic innervation of the rat hypothalamus: experimental biochemical and electron microscopic studies. *Brain Res* 191: 161-171, 1976.

40. **Roizen MF, Kobayashi RM, Muth EA, Jacobowitz DM.** Biochemical mapping of noradrenergic projections of axons in the dorsal noradrenergic bundle. *Brain Res* 104: 384-389, 1976.

41. **Conrad CA, Pfaff DW.** Efferents from medial-basal forebrain and hypothalamus in the rat. II. An autoradiographic study of the anterior hypothalamus. *J Comp Neurol* 169: 221-262, 1976.

42. **O'Donohue TL, Crowley WR, Jacobowitz DM.** Biochemical mapping of the noradrenergic ventral bundle projection sites: evidence for a noradrenergic-dopaminergic interaction. *Brain Res* 172: 87-100, 1979.

43. **Jacobowitz DM.** Fluorescence microscopic mapping of CNS norepinephrine systems in the rat forebrain, in *Anatomical Neuroendocrinology*, Stumpf WE, Grant LD. Eds. S. Karger, Basel, 1975, 368.

44. **Jacobowitz DM.** Monoaminergic pathways in the central nervous system, in *Psychopharmacology: A Generation of Progress*. Lipton MA, DiMascio A, Killam KF. Eds. Raven Press, New York, 1977, 119.

45. **Silver MA, Jacobowitz DM, Crowley WR, O'Donohue TL.** Retrograde transport of dopamine-beta-hydroxylase antibody (ADBH) by CNS noradrenergic neurons: hypothalamic noradrenergic innervations. *Anat Rec* 190: 541, 1978.

46. **Morrison JH, Molliver ME, Grzanna R, Coyle JT.** The intracortical trajectory of the coeruleo-cortical projection in the rat: a tangentially organized cortical afferent. *Neuroscience* 6: 139-158, 1981.

47. **Gatter KC, Powell TPS.** The projection of the locus coeruleus upon the neocortex in the macaque monkey. *Neuroscience* 2: 441-445, 1977.

48. **Freedman R, Foote SL, Bloom FE.** Histochemical characterization of a neocortical projection of the nucleus locus coeruleus in the squirrel monkey. *J Comp Neurol* 164: 209-232, 1975.

49. **Morrison JH, Molliver ME, Grzanna R.** Noradrenergic innervation of cerebral cortex: widespread effects of local cortical lesions. *Science* 205: 313-316, 1979.

50. **Austin JH, Takaori S.** Studies of connections between locus coeruleus and cerebral cortex. *Jpn J Pharmacol* 26: 145-160, 1976.

51. **Ossipov MH, Chatterjee TK, Gebhart GF.** Locus coeruleus lesions in the rat enhance the antinociceptive potency of centrally administered clonidine but not morphine. *Brain Res* 341: 320-330, 1985.

52. **Takagi H, Shiosaka S, Toyhama M, Senba E, Sakanaka M.** Ascending components of the medial forebrain bundle from the lower brain stem in the rat, with special reference to raphe and catecholamine cell groups. A study by the HRP method. *Brain Res* 193: 315-337, 1980.

53. **Speciale SG, Crowley WR, O'Donohue TL, Jacobowitz DM.** Forebrain catecholamine projections of the A5 cell group. *Brain Res* 154: 128-133, 1978.

54. **Crawley JN, Roth RH, Maas JW.** Locus coeruleus stimulation increases noradrenergic metabolite levels in rat spinal cord. *Brain Res* 166: 180-184, 1979.

55. **Commissiong JW, Hellstrom SO, Neff NH.** A new projection from locus coeruleus to the spinal ventral columns: histochemical and biochemical evidence. *Brain Res* 148: 207-213, 1978.

56. **Hancock MBD, Fougerousse CL.** Spinal projections from the nucleus locus coeruleus and nucleus subcoeruleus in the cat and monkey as demonstrated by the retrograde transport of horseradish peroxidase. *Brain Res Bull* 1: 229-234, 1976.

57. **Nygren L, Olson L.** A new major projection from locus coeruleus: the main source of noradrenergic nerve terminals in the ventral and dorsal columns of the spinal cord. *Brain Res* 132: 85-93, 1977.
58. **Fleetwood-Walker SM, Coote JH.** The contribution of brain stem catecholamine cell groups to the innervation of the sympathetic lateral cell column. *Brain Res* 205: 141-155, 1981.
59. **Coote JH, Fleetwood-Walker SM, Martin II.** The origin of the catecholamine innervation of the sympathetic lateral column. *J Physiol* 295: 57-58, 1979.
60. **Coote JH, McLeod VH.** The influence of bulbospinal monoaminergic pathways on sympathetic nerve activity. *J Physiol* 241: 453-475, 1974.
61. **Fleetwood-Walker, SM.** Catecholamine Systems Descending from the Lower Brainstem: Their Contribution to the Innervation of the Sympathetic Lateral Column. Ph.D. thesis, Birmingham University, 1979.
62. **Lowey AD, McKellar S, Saper CB.** Direct projections from the A5 catecholamine cell group to the intermediolateral cell column. *Brain Res* 174: 309-314, 1979.
63. **McKellar S, Lowey AD.** Spinal projections of norepinephrine-containing neurons in the rat. *Neurosci Abstr* 5:344, 1979.
64. **Neumayr RJ, Hare BS, Franz DN.** Evidence for bulbospinal control of sympathetic preganglionic neurons by monoaminergic pathways. *Life Sci* 14: 793-806, 1974.
65. **McNeill TH, Slader JR Jr.** Simultaneous monoamine histofluoroscence and neuropeptide immunocythochemistry. II. Correlative distribution of catecholamine varicosities and magnocellular neurosecretory neurons in the rat supraoptic and paraventricular nuclei. *J Comp Neurol* 193: 1023-1033, 1980.
66. **Klemfuss H, Seiden LS.** Water deprivation increases anterior hypothalamic norepinephrine metabolism in the rat. *Brain Res* 341: 222-226, 1985.
67. **Takigawa M, Mogenson GJ.** A study of inputs to antidromically identified neurons of the locus coeruleus. *Brain Res* 135: 217-230, 1977.
68. **Cedarbaum JM, Aghajanian GK.** Noradrenergic neurons of the locus coeruleus: inhibition by epinephrine and activation by the alpha-antagonist piperoxane. *Brain Res.* 112: 413-419, 1976.
69. **Nakamura S, Iwama K.** Antidromic activation of the rat locus coeruleus neurons from hippocampus, cerebral and cerebellar cortices. *Brain Res* 99: 372-376, 1975.
70. **Egan TM, North RA.** Acetylcholine acts on M_2-muscarinic receptors to excite rat locus coeruleus neurons. *Br J Pharmacol* 85: 733-735, 1985.
71. **Saper CB, Swanson LW, Cowan WM.** The efferent connections of the ventromedial nucleus of the hypothalamus. *J Comp Neurol* 169: 409-442, 1976.
72. **Hosoya Y.** Hypothalamic projections to the ventral medulla oblongata in the rat, with special reference to the nucleus raphe pallidus: a study using autoradiographic and HRP techniques. *Brain Res* 344: 338-350, 1985.
73. **Ciriello J, Caverson MM, Calaresu FR.** Lateral hypothalamic and peripheral cardiovascualr afferent inputs to ventrolateral medullary neurons. *Brain Res* 347: 173-176, 1985.
74. **Dampney RAL, Goodchild AK, Tan E.** Vasopressor neurons in the rostral ventrolateral medulla of the rabbit. *J Auton Nerv Syst* 14: 239-254, 1985.
75. **Swanson LW, Saper CB.** Direct neuronal inputs to locus coeruleus from basal forebrain. *Proc Soc Neurosci* 1: 683, 1975.
76. **Ward DG, Baertshi AJ, Gann, DS.** Activation of solitary nucleus neurons from the locus coeruleus and vicinity. *Proc Soc Neurosci* 1: 424, 1975.
77. **Ross CA, Ruggierod DA, Reis DJ.** Afferent projections to cardiovascular portions of the nucleus of the tractus solitarius in the rat. *Brain Res* 223: 402, 1981.
78. **Zandberg P, DeJong W.** Localization of catecholaminergic receptor sites in the nucleus tractus solitarii involved in the regulation of arterial blood pressure. *Prog Brain Res* 47: 117, 1977.
79. **Calaresu FR, Ciriello J.** Projection to the hypothalamus from buffer nerves and nucleus tractus solitarius in the cat. *Am J Physiol* 239: 126-129, 1980.
80. **Snyder DW, Nathan MA, Reis DL.** Chronic liability of arterial blood pressure produced by selective destruction of the catecholamine innervation of the nucleus solitarii in rats. *Circ Res* 43: 289-295, 1978.
81. **Yamane Y, Nakai M, Yamamoto J, Umeda Y, Ogino K.** Release of vasopressin by electrical stimulation of the intermediate portion of the nucleus of the tractus solitarius in rats with cervical spinal cordotomy and vagotomy. *Brain Res* 324: 358-360, 1984.
82. **Gavras H, Bain GT, Bland L, Vlahakos D, Gavras I.** Hypertensive response to saline microinjection in the area of the nucleus tractus solitarii of the rat. *Brain Res* 343: 113-119, 1985.
83. **Tucker DC, Saper CB.** Specificity of spinal projections from hypothalamic and brainstem areas which innervate sympathetic preganglionic neurons. *Brain Res* 360: 159-164, 1985.
84. **Vlahakos D, Gavras I, Gavras H.** Alpha-adrenoceptor agonists applied in the area of the nucleus tractus solitarii in the rat: effect of anesthesia on cardiovascular responses. *Brain Res* 347: 372-375, 1985.
85. **Calza I, Giardino L, Grimaldi R., Rigolf M, Steinbusch HWM, Tiengo M.** Presence of 5-HT-posititve neurons in the medial nuclei of the solitary tract. *Brain Res* 347: 135-139, 1985.

86. **Leslie RA.** Neuroactive substances in the dorsal vagal complex of the medulla oblongata: nucleus of the tractus solitarius, area postrema, and dorsal motor nucleus of the vagus. *Neurochem Int* 7: 191-211, 1985.

87. **Palkovits M.** Distribution of neuroactive substances in the dorsal vagal complex of the medulla oblongata. *Neurochem Int* 7: 213-219, 1985.

88. **Przuntek H, Philippu A.** Reduced pressor responses to stimulation of the locus coeruleus after lesion of the posterior hypothalamus. *Naunyn-Schmiedeberg's Arch Pharmcol* 276: 119-122, 1973.

89. **Ross RA, Reis DJ.** Effects of lesions of locus coeruleus on regional distribution of dopamine-beta-hydroxylase activity in rat brain. *Brain Res* 73: 161-166, 1974.

90. **Palkovits M, Leranth C, Zaborsky L, Brownstein MJ.** Electron microscopic evidence of direct connections from the lower brain stem to the median eminence. *Brain Res* 136: 339-344, 1977.

91. **Sakumoto T, Tohyama M, Satoh K, Kimoto T, Kinugasa T, Tamizawa O, Kurachi K, Shimizu N.** Afferent fiber connections from lower brain stem to hypothalamus studied by horseradish methods with special reference to noradrenaline innervation. *Exp Brain Res* 31: 81-94, 1978.

92. **Ward DG, Gunn CG.** Locus coeruleus complex: elicitation of a pressor response and a brain stem region necessary for its occurrence. *Brain Res* 107: 401-406, 1976.

93. **Bloch R, Feldman J, Bousquet P, Schwartz J.** Relationship between the ventromedullary clonidine-sensitive area and the posterior hypothalamus. *Eur J Pharmacol* 45: 55, 1977.

94. **Palkovits M, Zaborsky L.** Neuroanatomy of central cardiovascular control. Nucleus tractus solitarii: afferent and efferent neuronal connections in relation to the baroreceptor reflex arc, in *Hypertension and Brain Mechanisms. Progress in Brain Research*, Vol. 47. DeJong W, Provost AP, Shapiro AP. Eds., Elsevier, Amsterdam, 1977, 9.

95. **Sinha JN, Sharma DK, Gurtu S, Pant KK, Bhargava KP.** Nucleus locus coeruleus: evidence for alpha$_1$-adrenoceptor mediated hypotension in the cat. *Naunyn-Schmiedeberg's Arch Pharmacol* 326: 193-197, 1984.

96. **Gunn CG, Sevelius G, Puiggari J, Myers FK.** Vagal cardio-inhibitory mechanisms in the hind brain of the dog and cat. *Am J Physiol* 214: 258-262, 1968.

97. **Gurtu S, Sinha JN, Bhargava KP.** Receptors in the medullary cardio-inhibitory loci. Nucleus tractus solitarius: catecholaminergic modulation of baroreflex induced bradycardia. *Ind J Pharmacol* 14: 37, 1982.

98. **Gurtu S, Sinha JN, Bhargava KP.** Involvement of alpha-adrenoceptors of the nucleus tractus solitarius in baroreflex mediated bradychardia. *Naunyn-Schmiedeberg's Arch Pharmacol* 321: 38-43, 1982.

99. **Gurtu S, Sharma DK, Sinha JN, Bhargava KP.** Evidence of the involvement of alpha$_2$-adrenoceptors in the nucleus ambiguus in baroreflex mediated bradychardia. *Naunyn-Schmiedeberg's Arch Pharmacol* 323: 199-204, 1983.

100. **Gurtu S, Pant KK, Sinha JN, Bhargava KP.** Mechanism of the cardiovascular effects evoked by electrical stimulation of nucleus locus coeruleus and subcoeruleus in the cat. *Brain Res* 301: 59-64, 1984.

101. **Sakai K, Touret M, Salbert D, Leger L, Jouvet M.** Afferent projections to the cat locus coeruleus as visualized by the horseradish peroxidase technique. *Brain Res* 119: 21-41, 1977.

102. **Sharma DK, Gurtu S, Sinha JN, Bhargava KP.** Receptors in the medullary cardio-inhibitory loci. II. Nucleus ambiguus: changes in heart rate and blood pressure following microinjection of adrenergic and cholinergic agents. *Ind J Pharmacol* 14: d38, 1982.

103. **Shimizu N, Ohnishi S, Tohyama M, Maeda T.** Demonstration by degeneration silver method of the ascending projections from the locus coeruleus. *Exp Brain Res* 21: 181-192, 1974.

104. **Sawchenko PE, Swanson LW.** The organization of noradrenergic pathways from the brainstem to the paraventricular and supraoptic nuclei in the rat. *Brain Res Rev* 4: 275-325, 1982.

105. **Cirino M, Renaud P.** Influence of lateral septum and amygdala stimulation on the excitability of hypothalamic supraoptic neurons. An electrophysiological study in the rat. *Brain Res* 326: 357-361, 1985.

106. **Rogers RC, Nelson DO.** Neurons of the vagal division of the solitary nucleus activated by the paraventricular nucleus of the hypothalamus. *J Auton Nerv Syst* 10: 193-197, 1984.

107. **Kalia M, Mesulam MM.** Brainstem projections of sensory and motor components of the vagus complex in the cat. I. Cervical vagus and nodose ganglion. *J Comp Neurol* 193: 435-465, 1980.

108. **Novin D, Sundstein JW, Cross BA.** Some properties of antidromically activated units in the paraventricular nucleus of the hypothalamus. *Exp Neurol* 26:316-329, 1970.

109. **Sawchenko PE.** Central connections of the sensory and motor nuclei of the vagus nerve. *J Auton Nerv Syst* 9: 13-26, 1983.

110. **Beckstead RM, Morse JR, Norgren R.** The nucleus of the solitary tract in the monkey: projections to the thalamus and brainstem nuclei. *J Comp Neurol* 190: 259-282, 1980.

111. **Hosoya Y, Matsushita M.** Brainstem projections from the lateral hypothalamic area in the rat as studied with autoradiography. *Neurosci Lett* 24: 111-116, 1981.

112. **Loewy AD, Burton H.** Nuclei of the solitary tract: efferent projections to the lower brain stem and spinal cord of the cat. *J Comp Neurol* 181: 421-450, 1978.

113. **McKellar S, Lowey AD.** Efferent projections of the nucleus of the A1 catecholamine cell group in the rat: an autoradiographic study. *Brain Res* 241: 11-29, 1982.

114.   **Howe PRC.** Blood pressure control by neurotransmitters in the medulla oblongata and spinal cord. *J Auton Nerv Syst* 12: 95-115, 1985.

115.   **Granata AR, Kumada M, Reis DJ.** Sympathoinhibition by A1-noradrenergic neurons in the C1-area of the rostral medulla. *J Auton Nerv Syst* 14: 387-395, 1985.

116.   **Morest DK.** Experimental study of the projections of the nucleus of the tractus solitarius and the area prostrema in the cat. *J Comp Neurol* 130: 277-300, 1967.

117.   **Morgane PJ.** Historical and modern concepts of hypothalamic organization and function, in *Handbook of the Hypothalamus.* Vol. 1, Morgane PJ, Pannksepp J. Eds. Marcel Dekker, New York, 1979, 1.

118.   **Norgren R.** Projections from the nucleus of the solitary tract in the rat. *Neuroscience* 3: 207-218, 1978.

119.   **Renaud LP, Day TA.** Excitation of supraoptic putative vasopressin neurons following electrical stimulation of the A1 catecholamine cell group region of the rat medulla. *Soc Neurosci Abstr* 8: 422, 1982.

120.   **Ricardo JA, Koh ET.** Anatomical evidence of direct projections from the nucleus of the solitary tract to the hypothalamus, amygdala, and other forebrain structures in the rat. *Brain Res* 153: 1-26, 1978.

121.   **Sawchenko PE, Swanson LW.** Central noradrenergic pathways for the integration of hypothalamic neuroendocrine and autonomic responses. *Science* 214: 685-687, 1981.

122.   **Sawchenko PE, Swanson LW.** Anatomic relationships between vagal preganglionic neurons and aminergic and peptidergic neural systems in the brainstem of the rat. *Soc Neurosci Abstr* 8: 427, 1982.

123.   **Schawber JS, Kapp BS, Higgins GA, Rapp PR.** Amygdaloid and basal forebrain direct connections with the nucleus of the solitary tract and the dorsal motor nucleus. *J Neurosci* 2: 1424-1438, 1982.

124.   **Swanson LE, Sawchenko PE.** Hypothalamic integration: organization of the paraventricular and supraoptic nuclei. *Annu Rev Neurosci* 6: 269-324, 1983.

125.   **Swanson LW, Sawchenko PE, Berod A, Hartman BK, Helle KB, Van Orden DE.** An immunohisto-chemical study of the organization of catecholaminergic cells and terminal fields in the paraventricular and supraoptic nuclei of the hypothalamus. *J Comp Neurol* 196: 271-285, 1981.

126.   **Torvik A.** Afferent connections to the sensory trigeminal nuclei, the nucleus of the solitary tract and adjacent structures. An experimental study in the rat. *J Comp Neurol* 106: 51-141, 1956.

127.   **Neil JJ, Lowey AD.** Decreases in blood pressure in response to L-glutamate microinjections into the A5 catecholaminergic cell group. *Brain Res* 241: 271-278, 1982.

128.   **Loewy AD, Gregorie EM, McKellar S, Baker RP.** Electrophysiological evidence that the A5 catecholam-ine cell group is a vasomotor center. *Brain Res.* 178: 196-200, 1979.

129.   **Moore SD, Guyenet PG.** Effect of blood pressure on A2 noradrenergic neurons. *Brain Res* 338: 169-172, 1985.

130.   **Loewy AD, Neil JJ.** The role of descending monoaminergic systems in the central control of blood pressure. *Fed Proc Fed Am Soc Exp Biol* 40: 2778-2785, 1981.

131.   **Nauta WJH, Haymaker W.** Hypothalamic nuclei and fiber connections, in *The Hypothalamus.* Haymaker W, Anderson E, Nauta WJH. Eds. Charles C Thomas, Springfield, IL, 1969, 136.

132.   **McCall RB, Humphrey SJ.** Evidence for GABA mediation of sympathetic inhibition evoked from midline medullary depressor sites. *Brain Res* 339: 356-360, 1985.

133.   **Palkovits M.** Neuronal pathways and neurotransmitters in septum pellucidum of rat. *Endocrinol Exp* 19: 225-240, 1976.

134.   **Unnerstall JR, Kopajtic TA, Kuhard MJ.** Distribution of alpha$_2$-agonist binding sites in the rat and human central nervous system: analysis of some functional, anatomic correlates of the pharmacological effects of clonidine and related adrenergic agents. *Brain Res Rev* 7: 69-101, 1984.

135.   **Blessing WW, Reis DJ.** Inhibitory cardiovascular function of neurons in the caudal ventrolateral medulla of the rabbit; relationship to the area containing noradrenergic neurons. *Brain Res* 253: 161-171, 1982.

136.   **Blessing WW, West MJ, Chalmers J.** Hypertension, bradychardia and pulmonary edema in the conscious rabbit after brainstem lesions coinciding with the A1 group of catecholamine neurons. *Circ Res* 49: 949-958, 1981.

137.   **Elliot JM, Stead BH, West MJ, Chalmers J.** Cardiovascular effects of intracisternal 6-hydroxydopamine and of subsequent lesions of the ventrolateral medulla coinciding with the A1 group of noradrenaline cells in the rabbit. *J Auton Nerv Syst* 12: 117-130, 1985.

138.   **Coote JH, McLeod VH.** The spinal route of sympatho-inhibitory pathways descending from the medulla oblongata. *Pfleugers Arch Ges Physiol* 359: 335-347, 1975.

139.   **Day TA, Blessing WW, Willoughby JO.** Noradrenergic and dopaminergic projections to the medial preoptic area of the rat: a combined horseradish peroxidase/catecholamine fluorescence study. *Brain Res* 193: 543-548, 1980.

140.   **Moore SD, Guyenet PG.** An electrophysiological study of the forebrain projection of nucleus commis-suralis: preliminary identification of presumed A2 catecholaminergic neurons. *Brain Res* 263: 211-222d, 1983.

141.   **Miller AJ, McKoon M, Pinneau M, Silverstein R.** Postnatal synaptic development of the nucleus tractus solitarius (NTS) of the rat. *Dev Brain Res* 8: 205-213, 1983.

142. **Fleetwood-Walker SM, Coote JH, Gilbey MP.** Identification of spinally projecting neurons in the A1 catecholamine cell group of the ventrolateral medulla. *Brain Res* 273: 25-33, 1983.

143. **Amendt K, Czachurski J, Dembowsky K, Seller H.** Bulbospinal projections to the intermediolateral cell column; a neuroanatomical study. *J Autonom Nerv Syst* 1: 103-117, 1979.

144. **Kaba H, Saito H, Setu K, Kawakami M.** Antidromic identification of neurons in the ventrolateral part of the medulla oblongata with ascending projections to the preoptic and anterior hypothalamic area (POA/ AHA). *Brain Res* 234: 149-154, 1982.

145. **Guyenet PG** Baroreceptor-mediated inhibition of A5 noradrenergic neurons. *Brain Res* 303: 31-40, 1984.

146. **Chalmers JP, Blessing WW, West MJ, Howe PRC, Costa M, Furness JB.** Importance of new catecholamine pathways in control of blood pressure. *Clin Exp Hypertension* 3: 393-416, 1981.

147. **Hukuhara T, Tabeda R.** Neuronal organization of central vasomotor control mechanisms in the brain stem of the cat. *Brain Res* 87: 419-429, 1975.

148. **Day TA, Willoughby JO.** Noradrenergic afferents to median eminence: inhibitory role in rhythmic growth hormone secretion. *Brain Res* 202: 335-345, 1980.

149. **Byrum CHE, Stornetta R, Guyenet PG.** Electrophysiological properties of spinally-projecting A5 noradrenergic neurons. *Brain Res* 303: 15-29, 1984.

150. **Svensson TH, Thóren P.** Brain noradrenergic neurons in the locus coeruleus: inhibition by blood volume load through vagal afferents. *Brain Res* 172: 174-178, 1979.

151. **Maura G, Bonano G, Raiteri M.** Chronic clonidine induces functional down-regulation of presynaptic alpha$_2$-adrenoceptors regulating (3H) noradrenaline and (3H) 5-hydroxytryptamine release in the rat brain. *Eur J Pharmacol* 112: 105-110, 1985.

152. **Thoren P.** Role of cardiac vagal C-fibers in cardiovascular control. *Rev Physiol Biochem Pharmacol* 86: 2-94,1979.

153. **Reiner PB.** Clonidine inhibits central noradrenergic neurons in unanesthetized cats. *Eur J Pharmacol* 115: 249-257, 1985.

154. **Ward DG, Leftcourt AM, Gunn DS.** Responses of neurons in the locus coeruleus to hemodynamic changes. *Fed Proc Fed Am Soc Exp Biol* 37: 743, 1978.

155. **Kizer JS, Muth E, Jacobowitz DM.** The effect of bilateral lesions of the ventral noradrenergic bundle on endocrine-induced changes of tyrosine hydroxylase in the rat median eminence. *Endocrinology* 98: 886-893, 1976.

156. **Kostowski W, Jerlicz M, Bidzinski A, Hauptmann M.** Evidence for the existence of two oppostie noradrenergic brain systems controlling behavior. *Psychopharmacology* 59: 311-312, 1978.

157. **Engberg G, Elam M, Svensson TH.** Effect of adrenaline synthesis inhibition on brain noradrenaline neurons in locus coeruleus. *Brain Res* 223: 49-58, 1981.

158. **Bolme PH, Corrodi H, Fuxe K, Hökfelt T, Lidbrink P, Goldstein M.** Possible involvement of central adrenaline neurons in vasomotor and respiratory control. Studies with clonidined and its interaction with piperoxane and yohimbine. *Eur J Pharmacol* 28: 89-94, 1974.

159. **Fuxe K, Bolme P, Johnsson G, Agnati LF, Goldstein M, Hökfelt T, Schwarcz R, Engel J.** On the cardiovascular role of noradrenaline, adrenaline and peptide containing neuron systems in the brain, in *Nervous System and Hypertension.* Meyer P, Schmitt H. Eds. Flammarion Médecine-Science, Paris, 1979, 1.

160. **Hökfelt T, Fuxe K, Goldstein M, Johansson O.** Evidence for adrenaline neurons in the rat brain. *Acta Physiol Scand* 89: 286-288, 1973.

161. **Hökfelt T, Fuxe K, Goldstein M, Johansson O.** Immunohistochemical evidence for the existence of adrenaline neurons in the rat brain. *Brain Res* 66: 235-251, 1974.

162. **Sauter AM, Lew JY, Baba Y, Goldstein M.** Effect of phenylethanolamine-N-methyltransferase and dopamine-beta-hydroxylase inhibition on epinephrine levels in the brain. *Life Sci* 21: 261-266, 1977

163. **Saavedra JM, Grobeckerd H, Zivin J.** Catecholamines in the raphe nuclei of the rat. *Brain Res* 114: 339-345, 1976.

164. **Blondaux C, Juge A, Sordet F, Chouvet G, Jouvet M, Pujol JF.** Modification du métabolisme de la sérotonine (5-HT) cerebrale induite chez le rat par administration de 6 hydroxydopamine. *Brain Res* 50: 101-114, 1973.

165. **Johnson GA, Kim EG, Boukma SJ.** 5-Hydroxyindole levels in rat brain after inhibition of dopamine-beta-hydroxylase. *J Pharmacol Exp Ther* 180: 539-546, 1972.

166. **Kostowski W, Samanin R, Bareggi SR, Mare V, Garattini S, Valzelli L.** Biochemical aspects of the interaction between midbrain raphe and locus coeruleus in the rat. *Brain Res* 82: 178-182, 1974.

167. **Palkovits M, Brownstein M, Saavedra JM.** Serotonin content of the brain stem nuclei of the rat. *Brain Res* 80: 237-249, 1974.

168. **Gallager DW, Aghajanian GK.** Effect of antipsychotic drugs on the firing dorsal raphe cells. I. Role of adrenergic system. *Eur J Pharmacol* 39: 341-355, 1976.

169. **Couch J.** Responses of neurons in the raphe nuclei to serotonin, norepinephrine and acetylcholine and their correlation with an excitatory synaptic input. *Brain Res* 19: 137-150, 1970.

170. **Footed W, Sheard M, Aghajanian GK.** Comparison of effects of LSD and amphetamine on midbrain raphe units. *Nature* 222: 567, 1969.
171. **Gey K, Pletscher A.** Influence of chlorpromazine and chlorprothixene on the cerebral metabolism of 5-hydroxytryptamine, norepinephrine and dopamine. *J Pharmacol Exp Ther* 133: 18, 1961.
172. **Morgane P, Stern W, Berman E.** Inhibition of unit activity in the anterior raphe by stimulation of the locus coeruleus. *Anat Rec* 178: 42, 1974.
173. **Roizen M, Jacobowitz D.** Studies on the origin of innervation of the noradrenergic area bordering on the nucleus raphe dorsalis. *Brain Res* 101: 561, 1976.
174. **Gallager DW, Aghajanian GK.** Effect of antipsychotic drugs on the firing dorsal raphe cells. II. Reversal by picrotoxin. *Eur J Pharmacol* 39: 357-364, 1976.
175. **Sheard M, Zolovick A, Aghajanian GK.** Raphe neurons: effect of tricyclic antidepressant drugs. *Brain Res* 43: 690, 1972.
176. **Rochette L, Bralet J.** Effect of the norepinephrine receptor stimulating agent "clonidine" on the turnover of 5-hydroxtryptamine in some areas of the rat brain. *J Neural Trans* 37: 259-267, 1975.
177. **Johnson GA, Kim EG.** Increase of brain levels of tryptophan induced by inhibition of dopamine-beta-hydroxylase. *J Neurochem* 20: 1761-1764, 1973.
178. **Maj J, Baran L, Grabowska M, Sowinska H.** Effect of clonidine on the 5-hydroxytryptamine and 5-hydroxyindoleacetic acid brain levels. *Biochem Pharmacol* 22: 2679-2683, 1973.
179. **Baraban JM, Aghajanian GK.** Suppression of firing activity of 5-HT neurons in the dorsal raphe by alpha-adrenoceptor anatagonists. *Neuropharmacology* 19: 355-363, 1980.
180. **Svensson T, Bunney BS, Aghajanian GK.** Inhibition of both noradrenergic and serotonergic neurons in brain by the alpha-agonist cloinidine. *Brain Res* 92: 291-306, 1975.
181. **Reinhard JF Jr, Roth RH.** Noradrenergic modulation of serotonin synthesis and metabolism. I. Inhibition by clonidine in vivo. *J Pharmacol Exp Ther* 221: 541-546, 1981.
182. **Baraban JM, Aghajanian GK.** Noradrenergic innervation of serotonin neurons in the dorsal raphe: demonstration by electron microscopic autoradiography. *Brain Res* 204: 1-11, 1980.
183. **Papeschi R, Theiss P.** The effect of yohimbine on the turnover of brain catecholamines and serotonin. *Eur J Pharmacol* 33: 1-12, 1975.
184. **Reinhard JF Jr, Galloway MP, Roth RH.** Noradrenergic modulation of serotonin synthesis and metabolism. II. Stimulation by 3-isobutyl-1-methylxanthine. *J Pharmacol Exp Ther* 226: 764-766, 1983.
185. **Marwaha J, Aghajanian GK.** Relative potencies of alpha$_1$ and alpha$_2$ antagonists in the locus coeruleus, dorsal raphe and dorsal lateral geniculate nuclei: an electrophysiological study. *J Pharmacol Exp Ther* 222: 287-293, 1982.
186. **Trulson ME, Crisp T.** Role of norepinephrine in regulating the activity of serotonin-containing dorsal raphe neurons. *Life Sci* 35: 511-515, 1984.
187. **Dyr W, Kostowski W, Zacharski B, Bidzinski A.** Differential clonidine effects on EEG following lesions of the dorsal and median raphe nuclei in rats. *Pharmacol Biochem Behav* 19: 177-185, 1983.
188. **Kostowski W, Giacolone E, Garattini S, Valzelli L.** Studies on behavioral and biochemical changes in rats after lesions in midbrain raphe. *Eur J Pharmacol* 4: 371-374, 1968.
189. **Kostowski W.** Brain serotonergic and catecholaminergic systems. Facts and hypothesis, in *Current Developments in Psychopharmacology*, Vol. 1. Essman WB, Valzelli L. Eds., Spectrum, New York, 1975, 39.
190. **Kostowski W.** Noradrenergic interaction among central neurotransmitters, in *Neurotransmitters, Receptors and Drugs Action*. Essman WB. Ed. Spectrum, New York, 1980, 47.
191. **Kostowski W, Plaznik A, Pucilowski O, Bidzinski A, Hauptmann M.** Lesion of serotonergic neurons antagonizes clonidine-induced suppression of avoidance behavior and locomotor activity in rats. *Psychopharmacology* 73: 261-264, 1981.
192. **Leger L, Descarries L.** Serotonin nerve terminals in the locus coeruleus of adult rat: a radioautographic study. *Brain Res* 145: 1-13, 1978.
193. **Bobillier P, Seguin S, Petitjean F, Salvert D, Touret M, Jouvet M.** The raphe nuclei of the cat brain stem: a topographical atlas of their efferent projections as revealed by autoradiography. *Brain Res* 113: 449-486, 1976.
194. **Lewis BD, Renaud B, Buda M, Pujol JF.** Time-course variations in tyrosine hydroxylase activity in the rat locus coeruleus after electrolytic destruction of the nuclei raphe dorsalis or raphe centralis. *Brain Res* 108: 339-349, 1976.
195. **Pickel VM, Joh TH, Reis DJ.** A serotonergic innervation of noradrenergic neurons in nucleus locus coeruleus: demonstration by immunocytochemical localization of the transmitter specific enzymes tyrosine and tryptophan hydroxylase. *Brain Res* 131: 197-214, 1977.
196. **Pujol JF, Keane PE, McRae A, Lewis BD, Renaud B.** Biochemical evidence for serotonin control of the locus coeruleus, in *Interactions Among Putative Neurotransmitters in the Brain*. Garattini S, Pujol JF, Samanin R. Eds. Raven Press, New York, 1977.

197. **Pujol JF, Stein D, Blondaux Ch, Petitjean F, Froment JL, Jouvet M.** Biochemical evidences for interaction phenomena between noradrenergic and serotonergic systems in the cat brain, in *Frontiers in Catecholamine Research*. Usdin E, Snyder S. Eds. Pergamon Press, Elmsford, NY, 1973, 771.

198. **Renaud B, Buda M, Lewis BD, Pujol JF.** Effects of 5,6-hydroxytryptamine on tyrosine-hydroxylase activity in central catecholaminergic neurons of the rat. *Biochem Pharmacol* 24: 1739-1742, 1975.

199. **Saavedra JM.** 5-Hydroxy-L-tryptophan decarboxylase activity: microsassay and distribution in discrete rat brain nuclei. *J Neurochem* 26: 585-589, 1976.

200. **Saavedra JM, Axelrod J.** Effects of 5,7-dihydroxytryptamine on serotonin and tryptophan hydroxylase in discrete regions of the rat brain. *Neurosci Abstr* 1: 396, 1975.

201. **Elam M, Svensson TH, Thoren P.** Differentiated cardiovascular afferent regulation of locus coeruleus neurons and sympathetic nerves. *Brain Res* 358: 77-84, 1985.

202. **Taber Pierce E, Foote WE, Hobson JA.** The efferent connection of nucleus raphe dorsalis. *Brain Res* 107: 137-144, 1976.

203. **Bobillier P, Seguin S, Degueurce A, Lewis BD, Pujol JF.** The efferent connections of the nucleus raphe centralis superior in the rat as revealed by radioautography. *Brain Res* 166: 1-8, 1979.

204. **Conrad LCA, Leonard CM, Pfaff DW.** Connections of the median and dorsal raphe nuclei in the rat: an autoradiographic and degeneration study. *J Comp Neurol* 156: 179-206, 1974.

205. **Dickinson SL, Slatter P.** Effect of lesioning dopamine, noradrenaline and 5-hydroxytryptamine pathways on tremorine-induced tremor and rigidity. *Neuropharmacology* 21: 787-794, 1982.

206. **Gumulka W, Samanin R, Valzelli L, Consolo S.** Behavioral and biochemical effects following the stimulation of the nucleus raphe dorsalis in rats. *J Neurochem* 18: 533-535, 1971.

207. **Lorens SA, Guldberg HC.** Regional 5-hydroxytryptamine following selective midbrain raphe lesions in the rat. *Brain Res* 78: 45-56, 1974.

208. **Samanin R, Garattini S.** The serotonergic system in the brain and its possible functional connections with other aminergic systems. *Life Sci* 17: 1201-1207, 1975.

209. **Pazos A, Palacios JM.** Quantitative autoradiographic mapping of serotonin receptors in the rat brain. I. Serotonin-1 receptors. *Brain Res* 346: 205-230, 1985.

210. **Korsgaard S, Gerlach J, Christensson E.** Behavioral aspects of serotonin-dopamine interaction in the monkey. *Eur J Pharmacol* 118: 245-252, 1985.

211. **Nishikawa T, Scatton B.** Inhibitory influence of GABA on central serotonergic transmission. Involvement of the habenulo-raphe pathways in the GABAergic inhibition of ascending cerebral serotonergic neurons. *Brain Res* 331: 81-90, 1985.

212. **Rudorfer MW, Scheinin M, Karoum F, Ross RJ, Potter WZ, Linnoila M.** Reduction of norepinephrine turnover by serotonergic drug in man. *Biol Psychiatry* 19: 179, 1984.

213. **Laguzzi R, Talman WT, Reis DJ.** Serotonergic mechanisms in the nucleus tractus solitarius may regulate blood pressure and behavior in the rat. *Clin Sci* 63: 323-326, 1982.

214. **Baraban JM, Wang RY, Aghajanian GK.** Reserpine suppression of dorsal raphe neuronal firing: mediation by adrenergic system. *Eur J Pharmacol* 52: 27-36, 1978.

215. **Aghajanian GK, Wang RY.** Physiology and pharmacology of central serotonergic neurons, in *Psychopharmacology: A Generation of Progress*. Lipton MA, DiMascio A, Killam KF. Eds. Raven Press, New York, 1978, 171.

216. **Bunney BS, Walters J, Kuhar MJ, Roth RH, Aghajanian GK.** D and L amphetamine stereoisomers: comparative potencies in affecting the firing of central dopaminergic and noradrenergic neurons. *Psychopharmacol Commun* 1: 177, 1975.

217. **Echizen H, Freed CR.** Altered serotonin and norepinephrine metabolism in rat dorsal raphe nucleus after drug-induced hypertension. *Life Sci* 34: 1581-1589, 1984.

218. **Heym J, Trulson ME, Jacobs BL.** Effects of adrenergic drugs on raphe unit activity in freely moving cats. *Eur J Pharmacol* 74: 117-125, 1981.

219. **Trulson ME, Jacobs BL.** Raphe unit activity in freely moving cats: correlation with level of behavioral arousal. *Brain Res* 163: 135, 1979.

220. **Vandermaelen CP, Aghajanian GK.** Electrophysiological and pharmacological characterization of serotonergic dorsal raphe neurons recorded extracellularly and intracellulary in rat brain slices. *Brain Res* 289: 109-119, 1983.

221. **Steinbusch HWM.** Distribution of serotonin-immunoreactivity in the central nervous system — cell bodies and terminals. *Neuroscience* 6: 557-618, 1981.

222. **Vandermaelen CP, Aghajanian GK.** Noradrenergic activation of serotonergic dorsal raphe neurons recorded in vitro. *Soc Neurosci Abstr* 8: 482, 1982.

223. **Anden NE, Grabowska M.** Pharmacological evidence for a stimulation of dopamine neurons by noradrenaline neurons in the brain. *Eur J Pharmacol* 39: 275-282, 1976.

224. **Anden NE, Strömbom U.** Adrenergic receptor blocking agents: effects on central noradrenaline and dopamine receptors and on motor activity. *Psychopharmacologia* 38: 91, 1974.

225. **Persson T, Waldeck B.** Further studies on the possible interaction between dopamine and noradrenaline containing neurons in the brain. *Eur J Pharmacol* 11: 315, 1970.
226. **Pycock CJ, Donaldson LM, Marsden CD.** Circling behavior produced by unilateral lesions in the region of the locus coeruleus in rats. *Brain Res* 97: 317-329, 1975.
227. **Rochette L, Bralet J.** Effect of clonidine on the synthesis of cerebral dopamine. *Biochem Pharmacol* 24: 303, 1975.
228. **Eison MS, Stark AD, Ellison G.** Opposed effects of locus coeruleus and substantia nigra lesions on social behavior in rat colonies. *Pharmacol Biochem Behav* 7: 87-90, 1977.
229. **Antelman SM, Caggiula AR.** Norepinephrine-dopamine interactions and behavior. *Science* 181: 682-684, 1973.
230. **Lavielle S, Tassin JP, Thierry AM, Blanc G, Herve D, Barthelemy C, Glowinski J.** Blockade by benzodiazepines on the selective high increase in dopamine turnover induced by stress in mesocortical dopaminergic neurons of the rat. *Brain Res* 168: 585-594, 1978.
231. **Berger B, Tassin JP, Blanc G, Moyne MA, Thierry AM.** Histochemical confirmation for dopaminergic innervation of the rat cerebral cortex after destruction of the noradrenergic ascending pathways. *Brain Res* 81: 332-337, 1974.
232. **Mercuri N, Calabresi P, Stanzione P, Bernardi G.** Electrical stimulation of mesencephalic cell groups (A9-A10) produces monosynaptic excitatory potentials in rat frontal cortex. *Brain Res* 338: 192-195, 1985.
233. **Berger B, Thierry AM, Tassin JP, Moyne MA.** Dopaminergic innervation of the rat prefrontal cortex: a fluorescence histochemical study. *Brain Res* 106: 133-145, 1976.
234. **Bunney BS, Aghajanian GK.** Dopamine and norepinephrine innervated cells in the rat prefrontal cortex. Pharmacological differentiation using micro-iontophoretic techniques. *Life Sci* 19: 1783-1792, 1976.
235. **Thierry AM, Javoy F, Glowinski J, Kety SS.** Effects of stress on the metabolism of norepinephrine, dopamine and serotonin in the central nervous system of the rat. I. Modifications of norepinephrine turnover. *J Pharmacol Exp Ther* 163: 163-171, 1968.
236. **Carter CJ, Pycock CJ.** Behavioral and biochemical effects of dopamine and noradrenaline depletion within the medial prefrontal cortex of the rat. *Brain Res* 192: 163-176, 1980.
237. **Arnold GB, Molinoff PB, Rutledge CO.** The release of endogenous norepinephrine and dopamine from cerebral cortex by amphetamine. *J Pharmacol Exp Ther* 202: 544-557, 1977.
238. **Creese L, Iversen SD.** The pharmacological and anatomical substrates of the amphetamine response in the cat. *Brain Res* 83: 419-436, 1975.
239. **Randrup A, Munkvad I.** Role of catecholamines in the amphetamine excitatory response. *Nature* 211: 540, 1960.
240. **Scheel-Kruger J, Randrup A.** Stereotyped hyperactive behavior produced by dopamine in the absence of noradrenaline. *Life Sci* 6: 1389-1398, 1967.
241. **Haskins JT, Moyer JA, Muth EA, Sigg EB.** DMI, WY-45,030, WY-45,881 and ciramadol inhibit locus coeruleus neuronal activity. *Eur J Pharmacol* 115: 139-146, 1985.
242. **Saavedra JM, Zivin J.** Tyrosine hydroxylase and dopamine-beta-hydroxylase distribution in discrete areas of the rat limbic system. *Brain Res* 105: 517-524, 1976.
243. **Weinstock M, Zavadh AP, Muth EA, Crowley WR, O'Donohue TL, Jacobowitz DM, Kopin IJ.** Evidence that noradrenaline modulates the increase in striatal dopamine metabolism induced by muscarinic receptor stimulation. *Eur J Pharmacol* 68: 427-435, 1980.
244. **Cedarbaum JM, Aghajanian GK.** Catecholamine receptors on locus coeruleus neurons: pharmacological characterization. *Eur J Pharmacol* 44: 375, 1977.
245. **Maeda H, Mogenson GJ.** Electrophysiological responses of neurons of the ventral tegmental area to electrical stimulation of amygdala and lateral septum. *Neuroscience* 6: 367-376, 1981.
246. **German DC, Dalsass M, Kiser RS.** Electrophysiological examination of the ventral tegmental (A10) area in the rat. *Brain Res* 181: 191-197, 1980.
247. **Guyenet PG, Aghajanian GK.** Antidromic identification of dopaminergic and other output neurons of the substantia nigra. *Brain Res* 150: 69-87, 1978.
248. **Phillipson OT.** Afferent projections to the ventral tegmental area of Tsai and interfasicular nucleus: a horseradish peroxidase study in the rat. *J Comp Neurol* 187: 117-144, 1979.
249. **Simon H, LeMoal M, Calas A.** Efferents and afferents of the ventral tegmental-A10 region studies after local injection of (3H) leucine and horseradish peroxidase. *Brain Res* 178: 17-40, 1979.
250. **Yim CY, Mogenson GJ.** Electrophysiological studies of neurons in the ventral tegmental areas of Tsai. *Brain Res* 181: 303-313, 1980.
251. **Herve D, Blanc G, Glowinski J, Tassin JP.** Reduction of dopamine utilization in the prefrontal cortex but not in the nucleus accumbens after selective destruction of noradrenergic fibers innervating the ventral tegmental area in the rat. *Brain Res* 237: 510-516, 1982.
252. **Tassin JP, Lavielle S, Hervé D, Blanc G, Thierry AM, Alvarez C, Berger B, Glowinski J.** Collateral sprouting and reduced activity of the rat mesocortical dopaminergic neurons after selective destruction of the ascending noradrenergic bundles. *Neuroscience* 4: 1569-1582, 1979.

253. **Donaldson IMacG, Dolphin A, Jenner P, Marsden CD, Pycock C.** The roles of noradrenaline and dopamine in contraversive circling behavior seen after unilateral electrolytic lesions of the locus coeruleus. *Eur J Pharmacol* 39: 179-191, 1976.

254. **Pycock CJ.** Noradrenergic involvement in dopamine-dependent stereotyped and cataleptic responses in the rat. *Arch Pharmacol Weinheim* 298: 15-22, 1977.

255. **Worth WS, Collins J, Kett D, Austin JH.** Serial changes in norepinephrine and dopamine in rat brain after locus coeruleus lesions. *Brain Res* 106: 198-203, 1976.

256. **Plaznik A, Pucilowski O, Kostowski W, Bidzinski A, Hauptmann M.** Rotational behavior produced by unilateral ventral noradrenergic bundle lesions: evidence for a noradrenergic-dopaminergic interaction in the brain. *Pharmacol Biochem Behav* 17: 619-622, 1982.

257. **Jerlicz M, Kostowski W, Bidzinski A, Hauptmann M.** Effects of lesions in the ventral noradrenergic bundle on behavior and response to psychotropic drugs in rats. *Pharmacol Biochem Behav* 9: 721-724, 1978.

258. **Kostowski W, Jerlicz M, Bidzinski A, Hauptmann M.** Behavioral effects of neuroleptics, apomorphine and amphetamine after bilateral lesion to the locus coeruleus in rats. *Pharmacol Biochem Behav* 7: 289-293, 1977.

259. **Kostowski W, Jerlicz M.** Effects of lesions of the locus coeruleus and the ventral noradrenergic bundle on the antinociceptive action of clonidine in rats. *Pol J Pharmacol Pharm* 30: 647-651, 1978.

260. **Kostowski W, Plaznik A.** Effects of lesions of the ventral noradrenergic bundle on the two-way avoidance behavior in rats. *Acta Physiol Pol* 29: 509-514, 1978.

261. **Kostowski W.** Two noradrenergic systems in the brain and their interaction with other monoaminergic neurons. *Pol J Pharmacol Pharm* 31: 425-436, 1979.

262. **Donaldson IMacG, Dolphin A, Jenner P, Marsden CD, Pycock CJ.** Contraversive circling behavior produced by unilateral electrolytic lesions of the ventral noradrenergic bundle mimicking the changes seen with unilateral electrolytic lesions of the locus coeruleus. *J Pharm Pharmacol* 28: 329-330, 1976.

263. **Donaldson IMacG, Dolphin A, Jenner P, Marsden CD, Pycock CJ.** The involvement of noradrenaline in motor activity as shown by rotational behavior after unilateral lesion of the locus coeruleus. *Brain* 99: 427-446, 1976.

264. **Donaldson IMacG, Dolphin A, Jenner P, Pycock CJ, Marsden CD.** Rotational behavior produced by unilateral electrolytic lesions of the ascending noradrenergic bundles. *Brain Res* 138: 487-509, 1978.

265. **Harik SI.** Locus coeruleus lesion by local 6-hydroxydopamine infusion causes marked and specific destruction of noradrenergic neurons, long-term depletion of norepinephrine and the enzymes that synthetize it, and enhanced dopaminergic mechanisms in the ipsilateral cerebral cortex. *J Neurosci* 4: 699-707, 1984.

266. **Millan MH, Millan MJ.** Pimozide blocks the open-field hyperactivity produced by lesions of the ventral noradrenergic bundle in rats. *Pharmacol Biochem Behav* 20: 473-477, 1984.

267. **Hansen S, Stanfield EJ, Everitt BJ.** The effects of lesions of lateral tegmental noradrenergic neurons on components of sexual behavior and pseudopregnancy in female rats. *Neuroscience* 6: 1105-1112, 1981.

268. **Slopsema JS, Van der Gugten J, DeBruin JPC.** Regional concentrations of noradrenaline and dopamine in the frontal cortex of the rat: dopaminergic innervation of the prefrontal subareas and lateralization of prefrontal dopamine. *Brain Res* 250: 197-200, 1982.

269. **Palkovits M, Zaborszky L, Brownstein MJ, Fekete MIK, Herman JP, Kanyicska B.** Distribution of norepinephrine and dopamine in cerebral cortical areas of the rat. *Brain Res Bull* 4: 593-601, 1979.

270. **Geyer MA, Dawsey WJ, Mandell AJ.** Differential effects of caffeine, D-amphetamine and methylphenidate on individual raphe cell fluorescence: a microspectrofluorimetric demonstration. *Brain Res* 85: 135-139, 1975.

271. **Haigler HJ, Aghajanian GK.** LSD and serotonin: a comparison of effects on serotonergic neurons and neurons receiving a serotonergic input. *J Pharmacol Exp Ther* 188: 688-699, 1974.

272. **Aghajanian GK, Foote WE, Sheard MH.** Action of psychotogenic drugs on single midbrain raphe neurons. *J Pharmacol Exp Ther* 171: 178-187, 1970.

273. **Jacobs BL, Trimbach Ch, Eubanks EE, Trulson M.** Hippocampal mediation of raphe lesion- and PCPA-induced hyperactivity in the rat. *Brain Res* 94: 253-261, 1975.

274. **Jacobs BL, Wise WD, Taylor KM.** Differential behavior and neurochemical effects following lesions of the dorsal or median raphe nuclei in rats. *Brain Res* 79: 353-361, 1974.

275. **Geyer MA, Puerto A, Dawsey WJ, Knapp S, Bullard WP, Mandell AJ.** Histologic and enzymatic studies of the mesolimbic and mesostriatal serotonergic pathways. *Brain Res* 106: 241-256, 1976.

276. **Bobillier P, Petitjean F, Salvert D, Ligher M, Seguin S.** Differential projections of the nucleus raphe dorsalis and nucleus raphe centralis as revealed by autoradiography. *Brain Res* 85: 205-210, 1975.

277. **Brodal A, Taber E, Walberg F.** The raphe nuclei of the brain stem in the cat. II. Efferent connections. *J Comp Neurol* 114: 239-260, 1960.

278. **Fuxe K.** Further mapping of central 5-HT neurons: studies with the neurotoxic dihydroxytryptamines, in *Serotonin — New Vistas*, Vol. 2. Costa E, Gessa GL, Sandler M. Eds., Raven Press, New York, 1974, 1.

279. **Kuhar MJ, Aghajanian GK, Roth RH.** Tryptophan hydroxylase activity and synaptosomal uptake of serotonin in discrete brain regions after midbrain raphe lesions: correlations with serotonin levels and histochemical fluorescence. *Brain Res* 44: 165-176, 1972.

280. **Galindo-Mireles D, Meyer G, Castañeyra-Perdomo A, Ferres-Torres R.** Cortical projections of the nucleus centralis superior and the adjacent reticular tegmentum in the mouse. *Brain Res* 330: 343-348, 1985.

281. **Kuhar MJ, Roth RH, Aghajanian GK.** Selective reduction of tryptophan hydroxylase activity in rat forebrain after midbrain raphe lesions. *Brain Res* 35: 167-176, 1971.

282. **Geyer MA, Puerto A, Menkes DB, Segal DS, Mandell AJ.** Behavioral studies following lesions of the mesolimbic and mesostriatal serotonergic pathways. *Brain Res* 106: 257-270, 1976.

283. **Lorens SA, Sorensen JP, Yunger LM.** Behavioral and neurochemical effects of lesions in the raphe system of the rat. *J Comp Physiol Psychol* 77: 48-52, 1971.

284. **Neill DB, Grant LD, Grossman SP.** Selective potentiation of locomotor effects of amphetamine by midbrain raphe lesions. *Physiol Behav* 9: 655-657, 1972.

285. **Srebro B, Lorens SA.** Behavioral effects of selective midbrain raphe lesions in the rat. *Brain Res* 89: 303-325, 1975.

286. **Palkovits M, Brownstein M, Kizer JS, Saavedra JM, Kopin IJ.** Effect of stress on serotonin and tryptophan hydroxylase activity of brain nuclei, in *Catecholamines and Stress*. Usdin E, Kvetnansky R, Kopin IJ. Eds. Pergamon Press, New York, 1976, 51.

287. **Palkovits M, Brownstein M, Saavedra JM.** Serotonin content of the brain stem nuclei in the rat. *Brain Res* 80: 237, 1974.

288. **Saavedra JM, Palkovits M, Brownstein M, Axelrod J.** Serotonin distribution in the nuclei of the rat hypothalamus and preoptic region. *Brain Res* 77: 157-165, 1974.

289. **Costall B, Naylor RJ, Marsden CD, Pycock CJ.** Serotoninergic modulation of the dopamine response from the nucleus accumbens. *J Pharm Pharmacol* 28: 523-526, 1976.

290. **Wang RY, Aghajanian GK.** Inhibition of neurons in the amygdala by dorsal raphe stimulation: mediation through a direct serotonergic pathway. *Brain Res* 120: 85-102, 1977.

291. **Bjorklund A, Falck B, Stenevi U.** Classification of monoamine neurons in the rat mesencephalon: distribution of a new nomoamine neuron system. *Brain Res* 32: 269-285, 1971.

292. **Guilbaud G, Besson JM, Oliveras JL, Liebeskind LC.** Suppression by LSD of the inhibitory effect exerted by dorsal raphe stimulation on certain spinal cord interneurons in the cat. *Brain Res* 61: 417-422, 1973.

293. **Miller JJ, Richardson TL, Fibiger HC, McLennan H.** Anatomical and electrophysiological identification of a projection from the mesencephalic raphe to caudate-putamen in the rat. *Brain Res* 97: 133-138, 1975.

294. **Nakamura S.** Two types of inhibitory effects upon brain stem reticular neurons by low frequency stimulation of raphe nucleus in the rat. *Brain Res* 93: 140-144, 1975.

295. **Segal M.** Physiological and pharmacological evidence for a serotonergic projection to the hippocampus. *Brain Res* 94: 115-131, 1975.

296. **Key BJ, Krzywoskinski L.** Electrocortical changes induced by the perfusion of noradrenaline, acetylcholine and their antagonists directly into the dorsal raphe nucleus of the cat. *Br J Pharmacol* 61: 297-305, 1977.

297. **Garcia Ramos J.** Cortical effects of the electrical stimulation of the n. raphe dorsalis in the cat. *Acta Physiol Latinoam* 28: 83-95, 1978.

298. **Anderson CD, Pasquier DA, Forbes WB, Morgane PJ.** Locus coeruleus to dorsal raphe connections: electrophysiological and morphological studies. *Soc Neurosci Abstr* 2: 477, 1977.

299. **Morgane PJ, Forbes WB, Pasquier DA.** Retrograde transport studies of relations between raphe nuclei and locus coeruleus. *Neurosci Abstr*, Abstr. No. 1501, 1977.

300. **Mouren-Mathieu AM, Leger L, Descarries L.** Radioautographic visualization of central monoamine neurons after local instillation of tritiated serotonin and norepinephrine in adult cat. *Neurosci Abstr* 2(Abstr. No. 714), 1976.

301. **Palkovits M, Saavedra JM, Jacobowitz DM, Kizer JS, Zaborszky L, Brownstein MJ.** Serotonergic innervation of the forebrain: effect of lesions on serotonin and tryptophan hydroxylase levels. *Brain Res* 130: 121-134, 1977.

302. **Aghajanian GK, Bloom FE, Sheard MH.** Electron microscopy of degeneration within the serotonin pathway of rat brain. *Brain Res* 13: 266-273, 1969.

303. **Anden NE, Dahlström A, Fuxe K, Larsson K.** Mapping out of catecholamine and 5-hydroxytryptamine neurons innervating the telencephalon and diencephalon. *Life Sci* 4: 1275-1279, 1965.

304. **Anden NE, Fuxe K, Ungerstedt U.** Monoamine pathways to the cerebellar and cerebral cortex. *Experientia* 23: 838-839, 1967.

305. **Roizen MF, Jacobowitz DM.** Studies on the origin of innervation of the noradrenergic areas bordering on the nucleus raphe dorsalis. *Brain Res* 101: 561-568, 1976.

306. **Mosko SS, Haubrich D, Jacobs BL.** Serotonergic afferents to the dorsal raphe nucleus: evidence from HRP and synaptosomal uptake studies. *Brain Res* 119: 269-290, 1977.

307. **Aghajanian GK.** Chemical feedback regulation of serotonin-containing neurons in brain. *Ann NY Acad Sci* 193: 86-94, 1972.

308. **Aghajanian GK, Bloom FE.** Localization of tritiated serotonin in rat brain by electron microscopic autoradiography. *J Pharmacol Exp Ther* 156: 23-30, 1967.

309. **Aghajanian GK, Haigler HJ.** Direct and indirect actions of LSD, serotonin and related compounds on serotonin-containing neurons, in *Serotonin and Behavior.* Barchas J, Usdin F. Eds., Academic Press, New York, 1973, 263.

310. **Baumgarten HG, Bjorklund A, Lachenmayer L, Nobin A.** Evaluation of the effects of 5,7-dihydroxytryptamine on serotonin and catecholamine neurons in the rat CNS. *Acta Physiol Scand Suppl* 391: 1-22, 1973.

311. **Lorez HP, Richards JG.** Distribution of indolealkylamine nerve terminals in the ventricles of the rat brain. *Z Zellforsch* 144: 511-522, 1973.

312. **Olson L, Boreus LO, Seiger A.** Histochemical demonstration and mapping of 5-hydroxytryptamine and catecholamine-containing neuron systems in the human fetal brain. *Z Anat Entwickl Gesch* 139: d259-282, 1973.

313. **Mosko SS, Jacobs BL.** Electrophysiological evidence against negative neuronal feedback from the forebrain controlling midbrain raphe unit activity. *Brain Res* 119: 291-303, 1977.

314. **Reubi JC, Emson PC.** Release and distribution of endogenous 5-HT in rat substantia nigra. *Brain Res* 139: 164-168, 1978.

315. **Dray A, Gonye TJ, Oakley NR, Tanner T.** Evidence for the existence of a raphe projection to the substantia nigra in rat. *Brain Res* 113: 45-57, 1976.

316. **Fibiger HC, Miller JJ.** Raphe projections to the substantia nigra: a possible mechanism for integration between dopaminergic and serotonergic systems. *Proc Am Neurosci Meet* Abstr. 693, 1977.

317. **Pickel VM, Joh TH, Reis DJ.** Immunocytochemical demonstration of a serotonergic innervation of catecholamine neurons in locus coeruleus and substantia nigra. *Proc Am Neurosci Meet* Abstr. 496, 1975.

318. **Nojyo Y, Sano Y.** Ultrastructure of the serotonergic nerve terminals in the suprachiasmatic and interpeduncular nuclei of rat brains. *Brain Res* 149: 482-488, 1978.

319. **Lorez HP, Richards JG.** 5-HT nerve terminals in the fourth ventricle of the rat brain: their identification and distribution studied by fluorescence histochemistry and electron microscopy. *Cell Tissue Res* 165: 37-48, 1975.

320. **Giambalvo CT, Snodgrass SR.** Biochemical and behavioral effects of serotonin neurotoxins on the nigrostriatal dopamine system: comparison of injection sites. *Brain Res* 152: 555-556, 1978.

321. **Dray A, Davies J, Oakely NR, Tongroach P, Velluci S.** The dorsal and medial raphe projections to the substantia nigra in the rat: electrophysiological, biochemical and behavioural observations. *Brain Res* 151: 431-442, 1978.

322. **Aghajanian GK, Wang RY.** Habenular and other midbrain raphe afferents demonstrated by a modified retrograde tracing technique. *Brain Res* 122: 229-242, 1977.

323. **Bunney BS, Aghajanian GK.** The precise localization of nigral afferents in the rat as determined by a retrograde tracing technique. *Brain Res* 117: 423-435, 1976.

324. **Costall B, Naylor RJ.** Stereotyped and circling behaviour induced by dopaminergic agonists after lesions of the midbrain raphe nuclei. *Eur J Pharmacol* 29: 206-222, 1974.

325. **Dray A, Oakley NR.** Methiothepin and a 5-HT pathway to rat substantia nigra. *Experientia* 33: 1198-1199, 1977.

326. **Fahn S, Libsch LR, Cutler RW.** Monoamines in the human neostriatum topographic distribution in normals and in Parkinson's disease and their role in akinesia, rigidity, chorea and tremor. *J Neurol Sci* 14: 427-455, 1971.

327. **Grabowska M.** Influence of midbrain raphe lesions on some pharmacological and biochemical effects of apomorphine in rats. *Psychopharmacologia* 39: 315-322, 1974.

328. **Moore RY, Halaris AE.** Hippocampal innervation by serotonin neurons of the midbrain raphe in the rat. *J Comp Neurol* 164: 171-184, 1975.

329. **Pasquier DA, Anderson C, Forbes WB, Morgane PJ.** Horseradish peroxidase tracing of the lateral habenular-midbrain raphe nuclei connections in the rat. *Brain Res Bull* 1: 443-451, 1976.

330. **Pasquier DA, Reinoso-Suárez F.** Differential efferent connections of the brain stem to the hippocampus in the cat. *Brain Res* 120: 540-548, 1977

331. **Pasquier DA, Reinoso-Suárez F, Morganed PJ.** Effect of raphe lesions on brain serotonin in the cat. *Brain Res Bull* 1: 279-283, 1976.

332. **Pierce ET, Foote WE, Hobson JA.** The efferent connection of the nucleus raphe dorsalis. *Brain Res* 107: 137-144, 1976.

333. **Gallager DW, Pert A.** Afferents to brain stem nuclei (brain stem raphe, nucleus reticularis pontis caudalis and nucleus giganto-cellularis) in the rat as demonstrated by microinotophoretically applied horseradish peroxidase. *Brain Res* 144: 257-275, 1978.

334. **Casey KL.** Somatic stimuli, spinal pathways, and size of cutaneous fibers influencing unit activity in the medial medullary reticular formation. *Exp Neurol* 25: 35-36, 1969.

335. **Hamilton BL, Skultety FM.** Efferent connections of the periaqueductal gray matter in the cat. *J Comp Neurol* 139: 105-114, 1958.

336. **Jacquet YF, Lajtha A.** The periaqueductal gray: site of morphine analgesia and tolerance as shown by two-way cross tolerance between systemic and intracerebral injections. *Brain Res* 103: 501-513, 1976.

337. **Oliveras JL, Redjemi F, Guilbaud G, Besson JM.** Analgesia induced by electrical stimulation of the inferior centralis nucleus of the raphe in the cat. *Pain* 1: 139-145, 1975.

338. **Pert A, Yaksh T.** Sites of morphine induced analgesia in the primate brain: relation to pain pathways. *Brain Res* 80: 135-140, 1974.

339. **Proudfit HK, Anderson EG.** Morphine analgesia: blockade by raphe magnus lesions. *Brain Res* 98: 612-618, 1975.

340. **James TA, Starr MS.** Rotational behaviour elicited by 5-HT in the rat: evidence for an inhibitory role of 5-HT in the substantia nigra and corpus striatum. *J Pharm Pharmacol* 32: 196-200, 1980.

341. **Advis JP, Simpkins JW, Bennet J, Meites J.** Serotonergic control of prolactin release in male rats. *Life Sci* 24: 359-366, 1979.

342. **McCall RB, Aghajanian GK.** Serotonergic facilitation of facial motoneuron excitation. *Brain Res* 169: 11-27, 1979.

343. **Waldbillig RJ.** The role of the dorsal and median raphe in the inhibition of muricide. *Brain Res* 160: 341-346, 1979.

344. **Van de Kar LD, Lorens SA.** Differential serotonergic innervation of individual hypothalamic nuclei and other forebrain regions by the dorsal and median midbrain raphe nuclei. *Brain Res* 162: 45-54, 1979.

345. **Azmitia EC, Segal M.** An autoradiographic analysis of the differential ascending projections of the dorsal and median raphe nuclei in the rat. *J Comp Neurol* 179: 641-668, 1978.

346. **Beaudet A, Descarries L.** A serotonin–containing nerve cell group in rat hypothalamus. *Neurosci Abstr* 1: 678, 1976.

347. **Bloom FE, Hoffer PJ, Siggins GR, Barker JR, Nicoll RA.** Effects of serotonin on central neurons: microiontophoretic administration. *Fed Proc Fed Am Soc Exp Biol* 31: 97-106, 1972.

348. **Kellar KJ, Brown PA, Madrid J, Bernstein M, Verniko-Dannelis J, Mehler WR.** Origins of serotonin innervation of forebrain structures. *Exp Neurol* 56: 52-62, 1977.

349. **Jacobs BL, Asher R, Dement WC.** Electrophysiological and behavioral effects of electrical stimulation of the raphe nuclei in cats. *Physiol Behav* 11: 489-495, 1973.

350. **Chronister RB, DeFrancæ JF.** Organization of projection neurons of the hippocampus. *Exp Neurol* 66: 509-523, 1979.

351. **Geyer MA.** Both indoleamine and phenylethylamine hallucinogens increase serotonin in both dorsal and median raphe neurons. *Life Sci* 26: 431-434, 1980.

352. **Van der Kooy D, Hattori T.** Dorsal raphe cells with collateral projections to the caudate-putamen and substantia nigra: a fluorescent retrograde double labelling study in the rat. *Brain Res* 186: 1-7, 1980.

353. **Azmitia EC.** The serotonin-producing neurons of the midbrain median and dorsal raphe nuclei, in *Handbook of Psychopharmacology*, Vol. 9. Iversen LL, Snyder SH. Eds., Plenum Press, New York, 1978, 233.

354. **Bentivoglio M, Van der Kooy D, Kuypers HGJM.** The organization of the efferent projections of the substantia nigra in the rat. A retrograde fluorescent double labeling study. *Brain Res* 174: 1-17, 1979.

355. **Jacobs BL, Foote SL, Bloom FE.** Differential projections of neurons within the dorsal raphe nucleus of the rat: a horseradish peroxidase (HRP) study. *Brain Res* 147: 149-153, 1978.

356. **Moore RY, Halaris AE, Jones BE.** Serotonin neurons of the midbrain raphe: ascending projections. *J Comp Neurol* 180: 417-438, 1978.

357. **Lidov HGW, Grzanna R, Molliver ME.** The serotonin innervation of the cerebral cortex in the rat — an immunohistochemical analysis. *Neuroscience* 5: 207-227, 1980.

358. **Krulich L, Coppings RJ, Giachetti A, McCann SM, Mayfield MA.** Lack of evidence that the central serotoninergic system plays a role in the activation of prolactin secretion following inhibition of dopamine synthesis or blockade of dopamine receptors in the male rat. *Neuroendocrinology* 30: 133-138, 1980.

359. **Clemens JA, Roush ME, Fuller RW.** Evidence that serotonin neurons stimulate secretion of prolactin release factor. *Life Sci* 22: 2209-2214, 1978.

360. **Krulich L, Negro-Vilar A, Advis JP, Giachetti A.** On the dopaminergic-serotoninergic interactions in the regulation of prolactin secretion in male rats. *Abstr Fed Proc* 37: 1798, 1978.

361. **Richards GE, Holland FJ, Aubert ML, Ganong WF, Kaplan SL, Grumbach MM.** Regulation of prolactin and growth hormone secretion. *Neuroendocrinology* 30: 139-143, 1980.

362. **Fuller RW, Snoddy HD.** Effect of serotonin-releasing drugs on serum corticosterone concentration in rats. *Neuroendocrinology* 31: 96-100, 1980.

363. **Fuller RW, Snoddy HD, Molloy BB.** Pharmacologic evidence for a serotonin neural pathway involved in hypothalamus-pituitary-adrenal function in rats. *Life Sci* 19: 337-346, 1976.

364. **Popova NK, Maslova LN, Naumenko EV.** Serotonin and the regulation of the pituitary-adrenal system after deafferentation of the hypothalamus. *Brain Res* 47: 61-67, 1972.

365. **Parry O, Roberts MHT.** The responses of motoneurons to 5-hydroxytryptamine. *Neuropharmacology* 19: 515-518, 1980.

366. **Barasi S, Roberts MHT.** The modification of lumbar motoneurones excitability by stimulation of a putative 5-HT pathway. *Br J Pharmacol* 52: 339-348, 1974.

367. **Ternaux JP, Boireau A, Bourgoin S, Hamon M, Hery F, Glowinski J.** In vivo release of 5-HT in the lateral ventricle of the rat: effects of 5-hydroxytryptophan and tryptophan. *Brain Res* 101: 533-548, 1976.

368. **Henry JL, Calaresu FR.** Excitatory and inhibitory inputs from medullary nuclei projecting to spinal cardioacceleratory neurons in the cat. *Exp Brain Res* 20: 485-504, 1974.

369. **Neumayn RJ, Hare BD, Franz DN.** Evidence for bulbospinal control of sympathetic preganglionic neurons by monoaminergic pathways. *Life Sci* 14: 793-806, 1974.

370. **File SE.** Chemical lesions of both dorsal and median raphe nuclei and changes in social and aggressive behavior in rats. *Pharmacol Biochem Behav* 12: 855-859, 1980.

371. **Reader TA.** Distribution of catecholamines and serotonin in the rat cerebral cortex: absolute levels and relative proportions. *J Neural Transm* 50: 13-27, 1981.

372. **Massari VJ, Sanders-Bush E.** Synaptosomal uptake and levels of serotonin in rat brain areas after *p*-chloroamphetamine or B-9 lesions. *Eur J Pharmacol* 33: 419, 1975.

373. **Jones DL, Mogenson GJ, Wu M.** Injections of dopaminergic, cholinergic, serotoninergic and gabaergic drugs into the nucleus accumbens: effects on locomotor activity in the rat. *Neuropharmacology* 20: 29-37, 1981.

374. **Lyness WH, Moore KE.** Destruction of 5-hydroxytryptaminergic neurons and the dynamics of dopamine in nucleus accumbens septi and other forebrain regions of the rat. *Neuropharmacology* 20: 327-334, 1981.

375. **Carter CJ, Pycock CJ.** The effects of 5,7 dihydroxytryptamine lesions of extrapyradimal and mesolimbic sites of spontaneous motor behavior and amphetamine-induced stereotype. *Naunyn-Schmiedeberg's Arch Pharmacol* 208: 51-54, 1979.

376. **Lucki I, Harvey JA.** Increased sensitivity to D-amphetamine action after midbrain raphe lesions as measured by locomotor activity. *Neuropharmacology* 18: 243-249, 1979.

377. **Nicolaou NM, García-Muñoz M, Arbuthnott G, Eccleston D.** Interactions between serotonergic and dopaminergic systems in rat brain demonstrated by small unilateral lesions of the raphe nuclei. *Eur J Pharmacol* 57: 295-305, 1979.

378. **Samanin R, Quattrone A, Consolo S, Ladinsky H, Algeri S.** Biochemical and pharmacological evidence of the interaction of serotonin with other aminergic systems in the brain, in *Interactions Between Putative Neurotransmitters in the Brain.* Garattini S, Pujol JF, Samanin R. Eds. Raven Press, New York, 1978, 355.

379. **Parent A, Descarries L, Beaudet A.** Organization of ascending serotonin systems in the adult rat brain. A radioautographic study after intraventricular administration of (3H) 5-hydroxytryptamine. *Neuroscience* 6: 115-118, 1981.

380. **Bobillier P, Lewis BD, Seguin S, Pujol JF.** Evidence for direct anatomical connections between the raphe system and other aminergic groups of the central nervous system as revealed by radioautography, in *Interactions Between Putative Neurotransmitters in the Brain.* Garattini S, Pujol JF, Samanin R. Eds. Raven Press, New York, 1978, 1113.

381. **Jacobowitz DM, MacLean PD.** A brainstem atlas of catecholaminergic neurons and serotonergic perikarya in a pigmy primate (*Cebuella pygmaea*). *J Comp Neurol* 177: 397-416, 1978.

382. **Leger L, Wilkund L, Descarries L, Persson M.** Description of an indoleaminergic cell component in the cat locus coeruleus: a fluorescence histochemical and radioautographic study. *Brain Res* 168: 43-56, 1979.

383. **Steinbusch HWM.** Distribution of serotonin-immunoreactivity in the central nervous system of the rat-cell bodies and terminals. *Neuroscience* 6: 557-618, 1981.

384. **Calas A.** Radioautographic studies of aminergic neurons terminating in the median eminence. *Adv Biochem Psychopharmacol* 16: 79-88, 1977.

385. **Descarries L, Beaudet A.** The serotonin innervation of adult rat hypothalamus, in *Cell Biology of Hypothalamic Neurosecretion,* Vol. 80. Vincent JD, Waikins KC. Eds., Centre National de la Recherche Scientifique, Paris, 1978, 135.

386. **Levitt P, Moore RY.** Developmental organization of raphe serotonin neuron groups in the rat. *Anat Embryol* 154: 241-251, 1978.

387. **Schutz MTB, deAguiar JC, Graeff FG.** Anti-aversive role of serotonin in the dorsal periaqueductal gray matter. *Psychopharmacology* 85: 340-345, 1985.

388. **Ochi H, Shimizu K.** Occurrence of dopamine-containing neurons in the midbrain raphe nuclei of the rat. *Neurosci Lett* 8: 317-320, 1978.

389. **Saavedra JM.** Distribution of serotonin and synthesizing enzymes in discrete areas of the brain. *Fed Proc Fed Am Soc Exp Biol* 36: 2134-2141, 1977.

390. **Segu L, Calas A.** The topographical distribution of serotoninergic terminals in the spinal cord: quantitative radioautographic studies. *Brain Res* 153: 449-464, 1978.

391. **Sladek JR, Walker P.** Serotonin-containing neuronal perikarya in the primate locus coeruleus and subcoeruleus. *Brain Res* 134: 359-366, 1977.

392. **Gallager DW.** Spontaneous unit activity of neurons within the dorsal raphe nucleus of the neonatal rat. *Life Sci* 30: 2109-2113, 1982.

393. **Behbehani MM.** The role of acetylcholine in the function of the nucleus raphe magnus and the interaction of this nucleus with the periaqueductal gray. *Brain Res* 252: 299-307, 1982.

394. **Akil H, Liebeskind JC.** Monoaminergic mechanisms of stimulation-produced analgesia. *Brain Res* 94: 279-296, 1975.

395. **Basbaum AI, Fields HL.** The origin of descending pathways in the dorsolateral funiculus of the spinal cord of the cat and rat: further studies on the anatomy of pain modulation. *J Comp Neurol* 187: 513-532, 1979.

396. **Behbehani MM, Fields HL.** Evidence that an excitatory connection between the periaqueductal gray and nucleus raphe magnus mediates stimulation produced analgesia. *Brain Res* 170: 85-93, 1979.

397. **Behbehani MM, Pomeroy SL.** Effect of morphine injected in periaqueductal gray on the activity of single units in nucleus raphe magnus of the rat. *Brain Res* 149: 266-269, 1978.

398. **Behbehani MM, Pomeroy SL, Mack CE.** Interaction between central gray and nucleus raphe magnus: role of norepinephrine. *Brain Res Bull* 6: 361-364, 1981.

399. **Bradley PB, Dray A.** The effect of microiontophoretically applied morphine and transmitter substances in rats during chronic treatment and after withdrawal from morphine. *Br J Pharmacol* 51: 104-106, 1974.

400. **Carlton SM, Young EG, Leichnetz GR, Mayer DJ.** Nucleus raphe magnus afferents in the rat. A retrograde study using horseradish peroxidase gel implants and tetramethylbenzidine neurohistochemistry. *Neuroscience* 7: 229, 1981.

401. **Carstens E, Yokata T, Zimmermann M.** Inhibition of spinal neuronal responses to noxious skin heating by stimulation of mesencephalic periaqueductal gray in the cat. *J Neurophysiol* 42: 558-568, 1979.

402. **Castiglioni AJ, Gallaway MC, Coulter JD.** Spinal projections from the midbrain in monkey. *J Comp Neurol* 178: 329-346, 1978.

403. **Fields HL, Basbaum AI, Clanton CH, Anderson SD.** Nucleus raphe magnus inhibition of spinal cord dorsal horn neurons. *Brain Res* 126: 441-453, 1977.

404. **Kneisley LW, Biber MP, La Vail JH.** A study of the origin of brain stem projections to monkey spinal cord using the retrograde transport method. *Exp Neurol* 60: 116-139, 1978.

405. **Leichnetz GR, Watkins L, Griffin G, Murfin R, Mayer DJ.** The projection from nucleus raphe magnus and other brainstem nuclei to the spinal cord in the rat: a study using the HRP blue-reaction. *Neurosci Lett* 8: 119-124, 1978.

406. **Liebeskind JC, Guilhaud G, Besson JM, Oliveras JL.** Analgesia from electrical stimulation of the periaqueductal gray matter in the cat: behavioral observations and inhibitory effects on spinal cord interneurons. *Brain Res* 50: 441-446, 1973.

407. **Martin RF, Jordan LM, Willis WD.** Differential projections of cat medullary raphe neurons demonstrated by retrograde labelling following spinal cord lesions. *J Comp Neurol* 182: 77-88, 1978.

408. **Pomeroy SL, Behbehani MM.** Response of nucleus raphe magnus neurons to iontophoretically applied substance P in rats. *Brain Res* 202: 464-468, 1980.

409. **Shah Y, Dostrovsky JO.** Electrophysiological evidence for a projection of the periaqueductal gray matter to nucleus raphe magnus in cat and rat. *Brain Res* 193: 534-538, 1980.

410. **Zemlan FP, Pfaff DW.** Topographical organization of the medullary reticulospinal systems as demonstrated by the horseradish peroxidase technique. *Brain Res* 174: 161-166, 1979.

411. **Yezierski RP, Bowker RM, Kevetter GA, Westlund KN, Coulter JD, Willis WD.** Serotonergic projections to the caudal brain stem: a double label study using horseradish peroxidase and serotonin immunocytochemistry. *Brain Res* 239: 258-264, 1982.

412. **Abols IA, Basbaum AI.** Afferent connections of the rostral medulla of the cat: a neural substrate for midbrain-medullary interactions in the modulation of pain. *J Comp Neurol* 201: 285-297, 1981.

413. **Beitz AJ.** The origin of brain stem serotonergic and neurotensin projections to the rodent nucleus raphe magnus. *Neurosci Abstr* 7: 533, 1981.

414. **Besson JM, Oliveras JL, Chaouch A, Rivot JP.** Role of the raphe nuclei in stimulation producing analgesia, in *Advances in Experimental Biology and Medicine*, Vol. 133. Haber B, Gabay S, Issidorides MR, Alivistatos SGA. Eds., Plenum Press, New York, 1981, 153.

415. **Bowker RM, Steinbusch HWM, Coulter JD.** Serotonergic and peptidergic projections to the spinal cord demonstrated by a combined retrograde HRP histochemical and immunocytochemical staining method. *Brain Res* 211: 412-417, 1981.

416. **Briggs I.** Excitatory responses of neurons in rat bulbar reticular formation to bulbar raphe stimulation and to iontophoretically applied 5-hydroxytryptamine and their blockade by LSD_{25}. *J Physiol* 265: 327-340, 1977.

417. **Carstens E, Fraunhoffer M, Zimmermann M.** Serotonergic mediation of descending inhibition from midbrain periaqueductal gray, but not reticular formation, of spinal nociceptive transmission in the cat. *Pain* 10: 149-167, 1981.

418. **Fields HL, Anderson SD.** Evidence that raphe-spinal neurons mediate opiate and midbrain stimulation-produced analgesia. *Pain* 5: 333-349, 1978.

419. **Hubbard JE, DiCarlo V.** Fluorescence histochemistry of monoamine-containing cell bodies in the brain stem of the squirrel monkey (*Saimiri sciureus*). III. Serotonin-containing groups. *J Comp Neurol* 153: 385-398, 1974.

420. **Kuypers HGJM, Maisky VA.** Retrograde axonal transport of horseradish peroxidase from spinal cord to brain stem cell groups in the cat. *Neurosci Lett* 1: 9-14, 1975.

421. **Lovick TA, West DC, Wolstencroft JH.** Responses of raphe-spinal and other bulbar raphe neurons to stimulation of the periaqueductal gray in the cat. *Neurosci Lett* 8: 45-49, 1978.

422. **Messing RB, Lytle LD.** Serotonin-containing neurons: their possible role in pain and analgesia. *Pain* 4: 1-21, 1977.

423. **Schofield SPM, Everitt BJ.** The organization of indoleamine neurons in the brain of the rhesus monkey (*Macaca mulatta*). *J Comp Neurol* 197: 369-383, 1981.

424. **Yezierski RP, Kevetter GA, Bowker RM, Westlund KN, Coutler JD, Willis WD.** Midbrain projections to the caudal brainstem: a double label study using HRP and serotonin (5-HT) immunohistochemistry. *Anat Rec* 199: 284A, 1981.

425. **Laguzzi R, Talman WT, Reis DJ.** Serotonergic mechanisms in the nucleus tractus solitarius may regulate blood pressure and behaviour in the rat. *Clin Sci* 63: 323-326, 1982.

426. **Miura M, Reis DJ.** The role of the solitary and paramedian reticular nuclei in mediating cardiovascular reflex responses from carotid baro and chemo-receptors. *J Physiol* 223: 525-548, 1972.

427. **Basbaum AI, Clanton CH, Fields HL.** Three bulbospinal pathways from the rostral medulla of the cat: an autoradiographic study of pain modulating systems. *J Comp Neurol* 178: 209-224, 1978.

428. **Van de Kar LD, Bethea CL.** Pharmacological evidence that serotonergic stimulation of prolactin secretion is mediated via the dorsal raphe nucleus. *Neuroendocrinology* 35: 225-230, 1982.

429. **Curzon G.** Some behavioral interactions between 5-hydroxytryptamine and dopamine, in *Serotonin: Current Aspects of Neurochemistry and Function.* (Advances in Experimental Medicine and Biology Series, Vol. 133), Plenum Press, New York, 1981, 563.

430. **Van der Maelen CP, Aghajanian GK.** Serotonin-induced depolarization of rat facial motoneurons in vivo: comparison with amino acid transmitters. *Brain Res* 239: 139-152, 1982.

431. **Bloom FE, Hoffer BJ, Siggins GR, Barker JL, Nicoll RA.** Effects of serotonin on central neurons: microiontophoretic administration. *Fed Proc Fed Am Soc Exp Biol* 31: 97-106, 1972.

432. **Jones RSG, Broadbent J.** Further studies on the role of indoleamines in the responses of cortical neurons to stimulation of nucleus raphe medianus: effects of indoleamine precursor loading. *Neuropharmacology* 21: 1273-1277, 1982.

433. **Jones RSG.** Response of cortical neurons to stimulation of the nucleus raphe medianus: a pharmacological analysis of the role of indoleamines. *Neuropharmacology* 21: 511-520, 1982.

434. **Sastry BSR, Phillis JW.** Inhibition of cerebral cortical neurons by a 5-hydroxytryptaminergic pathway from the median raphe nucleus. *Can J Physiol Pharmacol* 55: 737-743, 1977.

435. **Van de Kar LD, Wilkinson CW, Skrobik Y, Brownfield MS, Ganong WF.** Evidence that serotonergic neurons in the dorsal raphe nucleus exert a stimulatory effect on the secretion of renin but not of corticosterone. *Brain Res* 235: 233-243, 1982.

436. **Willoughby JO, Menadue M, Jervois P.** Function of serotonin in physiologic secretion of growth hormone and prolactin: Action of 5,7-dihydroxytryptamine, fenfluramine and *p*-chlorophenylalanine. *Brain Res* 249: 291-299, 1982.

437. **Brown L, Rosellini RA, Samuels OB, Riley EP.** Evidence for a serotonergic mechanism of the learned helplessness phenomenon. *Pharmacol Biochem Behav* 17: 877-883, 1982.

438. **Culman J, Kvetnansky R, Kiss A, Mezey E, Murgaus K.** Interaction of serotonin and catecholamines in individual brain nuclei in adrenocortical activity during stress, in *Catecholamines and Stress: Recent Advances.* Usdin E, Kvetnansky R, Kopin IJ. Eds., Elsevier/North-Holland, New York, 1980, 69.

439. **Beart PM, McDonald D.** 5-Hydroxytryptamine and 5-hydroxytryptaminergic-dopaminergic interactions in the ventral tegmental area of rat brain. *J Pharm Pharmacol* 34: 591-593, 1982.

440. **Reisine TD, Soubrié P, Artaud F, Glowinski J.** Involvement of lateral habenula-dorsal raphe neurons in the differential regulation of striatal and nigral serotonergic transmission in cats. *J Neurosci* 2: 1062-1071, 1982.

441. **Felten DL, Harrigan P.** Dendrites bundle in nuclei raphe dorsalis and centralis superior of the rabbit. A possible substrate for local control of serotonergic neurons. *Neurosci Lett* 16: 275-280, 1980.

442. **Stern WC, Johnson A, Bronzino JD, Morgane PJ.** Neuropharmacology of the afferent projections from the lateral habenula and substantia nigra to the anterior raphe in the rat. *Neuropharmacology* 20: 974-979, 1981.

443. **Hery F, Ternaux JP.** Regulation of release processes in central serotonergic neurons. *J Physiol* 77: 287-301, 1981.

444. **Speciale SG, Neckers LM, Wyatt RJ.** Habenular modulation of raphe indoleamine metabolism. *Life Sci* 27: 2367-2372, 1980.

445. **Stern WC, Johnson A, Bronzino JD, Morgane PJ.** Effects of electrical stimulation of the lateral habenula on single-unit activity of raphe neurons. *Exp Neurol* 65: 326-342, 1979.

446. **Van der Kooy D, Hattori T.** Bilaterally situated dorsal raphe cell bodies have only unilateral forebrain projections in rat. *Brain Res* 192: 550-554, 1980.

447.   **Wang RY, Aghajanian GK.** Physiological evidence for habenula as major link between forebrain and midbrain raphe. *Science* 197: 89-91, 1977.

448.   **Heym J, Steinfels GF, Jacobs BL.** Activity of serotonin-containing neurons in the nucleus raphe pallidus of freely moving cats. *Brain Res* 251: 259-276, 1982.

449.   **Anderson EG, Proudfit HK.** The functional role of the bulbospinal serotonergic nervous system, in *Serotonin Neurotransmission and Behavior.* Jacobs BL, Gelperin A. Eds., MIT Press, Cambridge, MA, 1981, 307.

450.   **Brodal A, Taber E, Walberg F.** The raphe nuclei of the brainstem of the cat. II. Efferent connections. *J Comp Neurol* 114: 261-279, 1960.

451.   **Brodal A, Walberg F, Taber E.** The raphe nuclei of the brainstem of the cat. III. Afferent connections. *J Comp Neurol* 114: 261-279, 1960.

452.   **Coote JH, MacLeod VH.** The influence of bulbospinal monoaminergic pathways on sympathetic nerve activity. *J Physiol* 241: 453-475, 1974.

453.   **DeMontingy C, Aghajanian GK.** Preferential action of 5-methoxytryptamine and 5-methoxydimethyltryptamine on presynaptic serotonin receptors: a comparative iontophoretic study with LSD and serotonin. *Neuropharmacology* 16: 811-818, 1977.

454.   **Fox GQ, Pappas GD, Purpura DD.** Morphology and fine structure of the feline neonatal medullary raphe nuclei. *Brain Res* 101: 385-410, 1976.

455.   **Haigler HJ.** Morphine: effects on brainstem raphe neurons, in *Iontophoresis and Transmitter Mechanisms in the Mammalian Central Nervous System.* Ryal RW, Kelly JS. Eds., Elsevier/North-Holland, New York, 1978, 326.

456.   **Jacobs BL, Heym J, Steinfels GF.** Physiological and behavioral analysis of raphe unit activity, in *Handbook of Psychopharmacology,* Vol. 18. Iversen LL, Snyder SD, Snyder SH. Eds., Plenum Press, New York, 1981.

457.   **Jacobs BL, Heym J, Trulson ME.** Behavioral and physiological correlates of brain serotonergic unit activity. *J Physiol* 77: 431-436, 1981.

458.   **Lowey AD.** Raphe pallidus and raphe obscurus projections to the intermediolateral cell column in the rat. *Brain Res* 222: 129-133, 1981.

459.   **McGinty DJ, Harper RM.** Dorsal raphe neurons: depression of firing during sleep in cats. *Brain Res* 101:569-575, 1976.

460.   **Parent A.** The anatomy of serotonin-containing neurons across phylogeny, in *Serotonin Neurotransmission and Behavior.* Jacobs BL, Gelperin A. Eds., MIT Press, Cambridge, MA, 1981, 3.

461.   **Rogawski MA, Aghajanian GK.** Serotonin autoreceptors on dorsal raphe neurons: structure-activity relationships of tryptamine analogs. *J Neurosci* 1: 1148-1154, 1981.

462.   **Simon RP, Gershon MD, Brooks DC.** The role of the raphe nuclei in the regulation of ponto-geniculo-occipital wave activity. *Brain Res* 58: 313-330, 1973.

463.   **Tohayama M, Sakai K, Touret M, Salvert D, Jouvet M.** Spinal projections from the lower brainstem in the cat as demonstrated by the horseradish peroxidase technique. II. Projections from the dorsolateral pontine tegmentum and raphe nuclei. *Brain Res* 176: 215-231, 1979.

464.   **Trulson ME, Jacobs BL.** Dissociations between the effects of LSD on behavior and raphe unit activity in freely moving cats. *Science* 205: 515-518, 1979.

465.   **West DC, Wolstencroft JH.** Location and conduction velocity of raphespinal neurons in nucleus raphe magnus and raphe pallidus in the cat. *Neurosci Lett* 5: 147-151, 1977.

466.   **DeKloet ER, Kovacs GL, Telegdy G, Bohus B, Versteeg DHG.** Decreased serotonin turnover in the dorsal hippocampus of rat brain shortly after adrenalectomy: selective normalization after corticosterone substitution. *Brain Res* 239: 659-663, 1982.

467.   **Fuxe K, Hökfelt T, Ungerstedt U.** Localization of indoleamines in CNS, in *Advances in Pharmacology,* Vol. 6 (Part A). Garattinia S, Shore PA. Eds., Academic Press, New York, 1968, 235.

468.   **Thiebot MH, Hamon M, Soubrié P.** The involvement of nigral serotonin innervation in the control of punishment-induced behavioral inhibition in rats. *Pharmacol Biochem Behav* 19: 225-229, 1983.

469.   **Beckstead EM, Domesick VB, Nauta WJH.** Efferent connections of substantia nigra and ventral tegmental area in the rat. *Brain Res* 175: 191-217, 1979.

470.   **Carter CJ, Pycock CJ.** A study of the sites of interaction between dopamine and 5-hydroxtryptamine for the production of fluphenazine-induced catalepsy. *Naunyn Schmiedeberg's Arch Pharmacol* 304: 135-139, 1978.

471.   **Carter CJ, Pycock CJ.** The effects of 5,7-dihydroxytryptamine lesions of extrapyramidal and mesolimbic sites on spontaneous motor behavior and amphetamine-induced stereotype. *Naunyn-Schmiedeberg's Arch Pharmacol* 308: 51-54, 1979.

472.   **Rivot JP, Lamour Y, Ory-Lavollee L, Pointis D.** In vivo electro-chemical detection of 5-hydroxyindoles in rat somatosensory cortex: effect of the stimulation of the serotonergic pathways in normal and pCPA-pretreated animals. *Brain Res* 275: 164-168, 1983.

473.   **Aghajanian GK, Rosecrans JA, Sheard MH.** Serotonin: release in the forebrain by stimulation of midbrain raphe. *Science* 156: 402-403, 1967.

474. Fujiwara H, Uemoto M, Tanaka C. Stimulation of the rat dorsal raphe in vivo release labeled serotonin from the parietal cortex. *Brain Res* 216: 351-360, 1981.

475. LaMour Y, Rivot JP, Pointis D, Ory-Lavollée L. Laminar distribution of serotonergic innervation in rat somatosensory cortex as determined by in vivo electrochemical detection. *Brain Res* 259: 163-166, 1983.

476. Olpe HR. The cortical projection of the dorsal raphe nucleus: some electrophysiological and pharmacological properties. *Brain Res* 216: 61-71, 1981.

477. Rivot JP, Chiang CY, Besson JM. Increase of serotonin metabolism within the dorsal horn of the spinal cord during nucleus raphe magnus stimulation, as revealed by in vivo electrochemical detection. *Brain Res* 238: 117-126, 1982.

478. Soubrie P, Blas C, Ferron A, Glowinski J. Chlordiaxepoxide reduces in vivo serotonin release in the basal ganglia of éncephale isoléd but not anesthetized cats: evidence for a dorsal raphe site of action. *J Pharmacol Exp Ther* 226: 526-532, 1983.

479. Rosa M, Paillardo GP, Pasquier DA. Increase in activity of choline acetyltransferase in the dorsal raphe nucleus following habenular deafferentation. *Brain Res* 194: 578-582, 1980.

480. Llewleyn MB, Azami J, Roberts MHT. Effects of 5-hydroxytryptamine applied into nucleus raphe magnus on nociceptive thresholds and neuronal firing rate. *Brain Res* 258: 59-68, 1983.

481. Azami J, Llewelyn MB, Roberts MHT. The contribution of nucleus reticularis paragigantocellularis and nucleus raphe magnus to the analgesia produced by systemically administered morphine investigated with the microinjection technique. *Pain* 12: 229-246, 1982.

482. Azami J, Wright DM, Roberts MHT. Effects of morphine and naloxone on the responses to noxious stimulation of neurons in the nucleus reticularis paragigantocellularis. *Neuropharmacology* 20: 869-876, 1981.

483. Beitz AJ. The origin of brain stem serotonergic and neurotensin projections to the rodent nucleus raphe magnus. *Soc Neurosci Abstr* 7: 533, 1981.

484. Belcher G, Ryall RW, Shaffner R. The differential effects of 5-hydroxy-tryptamine, noradrenaline and raphe stimulation on nociceptive and non-nociceptive dorsal horn interneurons in the cat. *Brain Res* 151: 307-321, 1978.

485. Chance WT, Krynock GM, Rosecrans JA. Effects of medial raphe magnus lesions on the analgesic activity of morphine and methadone. *Psychopharmacologia* 56: 133-137, 1978.

486. Couch JR. Further evidence for a possible excitatory serotonergic synapse on raphe neurons of pons and lower midbrain. *Life Sci* 19: 761-768, 1976.

487. Llewelyn MB, Azami J, Roberts MHT. Effects of 5-hydroxytryptamine on nucleus raphe magnus studied by extracellular recording and nociceptive testing. *Pain* 1(Suppl.): S264, 1981.

488. Mohrland JS, Gebhart GF. Effects of focal electrical stimulation and morphine microinjection in the periaqueductal gray of the rat mesencephalon on neuronal activity in the medullary reticular formation. *Brain Res* 201: 23-37, 1980.

489. McRae-Deguerce A, Milon H. Serotonin and dopamine afferents to the rat locus coeruleus: a biochemical study after lesioning to the ventral mesencephalic tegmental-A10 region and raphe dorsalis. *Brain Res* 263: 344-347, 1983.

490. Leger L, McRae-Deguerce A, Pujol JF. Origine de l'innervation sérotoninerque du Locus Coeruleus chez le rat. *CR Acad Sci* 290: 807-810, 1980.

491. McRae-Deguerce A, Bérod A, Mermet A, Keller A, Chouvet G, Joh TH, Pujol JF. Alterations in tyrosine hydroxylase activity elicited by raphe nuclei lesions in the rat locus coeruleus: evidence for the involvement of serotonin afferents. *Brain Res* 235: 284-301, 1982.

492. Ochi J, Shimizu K. Occurrence of dopamine-containing neurons in the midbrain raphe nuclei of the rat. *Neurosci Lett* 8: 317-320, 1978.

493. Simon H, Lemoal M, Calas A. Efferents and afferents of the ventral tegmental A10 region studied after local injection of (3H) leucine and horseradish peroxidase. *Brain Res* 178: 17-40, 1979.

494. Frankfurt TM, Azmitia E. The effect of intracerebral injections of 5,7-dihydroxytryptamine and 6-hydroxydopamine on the serotonin-immunoreactive cell bodies and fibers in the adult rat hypothalamus. *Brain Res* 261: 91-99, 1983.

495. Segal M, Weinstock M. Differential effects of 5-hydroxytryptamine antagonists on behaviors resulting from activation of different pathways arising from the raphe nuclei. *Psychopharmacology* 79: 72-78, 1983.

496. Pasquier DA, Kemper TL, Forbes WB, Morgane PJ. Dorsal raphe substantia nigra and locus coeruleus: interconnections with each other and the neostriatum. *Brain Res Bull* 2: 323-329, 1977.

497. Picock CJ, Horton RW, Carter CJ. Interactions of 5-hydroxytryptamine with dopamine and gamma-aminobutyric acid. *Adv Biochem Psychopharmacol* 19: 323-341, 1978.

498. Silbergeld EK, Hruska RE. Lisuride and LSD: dopaminergic and serotonergic interaction in the "serotonin syndrome". *Psychopharmacology* 65: 233-237, 1979.

499. Oderfeld-Nowak B, Simon JP, Chang L, Aprison MH. Interactions of the cholinergic and serotonergic systems: Reevaluation of conditions for inhibition of acetylcholinesterase by serotonin and evidence for a new inhibition derived from this natural indoleamine. *Gen Pharmacol* 11: 37-45, 1980.

500. **Robinson SE.** Effect of specific serotonergic lesions on cholinergic neurons in the hippocampus, cortex and striatum. *Life Sci* 32: 345-353, 1983.

501. **Zemlan FP, Kow LM, Pfaff DW.** Spinal serotonin (5-HT) receptor subtypes and nociception. *J Pharmacol Exp Ther* 22: 477-485, 1983.

502. **Fink H, OelBner W.** LSD, mescaline and serotonin injected into medial raphe nucleus potentiate apomorphine hypermotility. *Eur J Pharmacol* 75: 289-292, 1981.

503. **Sandrew BB, Poletti CE.** Limbic influence on the periaqueductal gray: a single unit study in the awake squirrel monkey. *Brain Res* 303: 77-86, 1984.

504. **Grofova L, Ottersen OP, Rinvik F.** Mesencephalic and diencephalic afferents to the superior colliculus and periaqueductal gray substance demonstrated by retrograde axonal transport of horseradish peroxidase in the cat. *Brain Res* 146: 205-220, 1978.

505. **Herkenham M, Nauta WJH.** Efferent connections of the habenular nuclei in the rat. *J Comp Neurol* 187: 19-48, 1979.

506. **Krieger MS, Conrad LCA, Pfaff DW.** An autoradiographic study of the efferent connections of the ventromedial nucleus of the hypothalamus. *J Comp Neurol* 183: 785-816, 1979.

507. **Laemle LK.** Neuronal populations of the human periaqueductal gray, nucleus lateralis. *J Comp Neurol* 186: 93-108, 1975.

508. **Liu RPC, Hamilton BL.** Neurons of the periaqueductal gray matter as revealed by Golgi study. *J Comp Neurol* 189: 403-418, 1980.

509. **Mantyh PW.** The midbrain periaqueductal gray in the rat, cat and monkey: a Nissel, Weil and Golgi analysis. *J Comp Neurol* 204: 349-363, 1982.

510. **Mantyh PW.** Forebrain projections to the periaqueductal gray in the monkey with observations in the cat and rat. *J Comp Neurol* 206: 146-158, 1982.

511. **Sanders KH, Klein CE, Mayer TE, Heym CH, Handwerker HO.** Differential effects of noxious and non-noxious input on neurons according to location in ventral periaqueductal gray or dorsal raphe nucleus. *Brain Res* 186: 83-97, 1980.

512. **Morton CR, Duggan AW, Zhao AQ.** The effects of lesions of medullary midline and lateral reticular areas on inhibition in the dorsal horn produced by periaqueductal gray stimulation in the cat. *Brain Res* 301: 121-130, 1984.

513. **Carlton SM, Leichnetz GR, Young EG, Mayer DJ.** Supramedullary afferents of the nucleus raphe magnus in the rat: a study using the transcannula HRP gel and autoradiographic techniques. *J Comp Neurol* 214: 43-58, 1983.

514. **Chung JM, Kevetter GA, Yezierski RP, Haber LH, Martin RF, Willis WD.** Midbrain nuclei projecting to the medial medulla oblongata in the monkey. *J Comp Neurol* 214: 93-102, 1983.

515. **Hall JG, Duggan AW, Johnson SM, Morton CR.** Medullary raphe lesions do not reduce descending inhibition of dorsal horn neurons of the cat. *Neurosci Lett* 25: 25-29, 1981.

516. **Hall JG, Duggan AW, Morton CR, Johnson SM.** The location of brainstem neurons tonically inhibiting dorsal horn neurons of the cat. *Brain Res* 244: 215-222, 1982.

517. **Hamilton BL.** Projections of the nuclei of the periaqueductal gray matter in the cat. *J Comp Neurol* 152: 45-58, 1973.

518. **Hamilton BL, Skultety FM.** Efferent connections of the periaqueductal gray matter in the cat. *J Comp Neurol* 139: 105-114, 1970.

519. **Kneisley IW, Biber MPL, Lavail JH.** A study of the origin of brain stem projections to monkey spinal cord using the retrograde transport method. *Exp Neurol* 60: 116-139, 1978.

520. **Kuypers HGJM, Maisky VA.** Retrograde axonal transport of horseradish peroxidase from spinal cord to brain stem cell groups in the cat. *Neurosci Lett* 1: 9-14, 1975.

521. **Prieto GJ, Cannon JT, Liebeskind JC.** Nucleus raphe magnus lesions disrupt stimulation-produced analgesia from ventral but not dorsal midbrain areas in the rat. *Brain Res* 261: 53-57, 1983.

522. **Rose JD.** Projections to the caudatolateral medulla from the pons, midbrain, and diencephalon in the cat. *Exp Neurol* 72: 413-428, 1981.

523. **Sandkuhler J, Thalhammer JG, Geghart GF, Zimmermann M.** Lidocaine microinjected in the NRM does not block the inhibition by stimulation in the PAG of noxious-evoked responses of dorsal neurons in the cat. *Soc Neurosci Abstr* 8: 768, 1982.

524. **Senba E, Takagi H, Shiosaka S, Sakanaka M, Inagaki SN, Takatsuki K, Tohyama M.** On the afferent projections from some meso-diencephalic nuclei to nucleus raphe magnus of the rat. *Brain Res* 211: 387-392, 1981.

525. **Sandkuhler J, Gebhart GF.** Characterization of inhibition of a spinal nociceptive reflex by stimulation medially and laterally in the midbrain and medulla in the pentobarbital-anesthetized rat. *Brain Res* 305: 67-76, 1984.

526. **Dostrovsky JO, Hu JW, Sessle BJ, Sumino R.** Stimulation sites in periaqueductal gray, nucleus raphe magnus and adjacent regions effective in suppressing oral-facial reflexes. *Brain Res* 252: 287-297, 1982.

527. **Edeson RO, Ryall RW.** Systematic mapping of descending inhibitory control by the medulla of nociceptive spinal neurons in cats. *Brain Res* 271: 251-262, 1983.

528. **Gebhart GF, Sandkuhler J.** Lidocaine blockade of nucleus raphe magnus and the lateral medullary reticular formation indicates the descending pathways for inhibition of a spinal nociceptive reflex from the PAG are diffusely organized in the rat. *Soc Neurosci Abstr* 9: 787, 1983.

529. **Mayer DJ, Price DD.** Central nervous system mechanisms of analgesia. *Pain* 2: 379-404, 1976.

530. **Watkins LR, Griffin G, Leichnetz GR, Mayer DJ.** The somatotopic organization of the nucleus raphe magnus and surrounding brainstem structures as revealed by HRP slow release gels. *Brain Res* 181: 1-15, 1980.

531. **Sandkuhler J, Gebhart GF.** Relative contributions of the nucleus raphe magnus and adjacent medullary reticular formation to the inhibition by stimulation in the periaqueductal gray of a spinal nociceptive reflex in the pentobarbital-anesthetized rat. *Brain Res* 305: 77-87, 1984.

532. **Beall JE, Martin RF, Applebaum AE, Willis WD.** Inhibition of primate spinothalamic tract neurons by stimulation in the region of the nucleus raphe magnus. *Brain Res* 114: 328-333, 1976.

533. **Haber LH, Martin RF, Chatt AB, Willis WD.** Effects of stimulation in nucleus reticularis gigantocellularis on the activity of spinothalamic tract neurons in the monkey. *Brain Res* 153: 163-168, 1978.

534. **McCreery DB, Bloedel JR, Hames EG.** Effects of stimulating in raphe nuclei and in reticular formation on response of spinothalamic neurons to mechanical stimuli. *J Neurophysiol* 42: 166-182, 1979.

535. **Mohrland JS, McManus DO, Gebhart GF.** Lesions in nucleus reticularis gigantocellularis: effect on the antinociception produced by microinjection of morphine and local electrical stimulation in the periaqueductal gray matter. *Brain Res* 231: 143-152, 1982.

536. **Matsumura K, Nakayama T, Tsikawa Y.** Effects of median raphe electrical stimulation of the pre-optic thermo-sensitive neurons in the rat, in Int Symp Thermal Physiology, 29th IVPS Congress, Queensland, Australia, 1983.

537. **Veening JG, Swanson LW, Sawchenko PE.** The organization of projections from the central nucleus of the amygdala to brainstem sites involved in central autonomic regulation: a combined retrograde transport-immunohistochemical study. *Brain Res* 303: 337-357, 1984.

538. **Trulson ME, Trulson VM.** Activity of nucleus raphe pallidus neurons across the sleep-waking cycle in freely moving cats. *Brain Res* 237: 232-237, 1982.

539. **Costall B, Naylor RJ.** The behavioural effects of dopamine applied intracerebrally to areas of the mesolimbic system. *Eur J Pharmacol* 32: 87-92, 1975.

540. **Costall B, Naylor RJ.** The importance of the ascending dopaminergic systems to the extrapyramidal and mesolimbic brain areas for the cataleptic action of the neuroleptic and cholinergic agents. *Neuropharmacology* 13: 353, 1974.

541. **Horn AS, Cuello ACD, Miller RJ.** Dopamine in the mesolimbic system of the rat brain: endogenous levels and the effects of drugs on the uptake mechanism and stimulation of adenylate cyclase activity. *J Neurochem* 22: 265, 1974.

542. **Mora F, Sweeney KF, Rolls ET, Sanguinetti AM.** Spontaneous firing rate of neurons in the prefrontal cortex of the rat: evidence for a dopaminergic inhibition. *Brain Res* 116: 516-522, 1976.

543. **Fuxe K, Hökfelt T, Johansson O, Jonsson G, Lidbrink P, Ljungdahl A.** The origin of the dopamine nerve terminals in limbic and frontal cortex. Evidence for meso-cortical dopamine neurons. *Brain Res* 82: 349-355, 1974.

544. **Hökfelt T, Ljungdahl A, Fuxe K, Johansson O.** Dopamine nerve terminals in the rat limbic cortex: aspects of the dopamine hypothesis of schizophrenia. *Science* 184: 177-179, 1974.

545. **Lindvall O, Björklund A, Moore RY, Stenevi U.** Mesencephalic dopamine neurons projecting to neocortex. *Brain Res* 81: 325-331, 1974.

546. **Thierry AM, Blanc G, Sobel A, Glowinski J.** Dopaminergic terminals in the rat cortex. *Science* 182: 499-501, 1973.

547. **Brothers LA, Finch DM.** Physiological evidence for an excitatory pathway from entorhinal cortex to amygdala in the rat. *Brain Res* 359: 10-20, 1985.

548. **Thierry AM, Tassin JP, Blanc G, Glowinski J.** Topographic and pharmacological study of the mesocortical dopaminergic system, in *Brain Stimulation Reward*. Wuaquier A, Rolls ET. Eds., North-Holland, Amsterdam, 1976, 290.

549. **Kanazawa I, Marshall GR, Kelly JS.** Afferents to the rat substantia nigra studied with horseradish peroxidase, with special reference to fibres from the subthalamic nucleus. *Brain Res* 115: 485-491, 1976.

550. **Anden NE, Carlsson A, Dahlström A, Fuxe K, Hillarp NA, Larsson K.** Demonstration and mapping out of nigro-neostriatal dopamine neurons. *Life Sci* 3: 523-530, 1964.

551. **Grofova I.** The identification of striatal and pallidal neurons projecting to substantia nigra. An experimental study by means of retrograde axonal transport of horseradish peroxidase. *Brain Res* 91: 286-291, 1975.

552. **Grofova I, Rinvik E.** An experimental electron microscopic study on the striato-nigral projection in the cat. *Exp Brain Res* 11: 249-262, 1970.

553. **Hattori T, Fibiger HC, McGeer PL.** Demonstration of a pallido-nigral innervating dopaminergic neurons. *J Comp Neurol* 162: 487-504, 1975.

554. **Nauta WJH, Pritz MB, Lasek RJ.** Afferents to the rat caudato-putamen studied with horseradish peroxidase. An evaluation of a retrograde neuroanatomical research method. *Brain Res* 67: 219-238, 1974.

555. **Rinvik E.** Demonstration of nigrothalamic connections in the cat by retrograde axonal transport of horseradish peroxidase. *Brain Res* 90: 313-318, 1975.

556. **Kizer JS, Palkovits M, Brownstein MJ.** The projections of the A8, A9 and A10 dopaminergic cell bodies: evidence for a nigral-hypothalamic-median eminence dopaminergic pathway. *Brain Res* 108: 363-370, 1976.

557. **Bjorklund A, Lindvall O, Nobin A.** Evidence of an incerto-hypothalamic dopamine neuron system in the rat. *Brain Res* 89: 29-42, 1975.

558. **Woodruff GN, McCarthy PS, Walker RJ.** Studies on the pharmacology of neurons in the nucleus accumbens of the rat. *Brain Res* 115: 233-242, 1976.

559. **Cools AR, Van Rossum JM.** Excitation-mediating and inhibition-mediating dopamine receptors. *Psychopharmacologia* 45: 243-254, 1976.

560. **Kitai ST, Wagner A, Precht W, Ohno T.** Nigro-caudate and caudate-nigral relationship: an electrophysiological study. *Brain Res* 85: 44-48, 1975.

561. **Swanson LW, Cowan WM.** A note on the connections and development of the nucleus accumbens. *Brain Res* 92: 324-330, 1975.

562. **Unemoto H, Sasa M, Takaori S.** Inhibition from locus coeruleus of nucleus accumbens neurons activated by hippocampal stimulation. *Brain Res* 338: 376-379, 1985.

563. **Butcher SH, Butcher LL, Cho AK.** Modulation of neostriatal acetylcholine in the rat by dopamine and 5-hydroxytryptamine afferents. *Life Sci* 18: 733-744, 1976.

564. **Galley D, Simon H, LeMoal M.** Behavioral effects of lesions in the A10 dopaminergic area of the rat. *Brain Res* 124: 83-97, 1977.

565. **Koob GF, Fray PJ, Iversen SD.** Self-stimulation at the lateral hypothalamus and locus coeruleus after specific unilateral lesions of the dopamine system. *Brain Res* 146: 123-140, 1978.

566. **Clavier RM, Phillips AG, Fibiger HC.** Effects of unilateral nigrostriatal bundle lesions with 6-hydroxydopamine on self-stimulation from the A9 dopamine cell group. *Neurosci Abstr* 1: 479, 1975.

567. **Koob GF, Balcom GJ, Meyerhoff JL.** Increases in intracranial self-stimulation in the posterior hypothalamus following unilateral lesions in the locus coeruleus. *Brain Res* 101: 554-560, 1976.

568. **Lippa AS, Antelman SM, Fisher AE, Canfield DR.** Neurochemical mediation of reward: a significant role for dopamine. *Pharmacol Biochem Behav* 1: 23-28, 1973.

569. **Pellegrino LJ, Cushman AJ.** *A Stereotaxic Atlas of the Rat Brain.* Appleton-Century-Crofts, New York, 1967.

570. **LeMoal M, Stinus L, Galey D.** Radiofrequency lesion of the ventral mesencephalic tegmentum: Neurological and behavioral considerations. *Exp Neurol* 50: 521-535, 1976.

571. **Simon H, LeMoal M, Galey D, Cardo B.** Selective degeneration of central dopaminergic systems after injection of 6-hydroxydopamine in the ventral mesencephalic tegmentum of the rat. Demonstration by the Fink Heimer stain. *Exp Brain Res* 20: 275-384, 1974.

572. **Simon H, LeMoal M, Galey D, Cardo B.** Silver impregnation of dopaminergic systems after radiofrequency and 6-OHDA lesions of the rat ventral tegmentum. *Brain Res* 115: 215-331, 1976.

573. **Tassin JP, Stinus L, Simon H, Blanc G, Thierry AM, LeMoal M, Cardo B, Glowinski J.** Quantitative distribution of dopaminergic terminals in various areas of rat cerebral cortex. Implication of the dopaminergic mesocortical system in the so-called ventral tegmental area syndrome, in *Non-Striatal Dopaminergic Neurons.* Costa E, Gessa GL. Eds., Raven Press, New York, 1976.

574. **Maj J, Mogilnicka E, Klimek V.** Dopaminergic stimulation enhances the utilization of noradrenaline in the central nervous system. *J Pharm Pharmacol* 29: 569-570, 1977.

575. **Seeman P, Tedesco JL, Lee T, Chau-Wong M, Muller P, Bowles J, Whitaker PM, McManus C, Tittler M, Weinreich P, Friend WC, Brown GM.** Dopamine receptors in the central nervous system. *Fed Proc Fed Am Soc Exp Biol* 37: 128-137, 1978.

576. **Andrews DW, Patrick RL, Barchas JD.** The effects of 5-hydroxytryptophan and 5-hydroxytryptamine on dopamine synthesis and release in rat brain striatal synaptosomes. *J Neurochem* 30: 465-470, 1978.

577. **Tassin JP, Stinus L, Simon H, Blanc G, Thierry AM, LeMoal M, Cardo B, Glowinski J.** Relationship between the locomotor hyperactivity induced by A10 lesions and the destruction of the fronto-cortical dopaminergic innervation in the rat. *Brain Res* 141: 267-281, 1978.

578. **Beart PM, McDonald D.** Neurochemical studies of the mesolimbic dopaminergic pathway: somatodendritic mechanisms and gabaergic neurons in the rat ventral tegmentum. *J Neurochem* 34: 1622-1629, 1980.

579. **Aghajanian GK, Bunney BS.** Central dopaminergic neurons: neurophysiological identification and responses to drugs, in *Frontiers in Catecholamine Research.* Snyder SH, Usdin E. Eds., Pergamon Press, Elmsford, NY, 1973, 643.

580. **Carter DA, Fibiger HC.** Ascending projections of presumed dopamine-containing neurons in the ventral tegmentum of the rat as demonstrated by horseradish peroxidase. *Neuroscience* 2: 569-576, 1977.

581. **Glowinski J, Iversen LL.** Regional studies of catecholamines in the rat brain. I. The disposition of (3H) norepinephrine and (3H) DOPA in various regions of the brain. *J Neurochem* 13: 655-699, 1966.

582. **McGeer EG, Parkinson J, McGeer PL.** Neonatal enzymic development in the interpeduncular nucleus and surrounding ventral tegmentum. *Exp Neurol* 53: 109-114, 1976.

583. **Perez de la Mora M, Fuxe K.** Brain GABA, dopamine and acetylcholine interactions. I. Studies with oxotremorine. *Brain Res* 135: 107-122, 1977.

584. **Robinson SE, Malthe-Sorenssen D, Wood PL, Commissiong J.** Dopaminergic control of the septal-hippocampal cholinergic pathway. *J Pharmacol Exp Ther* 208: 476-479, 1979.

585. **Kalivas PW, Jennes L, Miller JS.** A catecholaminergic projection from the ventral tegmental area to the diagonal band of Broca: modulation by neurotensin. *J Pharmacol Exp Ther* 326: 229-238, 1985.

586. **Assaf SY, Miller JJ.** Excitatory action of the mesolimbic dopamine system on septal neurons. *Brain Res* 129: 353-360, 1977.

587. **Commissiong JW, Galli CL, Neff NH.** Differentiation of dopaminergic and noradrenergic neurons in rat spinal cord. *J Neurochem* 30: 1095-1099, 1978.

588. **Lindvall O.** Mesencephalic dopaminergic afferents to the lateral septal nucleus of the rat. *Brain Res* 87: 89-95, 1975.

589. **Nauta HJW.** A proposed conceptual reorganization of the basal ganglia and telencephalon. *Neuroscience* 4: 1875-1881, 1979.

590. **Carpenter MB, Nakano K, Kim R.** Nigrothalamic projections in the monkey demonstrated by autoradiographic technics. *J Comp Neurol* 165: 401-416, 1976.

591. **Cole M, Nauta WJH, Mehler WR.** The ascending projections of the substantia nigra. *Trans Am Neurol Assoc* 89: 74-78, 1964.

592. **Graybiel AM.** Organization of the nigrotectal connection: an experimental tracer study in the cat. *Brain Res* 143: 339-348, 1978.

593. **Hedreen JC.** Separate demonstration of dopaminergic and non-dopaminergic projections of substantia nigra in the rat. *Anat Rec* 169(Abstr.): 338, 1971.

594. **Nauta WJH, Smith GP, Faull RLM, Domesick VB.** Efferents connections and nigral afferents of the nucleus accumbens septi in the rat. *Neuroscience* 3: 385-401, 1978.

595. **Schwyk RC, Fox CA.** The primate substantia nigra: a Golgi and electron microscopic study. *J Hirnforsch* 15: 95-126, 1974.

596. **Hattori T, McGeer PL, McGeer EG.** Dendro axonic neurotransmission. II. Morphologic for the synthesis, binding and release of neurotransmitters in dopaminergic dendrites in the substantia nigra and cholinergic dendrites in the neostriatum. *Brain Res* 170: 71-83, 1979.

597. **Commissiong JW, Gentleman S, Neff NH.** Spinal cord dopaminergic neurons: evidence for an uncrossed nigrospinal pathway. *Neuropharmacology* 18: 565-568, 1979.

598. **Hökfelt T, Phillipson O, Goldstein M.** Evidence for a dopaminergic pathway in the rat descending from the A11 cell group to the spinal cord. *Acta Physiol Scand* 107: 393-395, 1979.

599. **Rosenfeld MR, Seeger TF, Sharpless NS, Gardner EL, Makman MJ.** Denervation supersensitivity in the mesolimbic system: involvement of dopamine-stimulated adenylate cyclase. *Brain Res* 173: 572-576, 1979.

600. **German DC, Dalsass M, Kiser RS.** Electrophysiological examination of the ventral tegmental (A10) area in the rat. *Brain Res* 181: 191-197, 1980.

601. **Dalsass M, German DC, Kiser RS.** Anatomical and electrophysiological examination of neurons in the nucleus A10 region of the rat. *Neurosci Abstr* 4: 422, 1978.

602. **Moore RY, Bloom FE.** Central catecholamine neuron systems: anatomy and physiology of the dopamine system. *Annu Rev Neurosci* 1: 129-169, 1978.

603. **Beckstead RM, Domesick VB, Nauta WJH.** Efferent connections of the substantia nigra and ventral tegmental area in the rat. *Brain Res* 175: 191-217, 1979.

604. **Afifi AK, Kaelbar WW.** Efferent connections of the substantia nigra in the cat. *Exp Neurol* 11: 474-482, 1965.

605. **Domesick VB, Beckstead RM, Nauta WJH.** Some ascending and descending projections of the substantia nigra and ventral tegmental area in the rat. *Neurosci Abstr* 11: 61, 1976.

606. **Fallon JH, Moore RY.** Catecholamine innervation of the basal forebrain. IV. Topography of the dopamine projection to the basal forebrain and neostriatum. *J Comp Neurol* 180: 545-580, 1978.

607. **Faull RLM, Carman JB.** Ascending projections of the substantia nigra in the rat. *J Comp Neurol* 132: 73-92, 1968.

608. **Faull RLM, Mehler WR.** Studies of the fiber connections of the substantia nigra in the rat using the method of retrograde transport of horseradish peroxidase. *Neurosci Abstr* 11: 62, 1976.

609. **Fox CA, Schmitz JT.** The substantia nigra and the entopeduncular nucleus in the cat. *J Comp Neurol* 80: 323-334, 1944.

610. **Graybiel AM, Sciascia TR.** Origin and distribution of nigrotectal fibers in the cat. *Neurosci Abstr* 1: 174, 1975.

611. **Hopkins DA, Niessen LW.** Substantia nigra projections to the reticular formation, superior colliculus and central gray in the rat, cat and monkey. *Neurosci Lett* 2: 253-259, 1976.

612. **Jayaraman A, Batton BR III, Carpenter MB.** Nigrotectal projections in the monkey: an autoradiographic study. *Brain Res* 135: 147-152, 1977.

613. **Mettler FA.** Nigrofugal connections in the primate brain. *J Comp Neurol* 138: 291-320, 1970.

614. **Moore RY, Bhatnagar RK, Heller A.** Anatomical and chemical studies of a nigro-neostriatal projection in the cat. *Brain Res* 30: 119-135, 1971.

615. **Nauta WJH, Domesick VB.** Crossroads of limbic and striatal circuitry: hypothalamo-nigral connections, in *Limbic Mechanisms.* Liningston KE, Hornykiewicz O. Eds., Plenum Press, London, 1978.

616. **Pasquier DA, Kemper TL, Forbes WB, Morgane PJ.** Dorsal raphe, substantia nigra and locus coeruleus: interconnections with each other and the neostriatum. *Brain Res Bull* 2: 323-339, 1977.

617. **Carpenter MB, McMasters RE.** Lesions of the substantia nigra in the rhesus monkey. Efferent fiber degeneration and behavioral observations. *Am J Anat* 114: 293-320, 1964.

618. **Deniau JM, Hammond C, Riszk A, Feger J.** Electrophysiological properties of identified output neurons of the rat substantia nigra (pars compacta and pars reticulata): evidences for the existence of branched neurons. *Exp Brain Res* 32: 409-422, 1978.

619. **Rinvik E, Grofova I, Ottersen OP.** Demonstration of nigrotectal and nigroreticular projections in the cat by axonal transport of proteins. *Brain Res* 112: 388-394, 1976.

620. **Vincent SR, Hattori T, McGeer EG.** The nigrotectal projection: a biochemical and ultrastructural characterization. *Brain Res* 151: 159-164, 1978.

621. **York DH, Faber JF.** An electrophysiological study of nigrotectal relationships: a possible role in turning behavior. *Brain Res* 130: 383-386, 1977.

622. **Chiodo LA, Antelman SM, Caggiula AR, Lineberry CG.** Sensory stimuli alter the discharge rate of dopamine (DA) neurons: evidence for two functional types of DA cells in the substantia nigra. *Brain Res* 189: 544-549, 1980

623. **Mulder AII, Stoof JC, Horns AS.** Activation of presynaptic alpha-noradrenaline receptors in rat brain by the potent dopamine-mimetic *N,N*-dipropyl-5,6-ADTN. *Eur J Pharmacol* 67: 147-150, 1980.

624. **Mulder AH, Wemer J, deLangen CDJ.** Presynaptic receptor-mediated inhibition of noradrenaline release from brain slices and synaptosomes by noradrenaline and adrenaline, in *Presynaptic Receptors* (Advances in Bioscience Series,Vol. 18). Langer SZ, Starke K, Dubocovich ML. Eds., Pergamon Press, Oxford, 1979, 219.

625. **Starke K, Taube HD, Borowski E.** Presynaptic receptor systems in catecholaminergic transmission. *Biochem Pharmacol* 26: 259-264, 1977.

626. **Taube HD, Starke K, Borowski E.** Presynaptic receptor systems on the noradrenergic neurons of rat brain. *Naunyn-Schmiedeberg's Arch Pharmacol* 299: 123-129, 1977.

627. **Plaznik A, Pucilowski O, Kostowski W, Bidzinski A, Hauptmann M.** Rotational behavior produced by unilateral ventral noradrenergic bundle lesions: evidence for a noradrenergic-dopaminergic interaction in the brain. *Pharmacol Biochem Behav* 17: 619-622, 1982.

628. **Rabey JM, Passeltiner P, Bystritsky A, Engel J, Goldstein M.** The regulation of striatal DOPA synthesis by alpha$_2$-adrenoceptors. *Brain Res* 230: 422-426, 1981.

629. **Antelman SM, Caggiula AR.** Norepinephrine dopamine interactions and behavior. *Science* 195: 646-653, 1977.

630. **Scatton B, Zivkovic B, Dedek J.** Antidopaminergic properties of yohimbine. *J Pharmacol Exp Ther* 215: 494-499, 1980.

631. **Anisman H, Ritch M, Sklar LS.** Noradrenergic and dopaminergic interactions in escape behavior: analysis of uncontrollable stress effects. *Psychopharmacology* 74: 263-268, 1981.

632. **Antelman SM, Black CA.** Dopamine-beta-hydroxylase inhibitors (DBHI) reverse the effects of neuroleptics under activating conditions: possible evidence for a norepinephrine (NE)-dopamine (DA) interaction. *Soc Neurosci* (Abstr.) 1977; cited in **Anisman H, Ritch M, Sklar LS.** *Psychopharmacology* 74: 263-268, 1981.

633. **Krayniak PF, Meibach RC, Siegel A.** A projection from the entorhinal cortex to the nucleus accumbens in the rat. *Brain Res* 209: 427-434, 1981.

634. **Oberlander C, Hunt PF, Dumont C, Boissier JR.** Dopamine independent rotational response to unilateral intranigral injection of serotonin. *Life Sci* 28: 2595-2601, 1981.

635. **Costall B, Domeney AM, Naylor RJ.** Persistent overstimulation of mesolimbic dopamine systems in the rat. *Neuropharmacology* 21: 327-335, 1982.

636. **Raiteri M, Marchi M, Maura G.** Presynaptic muscarinic receptors increase striatal dopamine release evoked by "quasi-psychological" depolarization. *Eur J Pharmacol* 83: 127-129, 1982.

637. **Giorguieff MF, LeFloc'h ML, Glowinski J, Besson MJ.** Involvement of cholinergic presynaptic receptors of nicotinic and muscarinic types in the control of the spontaneous release of dopamine from striatal dopaminergic terminals in the rat. *J Pharmacol Exp Ther* 200. 535-541, 1977.

638. **Giorguieff MF, Kemel ML, Glowinski J.** The presynaptic stimulating effect of acetylcholine on dopamine release is suppressed during activation of nigrostriatal dopaminergic neurons in the cat. *Neurosci Lett* 14: 177-181, 1979.

639. **Westfall TC.** Effect of muscarinic agonists on the release of ^{3}H-dopamine by potassium and electrical stimulation from rat brain slices. *Life Sci* 14: 1641-1648, 1974.

640. **Levine MS, Hull CD, Villablanca JR, Buchwald NA, García-Rill E.** Effects of caudate nuclear or frontal cortical ablation in neonatal kittens or adult cats on the spontaneous firing of forebrain neurons. *Dev Brain Res* 4: 129-138, 1982.

641. **Buchwald NA, Hull CD, Levine MS.** Neurophysiological and anatomical interrelationships of the basal ganglia, in *Brain Mechanisms in Mental Retardation*, Vol. 18. Buchwald NA, Brazier MA. Eds., Academic Press, New York, 1975, 187.

642. **Graybiel AM, Ragsdale CW Jr.** Fiber connections of the basal ganglia, in *Development and Chemical Specificity of Neurons, Progress in Brain Research*, Vol. 51. Cuenod E, Kreutzberg GW, Bloom FF. Eds., Elsevier/North-Holland, Amsterdam, 1979, 239.

643. **Strombon UH, Liedman B.** Role of dopaminergic neurotransmission in locomotor stimulation by dexamphetamine and ethanol. *Psychopharmacology* 78: 271-276, 1982.

644. **Carlsson A.** Dopaminergic autoreceptors, in *Chemical Tools in Catecholamine Research*, Vol. 2. Almgren O, Carlsson A, Engel J. Eds., North-Holland/American Elsevier, Amsterdam, 1975.

645. **Carlsson A, Kehr W, Lindquist M, Magnusson T, Atack CV.** Regulation of monoamine metabolism in the central nervous system. *Pharmacol Rev* 24(2), 1972.

646. **Skirboll LR, Grace AA, Bunney BS.** Dopamine auto- and postsynaptic receptors. Electrophysiological evidence for differential sensitivity to dopamine agonists. *Science* 206: 80-82, 1979.

647. **Strombon U.** On the functional role of pre- and postsynaptic catecholamine receptors in brain. Thesis, Gothenburg. *Acta Physiol Scand Suppl* 431: 1-43, 1976.

648. **Wang RY.** Dopaminergic neurons in the rat ventral tegmental area. Electrophysiological evidence for autoregulation. Annu Meet Soc Neurosci, Abstr. No. 88, Cincinnati, 1980. 3.

649. **Herman JP, Guilloneau D, Dantzer R, Scatton B, Semerdjian-Rouquier L, LeMoal M.** Differential effects of inescapable footshocks and of stimuli previously paired with inescapable footshocks on dopamine turnover in cortical and limbic areas of the rat. *Life Sci* 30: 2207-2214, 1982.

650. **Schmidt RH, Björklund A, Lindvall O, Loren I.** Prefrontal cortex: dense dopaminergic input in the newborn rat. *Dev Brain Res* 5: 222-228, 1982.

651. **Björklund A, Divac I, Lindvall O.** Regional distribution of catecholamines in monkey cerebral cortex, evidence for a dopaminergic innervation of the primate prefrontal cortex. *Neurosci Lett* 7: 115-119, 1978.

652. **Coyle JT, Molliver ME.** Major innervation of newborn rat cortex by monoaminergic neurons. *Science* 196: 444-447, 1977.

653. **Divac I, Björklund A, Lindvall O, Passingham RE.** Converging projections from the mediodorsal thalamic nucleus and mesencephalic dopaminergic neurons in the neocortex in three species. *J Comp Neurol* 180: 59-72, 1978.

654. **Fuster JM.** *The Prefrontal Cortex. Anatomy, Physiology and Neuropsychology of the Frontal Lobe*. Raven Press, New York, 1980.

655. **Thierry AM, Tassin JP, Blanc G, Glowinski J.** Selective activation of the mesocortical DA system by stress. *Nature* 263: 242-244, 1976.

656. **Costa E, Panula P, Thompson HK, Cheney DL.** The transynaptic regulation of the septal-hippocampal cholinergic neurons. *Life Sci* 32: 165-179, 1983.

657. **Childs JA, Gale K.** Neurochemical evidence for a nigrotegmental GABAergic projection. *Brain Res* 258: 109-114, 1983.

658. **Melis RM, Gale K.** Effect of dopamine agonists on gamma-aminobutyric acid (GABA) turnover in the superior colliculus: evidence that nigrotectal GABA projections are under the influence of dopaminergic transmission. *J Pharmacol Exp Ther* 226: 431, 1983.

659. **Nijima K, Yoshida M.** Electrophysiological evidence for branching nigral projections to pontine reticular formation, superior colliculus and thalamus. *Brain Res* 239: 279-282, 1982.

660. **Bannon MJ, Roth RH.** Pharmacology of mesocortical dopamine neurons. *Pharmacol Rev* 35: 53-68, 1983.

661. **Agnati LF, Fuxe K, Anderson K, Benfenati F, Cortelli P, D'Alessandro R.** The mesolimbic dopamine system: evidence for a high amine turnover and for a heterogeneity of the dopamine neuron population. *Neurosci Lett* 18: 45-51, 1980.

662. **Bannon MJ, Bunney EB, Roth RH.** Mesocortical dopamine neurons: rapid transmitter turnover compared to other brain catecholamine systems. *Brain Res* 218: 376-382, 1981.

663. **Bannon MJ, Chiodo LA, Roth RH, Bunney EB.** Mesocortical dopamine neurons. I. Electrophysiological and biochemical evidence for the absence of autoreceptors in a subpopulation. *Soc Neurosci Abstr* 8: 480, 1982.

664. **Bannon MJ, Michaud RL, Roth RH.** Mesocortical dopamine neurons: lack of autoreceptors modulating dopamine synthesis. *Mol Pharmacol* 19: 270-275, 1981.

665. **Bannon MJ, Reinhard JF Jr, Bunney EB, Roth RH.** Unique response to antipsychotic drugs is due to the absence of terminal autoreceptors in mesocortical dopamine neurons. *Nature* 296: 444-446, 1982.

666. **Bunney BS.** The electrophysiological pharmacology of midbrain dopaminergic systems, in *The Neurobiology of Dopamine*. Horn AS, Korf J, Westerink BHC. Eds. Academic Press, New York, 1979, 417.

667. **Carter DA, Fibiger HD.** Ascending projections of presumed dopamine containing neurons in the ventral tegmentum of the rat as demonstrated by horseradish peroxidase. *Neuroscience* 2: 569-576, 1977.

668. **Demerest KT, Moore KE.** Comparison of dopamine synthesis regulation in the terminals of nigrostriatal, mesolimbic, tuberoinfundibular and tuberohypophyseal neurons. *J Neural Transm* 46: 263-277, 1979.

669. **Deniau JM, Thierry AM, Feger J.** Electrophysiological identification of mesencephalic ventromedial tegmentum (VMT) neurons projecting to the frontal cortex, septum and nucleus accumbens. *Brain Res* 189: 315-326, 1980.

670. **Dichiara G, Porceddu ML, Fratta W, Gessa GL.** Postsynaptic receptors are not essential for dopaminergic feedback regulation. *Nature* 267: 270-272, 1977.

671. **Hadfield MG.** Mesocortical vs. nigrostriatal dopamine uptake in isolated fighting mice. *Brain Res* 222: 172-176, 1981.

672. **Herve D, Simon H, Blanc G, Lisoprawski A, LeMoal M, Glowinski J, Tassin JP.** Increased utilization of dopamine in the nucleus accumbens but not in the central cortex after dorsal raphe lesions in the rat. *Neurosci Lett* 15: 127-133, 1979.

673. **Herve D, Tassin JP, Bathelemy C, Blanc G, Lavielle S, Glowinski J.** Difference in the reactivity of the mesocortical dopaminergic neurons to stress in the BALB/C and C57BL/6 mice. *Life Sci* 25: 1659-1664, 1979.

674. **Phillipson OT.** Afferent projections to A10 dopaminergic neurons in the rat as shown by the retrograde transport of horseradish peroxidase. *Neurosci Lett* 9: 353-359, 1978.

675. **Pycock CJ, Carter CJ, Kerwin RW.** Effect of 6-hydroxydopamine lesions of the medial prefrontal cortex on neurotransmitter systems in subcortical sites in the rat. *J Neurochem* 34: 91-99, 1980.

676. **Pycock CJ, Kerwin RW, Carter CJ.** Effect of lesion of cortical dopamine terminals on subcortical dopamine receptors in rats. *Nature* 286: 74-77, 1980.

677. **Scatton B, Simon H, LeMoal M, Bischoff S.** Origin of dopaminergic innervation of the rat hippocampal formation. *Neurosci Lett* 18: 125-131, 1980.

678. **Swanson LW.** The projections of the ventral tegmental area and adjacent regions: a combined fluorescent retrograde and immunofluorescence study in the rat. *Brain Res Bull* 9: 321-354, 1982.

679. **Tassin JP, Bockaert J, Blanc G, Stinus L, Thierry AM, Lavielle S, Premont J., Glowinski J.** Topographical distribution of dopaminergic innervation and dopaminergic receptors of the anterior cerebral cortex of the rat. *Brain Res* 154: 241-251, 1978.

680. **Waldmeier PC.** Serotonergic modulation of mesolimbic and frontal cortical dopamine neurons. *Experientia* 36: 1092-1094, 1980.

681. **Wang RY.** Dopaminergic neurons in the rat ventral tegmental area. I. Identification and characterization. *Brain Res Rev* 3: 123-140, 1981.

682. **Anden NE, Grabowska-Anden M, Liljenberg B.** Demonstration of autoreceptors on dopamine neurons in different brain regions of rats treated with gammabutyrolactone. *J Neural Transm* 58: 143-152, 1983.

683. **Anden NE, Grabowska-Anden M, Liljenberg B.** On the presence of autoreceptors on dopamine neurons in different brain regions. *J Neural Transm* 57: 129-137, 1983.

684. **Björklund A, Lindvall O, Nobin A.** Evidence of an incerto-hypothalamic dopamine neuron system in the rat. *Brain Res* 89: 29-42, 1975.

685. **Moore KE, Wuerthele SM.** Regulation of nigrostriatal and tubero-infundibular-hypophyseal dopaminergic neurons. *Prog Neurobiol* 13: 325-359, 1979.

686. **White FJ, Wang RY.** A10 dopamine neurons: role of autoreceptors in determining firing rate and sensitivity to dopamine agonists. *Life Sci* 34: 1161-1170, 1984.

687. **White FJ, Wang RY.** Electrophysiological evidence for A10 dopamine autoreceptor subsensitivity following chronic D-amphetamine treatment. *Brain Res* 309: 283-292, 1984.

688. **Shepard PD, German DC.** A subpopulation of mesocortical dopamine neurons possesses autoreceptors. *Eur J Pharmacol* 98: 455-456, 1984.

689. **Loughlin SE, Fallon JH.** Substantia nigra and ventral tegmental area projections to cortex: topography and collateralization. *Neuroscience* 1984.

690. **Lee EHY, Geyer MA.** Dopamine autoreceptor mediation of the effects of apomorphine on serotonin neurons. *Pharmacol Biochem Behav* 21: 301-311, 1984.

691. **Lee EHY, Geyer MA.** Selective effects of apomorphine on dorsal raphe neurons: a cytofluorimetric study. *Brain Res Bull* 9: 719-725, 1982.

692. **Lee EHY, Geyer MA.** Indirect effects of apomorphine on serotonergic neurons in rats. *Neuroscience* 11: 437-442, 1984.

693. **Lee EHY, Geyer MA.** Similarities of the effects of apomorphine and 3-PPP on serotonin neurons. *Eur J Pharmacol* 94: 297-303, 1983.

694. **Pickel VM, Joh TH, Reis DJ.** Immunocytochemical demonstration of a serotonergic innervation of catecholamine neurons in locus coeruleus and substantia nigra. *Soc Neurosci Abstr* 1: 320, 1975.

695. **Stern WC, Johnson A, Bronzino JD, Morgane PJ.** Influence of electrical stimulation of the substantia nigra on spontaneous activity of raphe neurons in the anestehtized rat. *Brain Res Bull* 4: 561-565, 1979.

696. **Van Oene JC, deVries JB, Horn AS.** The effectiveness of yohimbine in blocking rat central dopamine autoreceptors in vivo. *Naunyn-Schmiedeberg's Arch Pharmacol* 327: 304-311, 1984.

697. **Goldstein M, Freedman LS, Backstrom T.** The inhibition of catecholamine biosynthesis by apomorphine. *J Pharm Pharmacol* 22: 715-716, 1970.

698. **Waldmeier RC, Ortmann R, Bischoff S.** Modulation of dopaminergic transmission by alpha-noradrenergic agonists and antagonists: evidence for antidopaminergic properties of some alpha antagonists. *Experientia* 38: 1168-1176, 1982.

699. **Westerink BHC.** Analysis of trace amounts of catecholamines and related compounds in brain tissue: a study near the detection limit of liquid chromatography with electrochemical detection. *J Liquid Chromatogr* 6: 2337-2351, 1983.

700. **Guertzenstein PG.** Blood pressure effects obtained by drugs applied to the ventral surface of the brain stem. *J Physiol* 229: 395-408, 1973.

701. **Armitage AK, Hall GH.** Further evidence relating to the mode of action of nicotine in the central nervous system. *Nature* 214:d 977-979, 1967.

702. **Bradley PB, Dhawan BN, Wolstencroft JH.** Pharmacological properties of cholinoceptive neurons in the medulla and pons of the cat. *J Physiol* 183: 658-674, 1966.

703. **Guertzenstein PG.** Vasodepressor and pressor responses to drugs topically applied to the ventral surface of the brain stem. *J Physiol* 224: 84-85P, 1972.

704. **Phillis JW, York DH.** Pharmacological studies on a cholinergic inhibition in the cerebral cortex. *Brain Res* 10: 297-306, 1968.

705. **Brezenoff HE, Rusin J.** Brain acetylcholine mediates the hypertensive response to physostigmine in the rat. *Eur J Pharmacol* 29: 262-266, 1974.

706. **Bartolini A, Bartolini R, Domino EF.** Effects of physostigmine on brain acetylcholine content and release. *Neuropharmacology* 12: 15-19, 1973.

707. **Brezenoff HE.** Centrally induced pressor responses to intravenous and intraventricular physostigmine evoked via different pathways. *Eur J Pharmacol* 23: 290-296, 1973.

708. **Rammelspacher H, Kuhar MJ.** Effect of lesions on the action of hemicholinium-3 on acetylcholine levels in rat brain. *Fed Proc Fed Am Soc Exp Biol* 33: 505-507, 1974.

709. **Kobayashi RM, Brownstein M, Saavedra JM, Palkovits M.** Cholineacetyltransferase content in discrete regions of the rat brain stem. *J Neurochem* 24: 637-640, 1975.

710. **Brownstein M, Kobayashi R, Palkovits M, Saavedra JM.** Choline acetyltransferase levels in diencephalic nuclei of the rat. *J Neurochem* 24: 35-38, 1975.

711. **Nistri A, DeBellis AM, Cammelli E.** Acetylcholine and acetylcholinesterase in six regions of the frog central nervous system. *Neuropharmacology* 14: 427-430, 1975.

712. **Pepeu G, Nistri A.** Effects of drugs on the regional distribution and release of acetylcholine: functional significance of cholinergic neurons, in *Psychopharmacology, Sexual Disorders and Drug Abuse*. Ban TA, Boissier JR, Gessa GJ, Heimann H, Hollister L, Lehman HE, Munkvad I, Steinberg H, Sulser F, Sundwall A, Vinar O. Eds., North-Holland, Amsterdam, 1973, 563.

713. **Bradley PB, Dray A.** Observations on the pharmacology of cholinoceptive neurons in the rat brain stem. *Br J Pharmacol* 57: 599-602, 1976.

714. **Bradley PB, Dray A.** Short latency excitation of brain stem neurons in the rat by acetylcholine. *Br J Pharmacol* 45: 372-374, 1972.

715. **Duggan AW, Headley PM, Lodge D.** Acetylcholine-sensitive cells in the caudal medulla of the rat: distribution, pharmacology and effects of pentobarbitone. *Br J Pharmacol* 54: 23-31, 1974.

716. **Jhamandas K, Sutak M.** Morphine-naloxone interaction in the central cholinergic system: the influence of subcortical lesioning and electrical stimulation. *Br J Pharmacol* 58: 101-107, 1976.

717. **Giorguieff MF, LeFloc'h ML, Glowinski J, Besson MJ.** Involvement of cholinergic presynaptic receptors of nicotinic and muscarinic types in the control of the spontaneous release of dopamine from striatal dopaminergic terminals in the rat. *J Pharmacol Exp Ther* 200: 535-544, 1977.

718. **Furchgott RF, Steinsland OS, Wakade TD.** Studies on prejunctional muscarinic and nicotinic receptors, in *Chemical Tools in Catecholamine Research*, Vol. 2. Almgren O, Carlsson A, Engels J. Eds., North-Holland, Amsterdam, 1975, 164.

719. **Guyenet PG, Agid Y, Yavoy F, Beaujouan JC, Rossier J, Glowinski J.** Effects of dopaminergic receptor agonists and antagonists on the activity of the neostriatal cholinergic system. *Brain Res* 84: 227-244, 1975.

720. **Hery F, Bourgoin S, Hamon M, Ternaux JP, Glowinski J.** The role of nicotinic and muscarinic cholinergic receptors in the control of the release of newly synthetized (3H)-5HT in rat hypothalamic slices. *Naunyn-Schmiedeberg's Arch Pharmacol* 296: 91-97, 1977.

721. **Ladinsky H, Consolo S, Bianchi S, Samanin R, Ghezzi D.** Cholinergic dopaminergic interaction in the striatum: the effect of 6-hydroxydopamine or pimozide treatment on the increased striatal acetylcholine levels induced by apomorphine, piribedil and D-amphetamine. *Brain Res* 84: 221-226, 1975.

722. **Muscholl E.** Cholinomimetic drugs and release of the adrenergic transmitter, in *New Aspects of Storage and Release Mechanism of Catecholamines*. Schumann HJ, Kroneberg G. Eds., Springer-Verlag, New York, 1970, 168.

723. **Westfall TC.** Effect of muscarinic agonists on the release of (3H)-norepinephrine and (3H)-dopamine by postassium and electrical stimulation from rat brain slices. *Life Sci* 14: 1641-1652, 1974.

724. **Westfall TC.** The effect of cholinergic agents on the release of (3H)-dopamine from rat striatal slices by nicotinic potassium and electrical stimulation. *Fed Proc Fed Am Soc Exp Biol* 33: 524, 1974.

725. **Yamamura HI, Snyder SH.** Muscarinic cholinergic binding in rat brain. *Proc Natl Acad Sci USA* 71: 1725-1729, 1974.

726. **Guyenet P, Euvrard C, Javoy F, Herbet A, Glowinski J.** Regional differences in the sensitivity of cholinergic neurons to dopaminergic drugs and quipazine in the rat striatum. *Brain Res* 136: 487-500, 1977.

727. **Agid Y, Guyenet P, Glowinski J, Beaujouan JC, Javoy F.** Inhibitory influence of the nigrostriatal dopamine system on the striatal cholinergic neurons in the rat. *Brain Res* 86: 488-492, 1975.

728. **Aquilonius SM, Eckernäs SA, Sundwall A.** Regional distribution of choline acetyltransferase in the human brain: changes in Huntington's chorea. *J Neurol Neurosurg Psychiatry* 38: 669-677, 1975.

729. **Agid Y, Javoy F, Guyenet P, Beaujouan JC, Glowinski J.** Effect of surgical and pharmacological manipulations of the dopaminergic nigro neostriatal neurons on the activity of the neostriatal cholinergic system in the rat, in *Neuropsychopharmacology, Proc. Coll. Int. Neuropsycho-pharmacologicum*. Boissier JR, Hippius H, Pichot D. Eds., Excerpta Medica, Amsterdam, 1975, 480.

730. **Koslow SH, Racagni G, Costa E.** Mass fragmentographic measurement of norepinephrine, dopamine, serotonin and acetylcholine in seven discrete nuclei of the rat telediencephalon. *Neuropharmacology* 13: 1123-1130, 1971.

731. **Schmidt DF.** Regional levels of choline and acetylcholine in rat brain following head focussed microwave sacrifice: effect of (+)amphetamine and (+)parachloro-amphetamine. *Neuropharmacology* 15: 77-84, 1976.

732. **Butcher LL.** Nature and mechanisms of cholinergic-monoaminergic interactions in the brain. *Life Sci* 21: 1207-1226, 1977.

733. **Kobayashi RM, Palkovits M, Hruska RE, Rothschild R, Yamamura HL.** Regional distribution of muscarinic cholinergic receptors in brain. *Brain Res* 154: 13-23, 1978.

734. **Beani L, Bianchi C, Giacomelli A, Tamberi F.** Noradrenaline inhibition of acetylcholine release from guinea-pig brain. *Eur J Pharmacol* 48: 179-193, 1978.

735. **Corrodi H, Fuxe K, Lidbrink P.** Interaction between cholinergic and catecholaminergic neurons in rat brain. *Brain Res* 43: 397-403, 1972.

736. **Dudar JD.** The effect of septal nuclei stimulation on the release of acetylcholine from the rabbit hippocampus. *Brain Res* 83: 123-126, 1975.

737. **Hadhazy P, Szerb JC.** The effect of cholinergic drugs on (3H)-acetylcholine release from slices of rat hippocampus, striatum and cortex. *Brain Res* 123: 311-316, 1975.

738. **Ho AKS, Singer G, Gershon S.** Biochemical evidence of adrenergic interaction with cholinergic function in the central nervous system of the rat. *Psychopharmacologia* 21: 238-245, 1971.

739. **Ho AKS, Tsai CS, Gershon S.** Adrenergic-cholinergic interaction in the central nervous system and amphetamine-induced behavior. *Drug Addict* 3: 259-263, 1975.

740. **Lewander T, Joh TH, Reis DS.** Prolonged activation of tyrosine hydroxylase in noradrenergic neurons of rat brain by cholinergic stimulation. *Nature* 258: 440-445, 1975.

741. **Mantovani P, Bartolini A, Pepeu G.** Interrelationship between dopaminergic and cholinergic systems in the cerebral cortex. *Adv Biochem Psychopharmacol* 16: 423-426, 1977.

742. **Shute CCD, Lewis PR.** The ascending cholinergic reticular system: neocortical, olfactory and subcortical projections. *Brain* 90: 497-500, 1967.

743. **Vizi ES.** Interaction between adrenergic and cholinergic system: presynaptic inhibitory effect of noradrenaline on acetylcholine release. *J Neural Transm Suppl* 9: 61-64, 1974.

744. **Ladinsky H, Consolo S, Bianchi S, Jori A.** Increase in striatal acetylcholine by picrotoxin in the rat: evidence for a gabaergic-dopaminergic-cholinergic link. *Brain Res* 108: 351-361, 1976.

745. **Woody CD, Swartz BE, Gruen E.** Effects of acetylcholine and cyclic GMP on input resistance of cortical neurons in awake cats. *Brain Res* 158: 373-395, 1978.

746. **Krnjevic K, Pumain R, Renaud L.** The mechanism of excitation by acetylcholine in the cerebral cortex. *J Physiol* 215: 247-268, 1971.

747. **Spehlmann R, Smathers CC.** The effects of acetylcholine and of synaptic stimulation on the sensorimotor cortex of cats. II. Comparison of the neuronal responses to reticular and other stimuli. *Brain Res* 74: 243-253, 1974.

748. **Cuello AC, Emson PC, Paxinos G, Jessell T.** Substance P containing and cholinergic projections from the habenula. *Brain Res* 149: 413-429, 1978.

749. **Kataoka K, Nakamura Y, Hassler R.** Habenulo-interpeduncular tract: a possible cholinergic neuron in rat brain. *Brain Res* 62: 264-267, 1973.

750. **Leranth CS, Brownstein MJ, Zaborsky L, Jaranyi ZS, Palkovits M.** Morphological and biochemical changes in the rat interpeduncular nucleus following the transection of the habenulo-interpeduncular tract. *Brain Res* 99: 127-128, 1975.

751. **Mitchell R.** Connections of the habenula and interpeduncular nucleus in the cat. *J Comp Neurol* 121: 441-457, 1963.

752. **Smaha LA, Kaelbar WW.** Efferent fiber projectory of the habenula and the interpeduncular nucleus. An experimental study in the opposum and cat. *Exp Brain Res* 16: 291-308, 1973.

753. **Kloog Y, Sokolovsky M.** Studies on muscarinic acetylcholine receptors from mouse brain: characterization of the interaction with antagonists. *Brain Res* 144: 31-48, 1978.

754. **Alberts P, Bartfi T.** Muscarinic acetylcholine receptor from rat brain. Partial purification and characterization. *J Biol Chem* 251: 1543-1547, 1976.

755. **Yamamura HI, Snyder SH.** Muscarinic cholinergic binding in rat brain. *Proc Natl Acad Sci USA* 71: 1725-1729, 1974.

756. **Guyenet P, Aghajanian GK.** ACh, substance P and met-enkephalin in the locus coeruleus: pharmacological evidence for independent sites of action. *Eur J Pharmacol* 53: 319-328, 1979.

757. **Bird SJ, Aghajanian GK.** The cholinergic pharmacology of hippocampal pyramidal cells: a microiontophoretic study. *Neuropharmacology* 15: 273-277, 1976.

758. **Butcher LL, Talbock K, Bilezikjan L.** Acetylcholinesterase neurons in dopamine-containing regions of the brain. *J Neural Transm* 37: 147-154, 1975.

759. **Butcher LL, Marchand R, Parent A, Poirier LJ.** Morphological characteristics of ACh-containing neurons in the CNS of DFP-treated monkeys. *J Neurol Sci* 32: 169-174, 1977.

760. **Cheney DL, Lefevre H, Racagni G.** Cholineacetyltransferase activity and mass fragmentographic measurement of ACh in specific nuclei and tracts of the rat brain. *Neuropharmacology* 14: 801-808, 1975.

761. **Jhamandas K, Sutak M.** Morphine-naloxone interaction in the central cholinergic system: the influence of subcortical lesioning and electrical stimulation. *Br J Pharmacol* 58: 101-112, 1976.

762. **Kuhar MJ, Atweh SF, Bird SJ.** Studies on cholinergic-monoaminergic interaction in rat brain, in *Cholinergic-Monoaminergic Interaction in the Brain.* Butcher LL. Ed., Academic Press, New York, 1978.

763. **Lewis PR, Schon FEG.** The localization of acetylcholinesterase in the locus coeruleus of the normal rat and after 6-hydroxydopamine treatment. *J Anat* 120: 373-380, 1975.

764. **Zsilla G, Cheney DL, Racagni G, Costa E.** Correlation between analgesia and the decrease of acetylcholine turn-over rate in cortex and hippocampus elicited by morphine, meperidine, viminol R$_2$ and azidomorphine. *J Pharmacol Exp Ther* 199: 662-669, 1976.

765. **Mason ST.** Central noradrenergic-cholinergic interaction and locomotor behavior. *Eur J Pharmacol* 56: 131-137, 1979.

766. **Amatruda TT, Black DA, McKenna TM, McCarley RW, Hobson JA.** Sleep cycle control and cholinergic mechanisms: differential effects of carbachol injections at pontine brain stem sites. *Brain Res* 98: 501-507, 1975.

767. **Bird SJ, Kuhar MJ.** Iontophoretic application of opiates to the locus coeruleus. *Brain Res* 122: 523-526, 1977.

768. **Mason ST, Fibiger HC.** Noradrenaline-acetylcholine interaction in brain: possible behavioral function in locomotor activity. *Neuroscience* 1979 (cited in **Mason ST, Fibiger HC.** *Nature* 277: 396-400, 1979).

769. **Mason ST, Fibiger HC.** Possible behavioral function for noradrenaline-acetylcholine interaction in brain. *Nature* 277: 396-400, 1979.

770. **Papp M, Bozsik G.** Comparison of cholinesterase activity in the reticular formation of the lower brain stem of the cat and rabbit. *J Neurochem* 13: 697-702, 1966.

771. **Pavlin R.** Cholinesterase in reticular nerve cells. *J Neurochem* 12: 515-518, 1966.

772. **Buccafusco JJ, Brezenoff HE.** Pharmacological study of a cholinergic mechanism within the rat posterior hypothalamic nucleus which mediates a hypertensive response. *Brain Res* 165; 295-310, 1979.

773. **Brezenoff HE, Wirecki TS.** The pharmacological specificity of muscarinic receptors in the posterior hypothalamus of the rat. *Life Sci* 9: 99-109, 1970.

774. **Bronk DW, Pitts RF.** Role of hypothalamus in cardiovascular regulation. *Res Publ Assoc Res Nerv Ment Dis* 20: 323-341, 1940.

775. **Brownstein M, Kobayashi R, Palkovits M, Saavedra JM.** Choline acetyltransferase levels in diencephalic nuclei of the rat. *J Neurochem* 24: 35-38, 1975.

776. **Faiers AA, Calaresu FR, Mogenson GJ.** Factor affecting cardiovascular responses to stimulation of the hypothalamus in the rat. *Exp Neurol* 51: 188-206, 1976.

777. **Phillipu A, Bohuschke N.** Hypothalamic superfusion with muscarinic drugs: their effects on pressor responses to hypothalamic stimulation. *Naunyn-Schmiedeberg's Arch Exp Pathol Pharmakol* 292: 1-7, 1976.

778. **Saelens JK, Simke JP, Allen MP, Conroy CA.** Some of the dynamics of choline and acetylcholine metabolism in rat brain. *Arch Int Pharmacodyn* 203: 305-312, 1973.

779. **Shute CCD, Lewis PR.** Cholinergic and monoaminergic pathways in the hypothalamus. *Br Med Bull* 22: 221-226, 1966.

780. **Stavinoha WB, Modak AT, Weintraub ST.** Rate of accumulation of acetylcholine in discrete regions of the rat brain after dichlorvos treatment. *J Neurochem* 27: 1375-1378, 1976.

781. **Stein L, Seifter J.** Muscarinic synapses in the hypothalamus. *Am J Physiol* 202: 751-756, 1962.

782. **Uchimura H, Saito M, Hirano M.** Regional distribution of choline acetyltransferase in hypothalamus of the rat. *Brain Res* 91: 161-164, 1975.

783. **Yammamura HI, Snyder SH.** Choline: high affinity uptake by rat brain synaptosomes. *Science* 187: 626-628, 1972.

784. **Costall B, Hui SCG, Naylor RJ.** Hyperactivity induced by injection of dopamine into the accumbens nucleus: actions and interactions of neuroleptic, cholinomimetic and cholinolytic agents. *Neuropharmacology* 18: 661-665, 1979.

785. **DeGroot J.** The rat forebrain in stereotaxic coordinates. *Verh K Ned Akad Wet* 52: 14-39, 1959.

786. **Fibiger HC, Lynch GS, Cooper HP.** A biphasic action of central cholinergic stimulation on behavioural arousal in the rat. *Psychopharmacologia* 20: 366-382, 1971.

787. **Hoover DB, Muth FA, Jacobowitz DM.** A mapping of distribution of acetylcholine, choline acetyltransferase and acetylcholinesterase in discrete areas of rat brain. *Brain Res* 153: 295-306, 1978.

788. **Kuhar MI, Yamamura HI.** Localization of cholinergic muscarinic receptor in rat brain by light microscopic radioautography. *Brain Res* 110: 229-243, 1976.

789. **Palkovits M, Saavedra JM, Kobayashi RM, Brownstein M.** Choline acetyltransferase content of limbic nuclei of the rat. *Brain Res* 79: 443-450, 1974.

790. **Uchimura H, Kim JS, Saito M, Hirano M, Ito M, Nakahara T.** Choline and acetyltransferase and acetylcholinesterase activities in limbic nuclei of the rat brain. *J Neurochem* 30: 269-272, 1978.

791. **Van Dongen PAM.** Locus coeruleus region: effects on behavior of cholinergic, noradrenergic and opiate drugs injected intracerebrally into freely moving cats. *Exp Neurol* 67: 52-78, 1980.

792. **Berman AL.** *The Brain Stem of Cat. A Cytoarchitectonic Atlas with Stereotaxic Coordinates*. University of Wisconsin Press, Milwaukee, 1968.

793. **Myers RD, Tytell M, Kawa A, Rudy T.** Microinjection of ^{3}H-acetylcholine, ^{14}C-serotonin and ^{3}H-norepinephrine into the hypothalamus of the rat: diffusion into tissue and ventricles. *Physiol Behav* 7: 743-751, 1971.

794. **Caputi AP, Rossi F, Carney K, Brezenoff HE.** Modulatory effect of brain acetylcholine on reflex-induced bradychardia and tachycardia in conscious rats. *J Pharmacol Exp Ther* 215: 309-316, 1980.

795. **Brezenoff HE.** Cardiovascular responses to intrahypothalamic injections of carbachol and certain cholinesterase inhibitors. *Neuropharmacology* 11: 637-644, 1972.

796. **Brezenoff HE, Jenden DJ.** Changes in arterial blood pressure after microinjections of carbachol into the medulla and IVth ventricle of the rat brain. *Neuropharmacology* 9: 341-348, 1970.

797. **Buccafusco JJ, Brezenoff HE.** Pharmacological study of a cholinergic mechanism within the rat posterior hypothalamic nucleus which mediates a hypertensive response. *Brain Res* 165: 295-310, 1979.

798. **Caputi AP, Brezenoff HE.** Cardiovascular effects produced by choline injected into the lateral cerebral ventricle of the unanesthetized rat. *Life Sci* 26: 1029-1036, 1980.

799. **Day MD, Roach AG.** Cardiovascular effects of carbachol and other cholinomimetics administered into the cerebral ventricle of the conscious cat. *Clin Exp Pharmacol Physiol* 4: 431-442, 1977.

800. **Doda M, Gyorgy L, Koltai MA.** Central cholinergic interactions in somato-vegetative reflexes. *Neuropharmacology* 16: 125-128, 1977.

801. **Helke CY, Muth EA, Jacobowitz DM.** Changes in central cholinergic neurons in spontaneously hypertensive rats. *Brain Res* 188: 425-436, 1980.

802. **Henning M, Trolin G.** Are spinal excitatory muscarinic receptors important for cardiovascular control? *J Pharm Pharmacol* 27: 452, 1980.

803. **Hoffman WE, Phillips HI.** A pressor response to intraventricular injections of carbachol. *Brain Res* 105: 157-162, 1976.

804. **Lang WJ, Rush ML.** Cardiovascular responses to injections of cholinomimetic drugs into the cerebral ventricles of unanesthetized dogs. *Br J Pharmacol* 47: 196-200, 1973.

805. **Ozawa H, Uematsu T.** Centrally mediated cardiovascular effects of intracisternal application of carbachol in anesthetized rats. *Br J Pharmacol* 26: 339-346, 1976.

806. **Tangri KK, Jain JP, Bhargava KP.** Role of central cholinoceptors in cardiovascular regulation. *Prog Brain Res* 47: 123-129, 1977.

807. **Zandberg P, DeJong W.** Localization of catecholaminergic receptor sites in the nucleus tractus solitarii involved in the regulation of arterial blood pressure. *Prog Brain Res* 47: 117-122, 1977.

808. **Hoffman WE, Schmid PG, Phillips MI.** Central cholinergic and noradrenergic stimulation in spontaneously hypertensive rats. *J Pharmacol Exp Ther* 206: 644-651, 1978.

809. **Masserano JM, King C.** Effects on sleep of acetylcholine perfusion of the locus coeruleus of cats. *Neuropharmacology* 21: 1163-1167, 1982.

810. **Hobson JA, Brazier MA.** *The Reticular Formation Revisited: Specifying Functions for a Nonspecific System.* Raven Press, New York, 1980, 552.

811. **Karczmar AG.** Brain acetylcholine and animal electrophysiology, in *Brain Acetylcholine and Neuropsychiatric Disease.* Davis KL, Berger PA. Eds., Plenum Press, New York, 1979, 265.

812. **Matzukai M, Okada Y, Shuto S.** Cholinergic actions related to paradoxical sleep induction in the mesencephalic cat. *Experientia* 23: 1029-1030, 1967.

813. **McCarley RW, Hobson JA.** Neuronal excitability modulation over the sleep cycle: a structural and mathematical model. *Science* 189: 58-60, 1975.

814. **Sitaram N, Moore AM, Gillin JC.** The effect of physostigmine on normal human sleep and dreaming. *Arch Gen Psychiatry* 35: 1239-1243, 1978.

815. **Sitaram N, Moore AM, Gillin JC.** Experimental acceleration and slowing of REM sleep ultradian rythm by cholinergic agonist and antagonist. *Nature* 274: 490-492, 1978.

816. **Sitaram N, Gillin JC.** Development and use of pharmacologic probes of the CNS in man: evidence of cholinergic abnormality in primary affective illness. *Biol Psychiatry* 15: 925-955, 1980.

817. **Krstic MK.** A further study of the cardiovascular responses to central administration of acetylcholine in rats. *Neuropharmacology* 21: 1151-1162, 1982.

818. **Bhargava KP, Jain IP, Saxena AK, Sinha JN, Tangri KK.** Central adrenoceptors and cholinoceptors in cardiovascular control. *Br J Pharmacol* 63: 7-15, 1978.

819. **Krstic MK.** Adrenergic activation by intracerebroventricular administration of acetylcholine and 5-hydroxytryptamine in rats, in *Catecholamines: Basic and Clinical Frontiers.* Usdin F, Kopin IJ, Barchas J. Eds., Pergamon Press, Elmsford, NY, 1979, 1473.

820. **Kubo T, Misu Y.** Changes in arterial blood pressure after microinjections of nicotine into the dorsal area of the medulla oblongata of the cat. *Neuropharmacology* 20: 521-524, 1981.

821. **Robinson SE.** Interaction of the median raphe nucleus and hypothalamic serotonin with cholinergic agents and pressor responses in the rat. *J Pharmacol Exp Ther* 223: 662, 1982.

822. **Buccafusco JJ, Spector S.** Role of central cholinergic neurons in experimental hypertension. *J Cardiovasc Pharmacol* 2: 347-355, 1980.

823. **Finberg JPM, Buccafusco JJ, Spector S.** Regional brain acetylcholine kinetics. Effects of reserpine. *Life Sci* 25: 147-156, 1979.

824. **Vocci FJ, Karbowski MJ, Dewey WL.** Apparent in vivo acetylcholine turnover rate in whole mouse brain: evidence for a two compartment model by two independent kinetic analysis. *J Neurochem* 32: 1417-1422, 1979.

825. **Sitaram N, Nurnberger JI, Gershon ES, Gillin JC.** Cholinergic regulation of mood and REM sleep: potential model and marker of vulnerability of affective disorder. *Am J Psychiatry* 139: 571-576, 1982.

826. **Hershkowitz M, Eliash S, Cohen S.** The muscarinic cholinergic receptors in the posterior hypothalamus of hypertensive and normotensive rats. *Eur J Pharmacol* 86: 229-236, 1983.

827. **Buñag RD, Eferakeya A, Langdon D.** Enhancement of hypothalamic pressor responses in spontaneous hypertensive rats. *Am J Physiol* 228: 217-223, 1975.

828. **Cantor EH, Abraham S, Spector S.** Central neurotransmitter receptors in hypertensive rats. *Life Sci* 28: 519-524, 1981.

829. **Caputi AP, Camilleri BH, Brezenoff HE.** Age-related hypotensive effect of atropine in unanesthetized spontaneously hypertensive rats. *Eur J Pharmacol* 66: 103-110, 1980.

830. **Folkow BUG, Hallback MIL.** Physiopathology of spontaneous hypertension in rats, in *Hypertension.* Genest J, Koiw E, Kuchel O. Eds. McGraw-Hill, New York, 1977, 513.

831. **Juskevich J, Robinson D, Whitehorn D.** Effect of hypothalamic stimulation in spontaneously hypertensive and Wistar-Kyoto rats. *Eur J Pharmacol* 51: 249-260, 1978.

832. **Haring JH, Davis JN.** Acetylcholinesterase neurons in the lateral hypothalamus project to the spinal cord. *Brain Res* 268: 275-283, 1983.

833. **Albanese A, Butcher LL.** Acetylcholinesterase and catecholamine distribution in the locus coeruleus of the rat. *Brain Res Bull* 5: 127-134, 1980.

834. **Bagnali P, Beaudet A, Stella M, Cuenod M.** Selective retrograde labeling of cholinergic neurons with (3H) choline. *J Neurosci* 1: 691-695, 1981.

835. **Crutcher KA.** Cholinergic denervation of rat neocortex results in sympathetic innervation. *Exp Neurol* 324-329, 1981.

836. **Crutcher KA, Brothers L, Davis JN.** Sympathetic noradrenergic sprouting in response to central cholinergic denervation: a histochemical study of neuronal sprouting in the rat hippocampal formation. *Brain Res* 210: 115-128, 1981.

837. **Divac I.** Magnocellular nuclei of the basal forebrain project to neocortex, brain stem and olfactory bulb: review of some functional correlates. *Brain Res* 93: 385-398, 1975.

838.   **Jones EG, Burton H, Saper CB, Swanson IW.** Midbrain, diencephalic and cortical relationships of the basal nucleus of Meynert and associate structures in primates. *J Comp Neurol* 167: 385-420, 1976.

839.   **Karczmar AG.** Cholinergic influences on behavior, in *Cholinergic Mechanisms*. Waser PG. Ed., Raven Press, New York, 1975, 501.

840.   **Kievit J, Kuypers HGJM.** Basal forebrain and hypothalamic connections to frontal and parietal cortex in the rhesus monkey. *Science* 187: 660-662, 1974.

841.   **Kimura H, McGeer PL, Peng JH, McGeer EG.** Choline acetyltransferase containing neurons in rodent brain demonstrated by immunohistochemistry. *Science* 208: 1057-1059, 1981.

842.   **Kimura H, McGeer PJ, Peng JH, McGeer EG.** The central cholinergic system studied by choline acetyltransferase immunohistochemistry in the cat. *J Comp Neurol* 200: 151-201, 1981.

843.   **Nagai T, Kimura H, Maeda T, McGeer PI, Peng F, McGeer EG.** Cholinergic projections from the basal forebrain of the rat to the amygdala. *J Neurosci* 2: 513-520, 1982.

844.   **Ribak CE, Kramer WG III.** Cholinergic neurons in the basal forebrain of the cat have direct projections to the sensorimotor cortex. *Exp Neurol* 75: 453-465, 1982.

845.   **Wenk H, Bigal V, Meyer U.** Cholinergic projections from magnocellular nuclei of the basal forebrain to cortical areas in rats. *Brain Res Rev* 2: 295-316, 1980.

846.   **Wyss JM, Swanson LW, Cowan WM.** A study of subcortical afferents to the hippocampal formation in the rat. *Neuroscience* 4: 463-476, 1979.

Chapter 2

CENTRAL NEURONAL PATHWAYS INVOLVED IN ANXIETY BEHAVIOR: EXPERIMENTAL FINDINGS

Fuad Lechin, Bertha van der Dijs, José Amat, and Scarlet Lechin

TABLE OF CONTENTS

I. INTRODUCTION

The word anxiety is an umbrella covering an array of psychic symptoms and their shades of linguistic meaning. Although the various concepts differ, their definitions have sufficient common elements to guarantee conventional acceptance.

In this sense, anxiety is understood to be a state of mental/emotional stress in which the individual no longer feels to be in a safe, stable, and protected state for conducting vital functions.

Words such as anguish, fear, disquiet, restlessness, impatience, acceleration, agitation, irritability, phobia, etc. are commonly accepted as equivalents for anxiety. From a physical viewpoint, the psychic symptoms are accompanied by corporal, or somatic, manifestations including tachycardia, tachypnea, sweating, disorders of skin and mucous linings, increase of voluntary motor activity, and insomnia. These observable signs have become valuable research tools for the study of animal behavior in experimental psychology laboratories. For example, psychomotor agitation and tremors are thought to betray anxiety in an animal incapable of verbally expressing emotions.

Complex models of animal behavior have been designed to correspond with different types of stimulus (pleasurable or aversive) in order to infer from responses obtained whether the animal experiences fear or not, i.e., whether it is a state of anxiety. For example, if an animal procures a reward (food) despite punishment, we suppose that it is not experiencing anxiety. However, if the animal fails to respond to a stimulus usually associated with reward, we assume that this inactivity is caused by fear of punishment previously applied in association with the reward. Subjected to internal conflict, the animal is said to be in a state of anxiety.

Although extrapolation of anxiety situations from animal models to the human, whose psyche is much more complex, may appear to be bold if not foolhardy, therapeutic trials made with different kinds of psychoactive drugs (anxiolytic) have demonstrated that animal models are of real use in making inferences regarding human behavior. This, in turn, has allowed clinical and therapeutical investigations to contribute additional data to those from experiments with animals.

Several brain structures and pathways have been involved in the mechanisms underlying anxiety. Below, the main evidence concerning these structures is reviewed.

II. NORADRENERGIC SYSTEM AND ANXIETY-LIKE BEHAVIOR

A. Locus Coeruleus Nucleus

A role for the locus coeruleus (LC) (A6 nucleus) in anxiety was first postulated by Redmond et al.[1-4] and Foote et al.[5,6] on the basis of lesion and electrical stimulation of this nucleus. They found that electrical stimulation of LC in the awake restrained monkey (*Macaca arctoides*) elicited a behavioral response of scratching and grimacing which, in the wild, is usually observed in fearful or threatening situations. Conversely, electrolytic LC lesion in these primates eliminated grimacing. In humans, electrical stimulation of LC is reported to induce feelings of panic and fear of death, while a small LC lesion has a calming effect.[7] Manipulations of LC firing activity using drugs appears to support this hypothesis.[8-10] For instance, the α-antagonist drugs piperoxane and yohimbine, both of which have been found in microiontophoretic studies to increase the firing of LC neurons,[11] elicited a corresponding increase in fear-associated behavior patterns in the restrained monkey and reportedly induced anxiety in human volunteers. Conversely, drugs such as clonidine, desipramine, morphine, and the barbiturates whose actions have been found, in microiontophoretic studies, to reduce the firing rate of LC cells, diminish the incidence of fear behavior in the monkey and are well known clinically for their antianxiety effects.[12-18] The use of propranolol to further block the postsynaptic effect of norepinephrine (NE) released as a result of LC activity also reduces fear behavior in experimental situations; this drug (a β blocker) has been claimed clinically to possess antianxiety properties.[19,20]

Considerable pharmacological evidence suggests that NE system may be involved in fear and anxiety. Clonidine, an α_2-agonist which inhibits NE synthesis and release by stimulating α_2 autoreceptors on NE neurons, is reported to induce sedation and to reduce the signs of fear both in animals and humans. The drug is known also to increase lever-pressing behavior during the punishment component of Geller-Seifter operant conflict. It has also been proposed that clonidine may be clinically useful in controlling certain types of anxiety. Conversely, several clinical reports have shown that yohimbine and piperoxane (α_2-antagonists) can provoke anxiety and panic attacks. Yohimbine accentuates signs of fear induced by threat in monkeys and impedes the effects of clonidine in potentiated startle and punishment responses. The anxiogenic effects of yohimbine in human subjects has been found recently to be annulled by both clonidine and diazepam, a benzodiazepine derivate which reduces the firing of NE-LC neurons.[21,22] The effects of the above-mentioned and other α-adrenoceptor antagonists in a maze-exploration model of fear-motivated behavior provide further evidence for NE involvement.[23,24]

Among LC projections, the coeruleocortical fiber system is the structure most often postulated as being involved in aversion- or fear-motivated behavior. This postulation is based on the following findings: (1) NE turnover is increased by stress,[25,26] (2) the anxiolytic benzodiazepines reduce NE turnover,[27-31] and (3) destruction of the coeruleocortical NE fiber system with 6-OHDA impairs behavioral responses to nonreward.[32,33]

Evidence suggests that stimulation of NE-LC system plays an inhibitory role in rat mouse-killing (muricidal) behavior; thus LC stimulation simultaneously induces fear behavior and the inhibition of aggression.[3,23] With respect to this, anxiolytic drugs (benzodiazepines), all of which reduce the firing rate of NE-LC neurons, are capable of antagonizing LC-mediated suppression of muricidal aggression.[34] In effect, muricide is inhibited by microinjections of NE into the amygdala, a brain structure innervated by NE-LC axons, while destruction of the dorsal NE bundle (DNB) which includes these axons, potentiates muricidal aggression.[35-37] Another line of evidence shows that there are raised levels of NE and its metabolite (3-methoxy-4-hydroxyphenylethyleneglycol, MHPG) in the plasma of subjects affected by panic or anxiety. A positive correlation is found between NE plasma levels and psychic-anxiety behavior (Hamilton Depression Rating Scale) of hyposomnic-agitated patients (anxious depression). These labels refer to depressed subjects who show hypersympathetic activity.[38,39] Further, reduction in NE plasma levels strongly correlates with anxiety relief.[40-43] These findings are consistent with others involving "rapid-cycling" patients which show that plasma MHPG concentrations are significantly higher during mania than during depression.[44-46]

Several lines of investigation implicate sympathetic hyperactivity and β-adrenergic stimulation in panic attacks. Racing and pounding heartbeat, shortness of breath, sweating, rapid deep breathing, and chest pains which occurs during panic attacks suggest β-adrenergic activation. Tricyclic antidepressants prevent panic attacks, down regulate β-adrenoceptors, and decrease β-adrenergic cyclic adenosine monophosphate (cAMP) production. The gradual decrease in receptor number during antidepressant therapy is temporally related to clinical improvement.[18,24,47] Isoproterenol hydrochloride, a β-adrenergic agonist, causes normal subjects to experience many somatic manifestations of panic. Moreover, patients with "hyperdynamic β-adrenergic states" or acute "anxiety-like" attacks experience these and other symptoms following low doses of isoproterenol. Propranolol, a β-adrenergic blocking agent, relieves the somatic signs of anxiety, and can block isoproterenol-induced anxiety-like symptoms.[19,20] Finally, if heavy doses of propranolol are stopped suddenly, panic symptoms occur.[19,20,48] The fact that patients with panic disorder have elevated plasma catecholamine levels is consistent with the above hypersympathetic activity found in hyposomnic-agitated depressed patients.[38]

Other investigations show that in depressed patients there is a significant positive correlation between (1) rated anxiety and CSF levels of NE[49] and (2) changes in anxiety state and urinary

MHPG level.[44,50-53] Further studies with panic disorder patients show that structured exposure to a phobic stimulus produces an increase in plasma-free MHPG level which correlates significantly with increases in anxiety ratings.[45] Finally, hyperactivity of NE system is found in the distal colon motility of those patients showing anxious depression and raised NE plasma levels.[38,39,54-56]

Although the magnitude of central NE activity cannot be assessed directly in those patients showing raised NE plasma levels, it has been found that peripheral NE plasma levels correlate highly with central NE levels derived from CSF of the same subjects.[57] In resting subjects, the circulating norepinephrine represents the portion of neurotransmitter released from sympathetic nerves which is neither withdrawn by reuptake nor catabolized in surrounding tissues and which diffuses into the vascular compartment.

The principal brain metabolite of NE is MHPG. Up to 60% of plasma MHPG is derived from brain metabolism, indicating that plasma MHPG reflects central NE activity. Highly significant correlations are found in concentrations of MHPG among the various brain areas, between plasma and CSF, between plasma and brain areas, and between CSF and brain areas.[42,45,52,57-59]

NE projections to many sites in the brain and spinal cord come from the LC. Electrical stimulation of LC increases MHPG content of many brain regions, and lesions of NE cells in the LC or their axons result in a reduction of brain MHPG concentration proportional to the extent of destruction. Noxious stimuli such as electrical footshock augment NE turnover and MHPG concentration. Conversely, clonidine, an α_2 agonist, decreases NE turnover and MHPG formation. Therefore, it is evident that brain MHPG concentration is a good index of NE activity.[44,59-63]

Taken together, these findings concerning peripheral parameters of hypersympathetic activity in anxiety-like syndrome, lead us to suppose that both hypersympathetic activity and anxiety reflect central NE overactivity, depending in large degree on the activity of LC neurons.[19,20] Yet, there are other NE nuclei which could also be involved in anxiety-like behavior.

B. Noncoeruleus (A1, A2, A5) Nuclei

According to findings mentioned earlier, anxiety-like symptoms are closely related to sympatho-excitatory effects (rising blood pressure, pulse increase, elevated catecholamine plasma levels, etc.); thus only those NE nuclei which exert sympatho-excitatory effects should be considered possible candidates for sharing the anxiogenic properties of NE-LC nucleus (A6 cell group). Norepinephrine released at preganglionic sympathetic neurons (medullar reticular lateralis nucleus and cells of spinal intermediolateral horn) exerts a sympatho-inhibitory effect. Taking into account that this norepinephrine is released by axons provided by A1 and A5 nucleus, it is unlikely that these NE nuclei would display anxiogenic effects. On the contrary, NE-A2 nucleus, located in the dorsal paramedian region of medulla oblongata, supplies NE axons to parasympathetic preganglionic vagal neurons (dorsal motor nucleus vagii, nucleus ambiguus) at which level NE release might exert a parasympathetic inhibitory effect and, consequently, an indirect sympatho-excitatory action. Therefore, NE-A2, but not NE-A1 or NE-A5 nucleus, might cooperate with NE-A6 (LC) nucleus. The fact that both A6 and A2 nuclei send NE axons to posterior hypothalamus reinforces this hypothesis since norepinephrine released or microinjected at this central region exerts sympatho-excitatory effects. Conversely, NE released or injected in the ventromedial and anterior hypothalamus provokes inhibition of the peripheral sympathetic system. Most NE terminals reaching these hypothalamic areas arise from A1 and A5 cell groups.[64-69]

Some investigators have questioned the anxiogenic role assigned to LC, objecting that (1) findings from studies of the rat are not as conclusive as those made in primates; (2) rats with lesioned LC do not show learning incapacity or an increase in distractibility when subjected to irrelevant stimuli, neither do they show greater propensity for forgetting learned tasks; and (3) neither lesioning of anterior and lateral regions bordering LC, nor total depletion of brain NE through administration of drugs or neurotoxins (6-OHDA), produces results coherent with those attained through stimulation and discrete lesioning of this nucleus.[70-72]

From the authors' standpoint, the above objections overlook the fact that NE-LC is just one nucleus in a complex neuronal circuit, the circuit responsible for anxiety-like behavior. It is difficult to conceive that feelings as complex as fear, anxiety, and other high-level brain function such as learning and memory can be investigated without taking into account the changes which may be occurring on other brain levels, e.g., the prefrontal cortex. Dopaminergic (DA) axons of mesocortical DA system project to the deepest layers of prefrontal cortex, while NE axons of LC and serotonergic (5HT) axons of the median raphe (MR) project throughout cortical layers.[69,73-76]

When NE diminishes in prefrontal cortex, an increase of DA as much as 40% is registered in this region of the brain.[77] Likewise, we know that cortical DA acts as a brake on DA in subcortical mesolimbic structures, so much so that when cortical DA is suppressed, an increase is observed in distractibility, attention deficit, lack of concentration, learning incapacity, memory deficiency, and aggressiveness. On the other hand, DA cortical predominance over subcortical mesolimbic DA leads to motor hypoactivity, anxiety increase, lessening of aggressiveness, greater learning capacity, improved attention, etc. Consequently, it should come as no surprise that a shortage of NE at cortical level, prolonged over time, would favor increased anxiety rather than its expected reduction.[78-83] Thus, for example, the increase in memory and learning gained through electrical stimulation of LC is registered only after suspension of stimuli.[84] Then again, it must be remembered that, when lesioning a brain structure which forms part of complex multiple neuronal circuits, physiological changes are produced which tend to compensate for the experimentally provoked flaw, even to the point of exceeding the original function. For example, it is now firmly established that NE neurons of LC (A6) interchange axons with A1 group of the medulla oblongata.[65] Norepinephrine liberated by both groups of axons would interchange inhibitory influences through stimulation of α_2 adrenoceptors existing in the NE neuronal somatodendritic area. Therefore, destruction of LC should lead to disinhibition of NE-A1 neurons which send ample innervation to all 5HT and dopaminergic nuclei, as well as to their projection zones (hypothalamus, mesolimbic structures, etc.). There is clear evidence that ventral NE bundle (VNB), through which pass NE-A1 axons, innervates 5HT-MR which, in turn, has an inhibitory influence on mesocortical DA system.[85-92] Consequently, it is possible to speculate that if NE-A1 group is liberated of the brake provided by LC, the inhibition exerted by A1 neurons over 5HT-MR would be increased. The greater inhibition of 5HT activity arising from MR would permit disinhibition of mesocortical DA activity, which has been positively correlated with degrees of experienced anxiety. It becomes apparent that a lesion of the LC which, in the first sharp stage, reduces anxiety level may later favor the opposite behavior. Similar considerations could be made about changes in learning, memory, distractibility, motor activity, etc. Furthermore, the repercussions which LC lesion may have on 5HT-MR system and DA-ventral tegmental area (VTA) system must be taken into account.

Briefly, it is possible to interpret experimental results of neurophysiological and neuropharmacological investigations using "systemic" thinking, but not "lineal" thinking. Moreover, it is difficult to discuss the results of such experimental studies without having available the maximum information stemming from anatomical, physiological, pharmacological, and animal behavior findings made by the entire research community working in these areas.

III. NONNORADRENERGIC SYSTEMS AND ANXIETY-LIKE BEHAVIOR

In experimental animals, inhibition of normal behavior by novelty, punishment, or aversive stimuli is interpreted as an anxiety effect elicited by these stimuli. This inhibition of normal behavior may be strong enough to suppress the normal response of the animal to the presentation of a reward, i.e., food, in what is called punishment activity. Therefore, punishment behavior is considered as a conflict generated by fear or anxiety. This animal model of anxiety has been used to investigate the anxiogenic or anxiolytic effects of pharmacological manipulations and discrete brain lesions or stimulations.

Besides involvement of LC and other NE nuclei in anxiety-like behavior discussed above, evidence from animal experiments is accumulating to implicate 5HT and DA systems in control of behavior suppressed by aversive stimulation. However, experimental results are sometimes contradictory and shed doubts about conclusions emanating from the great bulk of evidence which apparently points to a specific direction. Under these circumstances, only the integration of sufficient related information can shed light on apparent disagreements.

A. Serotonergic and Dopaminergic Systems and Anxiety-Like Behavior

Consistent experimental evidence indicates that activation of ascending 5HT pathways mediates the suppression response induced by punishment (anxiogenic behavior).[93,94] The efficacy of 5HT receptor antagonists in releasing punished behavior (anxiolytic behavior) provides some of the strongest evidence implicating brain 5HT in punishment (anxiety).[95-97] In spite of controversy about the ability of some peripheral 5HT antagonists to block central 5HT receptors in electrophysiological studies, antipunishment of anxiolytic action of anti-5HT drugs is remarkably consistent with the effect of other manipulations that decrease 5HT activity in the brain, such as PCPA or α-propyl-dopacetamide administration, and neurotoxin lesion of ascending 5HT pathways in the rat midbrain.[98-103] In view of solid experimental evidence showing that methergoline blocks 5HT receptors in the brain, the marked antipunishment action of the drug demonstrated by some investigators further supports the mediating role of brain 5HT in punishment-induced response suppression.[104,105] Nevertheless, concurrent participation of other brain neuronal systems in this phenomenon should be taken into account.

Strong evidence implicates the DR in punishment. In effect, infusion of 5,7-dyhydroxy-tryptamine (5,7-DHT), a 5HT neurotoxin, into the DR causes attenuation of punishment-induced suppression (anxiety).[106,107] In addition, infusion of this neurotoxin into substantia nigra (SN) (bilaterally) lessens experimentally induced anxiety. Although SN has never been claimed to be crucially implicated in control of emotional responses, this role might be expected in the light of anatomical data showing SN to be in a position to influence serotonergic cells through its projection to DR.[108] However, DR neurons projecting to amygdaloid complex and to nucleus accumbens do not seem to be involved in anti-anxiety effects since marked destruction of 5HT innervation of these forebrain structures fails to release punishment behavior.[109,110]

The question arises as to whether attenuation of punished behavior following damage of nigral-serotonergic terminals is really associated with lessened fear and anxiety. Indeed, nigral-serotonergic innervation is essential for control of motor components of behavior and, in particular, for expression of fear and anxiety through inhibition of ongoing responses. From abundant data supporting the hypothesis that 5HT innervation in SN inhibits DA neurons, it can be assumed that this inhibition is essential for response suppression. Further, the fact that SN sends DA axons to DR nucleus, releasing 5HT-DR neurons from the inhibitory influence of SN nuclei, may be an additional factor in experimentally induced anxiety.[110-114]

Other dopaminergic neurons, such as those located within VTA or DA-A10 cell group, can be influenced by intranigral 5,7-DHT. Thus, they can participate in observed behavioral changes.[115-118] Indeed, DA-VTA neurons project to and control frontal cortex, the ablation of which has been reported to attenuate punished behavior. Nevertheless, although experiments suggest that 5,7-DHT infusion into VTA fails to modify punishment-induced suppression, it should be remembered that in this area there exist two different DA neurons which give rise to two self-controlling DA systems, one subcortical mesolimbic and the other mesocortical.[119-121] Predominance of DA mesolimbic system is accompanied by motor hyperactivity and loss of fear and anxiety, whereas DA mesocortical predominance is accompanied by suppression of motor activity and anxiety-like behavior.[88,116,122-125] Thus, lesioning 5HT terminals into VTA region would release from serotonergic bridle the two types of DA neurons.[85,89,90,126-131] In this case no predominance would be produced between the two DA systems.

Sufficient evidence shows that both 5HT-DR and DA-SN neurons exert inhibitory influences on DA mesolimbic system but not on DA mesocortical system.[99,101,106,107,109,110, 115,132-137] Moreover, experimental studies show that 5HT neurons of MR send inhibitory projections to DA mesocortical neurons.[78,99,101,107,109,135-139] Taken together, the above data suggest that 5HT-DR system displays an axiogenic role which may be mediated through inhibition of subcortical mesolimbic DA system, whereas 5HT-MR system would display an anxiolytic role through inhibition of DA mesocortical system. The fact that 5HT-MR neurons send inhibitory axons to 5HT-DR neurons may constitute an additional mechanism favoring the anxiolytic role of 5HT-MR system.[140] Finally, the finding that DA turnover in both nigral and frontal cortex is enhanced during the first stage (alert) of stress situations is coherent with the hypothesis associating anxiety with hyperactivity of DA mesocortical system and nonanxiety with hypoactivity of this DA system.[25,79,93,117,118,141-145]

Other experimental studies support the postulation that 5HT-MR neurons display an anxiolytic role.[86,107,111,136-138,146-149] Namely, 5HT agonists and 5-hydroxytryptophane (5HTP, a serotonin precursor) are proven to exert an antipunishment or anxiolytic action which cannot be exerted through 5HT-DR system, since this 5HT system is known to be associated with anxiogenic activity.[98-102,104-107,110,123,134-137,146,147,150-166] Moreover, the proven rewarding (antipunishment) effect of habenular stimulation induces decrease in the firing rate of 5HT-DR neurons.[105,167] Finally, strong positive reward response has been produced by stimulation of 5HT-MR nucleus. This MR-induced reward is inhibited by PCPA, a serotonin synthesis inhibitor.[98,100-103,107,136,139,161,163,164]

For more information regarding monoaminergic circuits involved in anxiety-like behavior, it is convenient to recall studies showing that NE neurons from LC exert an excitatory effect not only on 5HT-DR, but also on DA mesocortical neurons. In effect, destruction of NE innervation arising from LC and reaching DA-VTA neurons provoked a reduction of DA utilization in prefrontal cortex but not in the subcortical mesolimbic structures. Taking into account the positive correlation demonstrated between NE-LC activity, and between DA mesocortical activity and anxiety, as well as the fact that, NE-LC neurons exert a tonic excitatory effect on 5HT-DR neurons, the above discussions fit well with this complex circuit.[64,66,77,79,86,91,92,115,143,147,168-181]

Anxiolytic or anticonflict drugs such as benzodiazepine have used frequently in experimental studies addressing this matter. Although many of these studies point to the hypothesis that the effects of the drugs are exerted through inhibition of ascending 5HT pathways, recent studies demonstrate that the anxiolytic effect of benzodiazepines is not dependent on 5HT activity. In effect the anxiolytic action of these drugs is obtained with doses much lower than those required to inhibit 5HT neurons. In addition, the blockade of GABA activity with GABA antagonists injected into the DR did not eliminate the anxiolytic effect of benzodiazepines. It should be noted here that the GABA-mimetic effect of benzodiazepines has been widely demonstrated at the LC, DR, and VTA nuclei, all of which receive important GABA inputs.[27,28,30,31,34,37,94,104,106,111,123,141,148-151,153,155,156,159,165,167,182-184]

Although benzodiazepines decrease the firing rate of NE-LC neurons, other anxiolytic drugs such as buspirone are not able to do this. Since buspirone does not block 5HT receptors its anxiolytic property cannot be attributed to this mechanism. However, recent studies demonstrated that buspirone blocks DA autoreceptors and increases DA metabolism.[185-193] Taking into account that DA mesocortical system lacks autoreceptors, buspirone would enhance dopamine release and turnover in DA mesolimbic and nigrostriatal but not DA mesocortical system. This predominance of the former over the second DA system would explain the anxiolytic effect of the drug.[185-193]

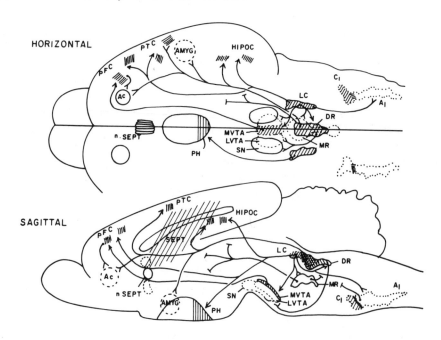

FIGURE 1. Anxiety–like behavior. Direct and indirect experimental evidence shows that during states of anxiety hyperactivity of noradrenergic (NE) neurons in the locus coeruleus (LC), serotonergic (5HT) neurons in the dorsal raphe nucleus (DR), dopaminergic (DA) neurons in the medial-anterior part of the ventral tegmental area (MVTA), and adrenergic (A) neurons in the C1 cell group is registered. Peripheral sympathetic hyperactivity registered in anxiety-like behavior would be due to stimulation by NE-LC axons of sympathetic preganglionic neurons located in posterior hypothalamus (PH) and pontine reticular formation. NE-LC axons also exert inhibitory influence on sympathoinhibitory NE-A1 neurons. In turn, adrenaline (A) neurons, freed from the NE-A1 bridle, contribute sympatho-excitatory activity by stimulating lateral medullary reticular formation and inhibiting preganglionic parasympathetic neurons located in the medulla oblongata (dorsal motor nucleus of the vagii and nucleus ambiguus) and in the lateral hypothalamus. DA-MVTA neurons receive excitatory input from the hyperactive NE-LC neurons. Hence, dopamine would be released in the projection areas of this DA mesocortical system, i.e., prefrontal cortex (PFC). With respect to this, it is known that DA mesocortical hyperactivity results in DA mesolimbic inhibition of n. accumbens (Ac), amygdala (Amig), septum (SEPT), etc. Such increase of DA mesocortical/DA mesolimbic ratio is consistent with heightened state of alert and lowered motor activity. However, the prolonging of anxiety state leads to exhaustion of DA mesocortical activity and a progressive increase in DA mesolimbic activity which in turn results in enhanced motor activity. The hyperactivity of 5HT-DR neurons is secondary to excitatory input from NE-LC neurons. DR neurons send inhibitory axons to DA neurons located in the lateral part of ventral tegmental area (LVTA) = DA mesolimbic neurons. Further, DR axons antagonize DA activity at the level of mesolimbic structures (n. accumbens, amygdala, septum, etc.) which receive these 5HT-DR axons. NE-LC hyperactivity would bridle median raphe (MR) serotonergic nucleus, since NE-LC axons project to MR, exerting inhibitory influence through postsynaptic α_1 receptors. PTC = Parieto temporal cortex, HIPOC = hippocampus, (/////////////) = hyperactive nuclei, (——>) = excitatory input, (——<) = inhibitory input, and (———|) = annulled output. (Reproduced with kind permission from PJD Publications Limited, Westbury, N.Y., 11590, U.S.A., from *Res Commun Psychol Psychiatr Behav*, 11: 113-143, 1986. Copyright © by PJD Publications Ltd.)

Summarizing, despite some inconsistent experimental results, there is sufficient evidence to state that NE-LC and 5HT-DR nuclei are positively correlated with anxiety. On the other hand, the activity of DA-SN nucleus is inversely correlated with anxiety. With respect to this, it is highly significant that while NE-LC stimulates 5HT-DR firing rate, DA-SN activity displays the opposite effect.

Other evidence strongly suggests that DA-VTA mesolimbic and DA-VTA mesocortical nuclei are negatively and positively correlated with anxiety, respectively. Since NE-LC correlates positively with DA-mesocortical, while 5HT-DR correlates negatively with DA-mesolimbic system, it is possible to attempt a theoretical approach which would integrate all those systems within a coherent model. Finally, the finding that 5HT-MR nucleus exerts inhibitory influences on 5HT-DR, NE-LC, and DA-mesocortical systems, as well as its strong reward property, gives support to the anti-anxiety role of 5HT-MR nucleus (see Figure 1).

REFERENCES

1. **Redmond DE Jr, Huang YH, Snyder DR, Maas JW, Baulu J.** Behavioral changes following lesions of the locus coeruleus in *Macaca arctoides. Neurosci Abstr* 1: 472, 1976.
2. **Redmond DE Jr.** Alterations in the function of the nucleus locus coeruleus: a possible model for studies of anxiety, in *Animal Models in Psychiatry and Neurology.* Hannin I, Usdin E. Eds., Pergamon Press, Elmsford, NY, 1977, 293.
3. **Huang YH, Redmond DE Jr, Snyder DR, Maas JW.** Loss of fear following bilateral lesions of the locus coeruleus in the monkey. *Neurosci Abstr* 2: 573, 1976.
4. **Gold MS, Redmond DE Jr.** Pharmacological activation and inhibition of noradrenergic activity alter specific behaviors in nonhuman primates. *Neurosci Abstr* 3: 250, 1977.
5. **Foote S, Bloom FE.** Activity of locus coeruleus neurons in the unanesthetized squirred monkey, in *Catecholamines: Basic and Clinical Frontiers.* Usdin E. Ed., Pergamon Press, Elmsford, NY, 1979, 625.
6. **Foote SL, Aston-Jones G, Bloom FE.** Impulse activity of locus coeruleus neurons in awake rats and monkeys is a function of sensory stimulation and arousal. *Proc Natl Acad Sci* 77: 3033-3039, 1980.
7. **Charney DS, Heninger GR, Redmond DE Jr.** Noradrenergic function and human anxiety, paper presented at the 136th Annu. Meet. American Psychiatric Association, New York, May 1983.
8. **Olpe H-R, Jones RSG, Steinmann MW.** The locus coeruleus: actions of psychoactive drugs. *Experientia* 39: 242-249, 1983.
9. **Hanson HM, Witoslawski JJ, Campbell EA.** Drug effects in squirrel monkeys trained on a multiple schedule with a punishment contingency. *J Exp Anal Behav* 10: 565-569, 1967.
10. **Charney DS, Heninger GR, Breier A.** Noradrenergic function in panic anxiety. Effects of yohimbine in healthy subjects and patients with agoraphobia and panic disorder. *Arch Gen Psychiatry* 41: 751-763, 1984.
11. **Charney DS, Heninger GR, Redmond DE Jr.** Yohimbine induced anxiety and increased noradrenergic functions in humans: effects of diazepam and clonidine. *Life Sci* 33: 19-30, 1983.
12. **McNair DM, Kahn RJ.** Imipramine compared with a benzodiazepine for agoraphobia, in *Anxiety: New Research and Changing Concepts.* Klein DF, Rabkin J. Eds., Raven Press, New York, 1981, 69.
13. **Geyer MA, Lee EHY.** Effects of clonidine, piperoxane and locus coeruleus lesion on the serotonergic and dopaminergic systems in raphe and caudate nucleus. *Psychopharmacology* 33: 3399-3404, 1984.
14. **Zitrin GM, Klein DF, Woerner MG, Ross DC.** Treatment of phobias: comparison of imipramine and placebo. *Arch Gen Psychiatry* 40: 125-138, 1983.
15. **Uhde TW, Siever LJ, Post TM.** Clonidine: acute challenge and clinical trial paradigms for the investigation and treatment of anxiety disorders, affective illness, and pain syndromes, in *Neurobiology of the Mood Disorders.* Post RM, Ballenger JC. Eds., Williams & Wilkins, Baltimore, 1984.
16. **Hoehn-Saric R, Merchant AF, Keyser MC, Smith VK.** Effects of clonidine on anxiety disorders. *Arch Gen Psychiatry* 38: 1278-1286, 1981.
17. **Nybaeck HV, Walters JR, Aghajanian GK.** Tricyclic antidepressants: effects on the firing rate of brain noradrenergic neurons. *Eur J Pharmacol* 32: 302-312, 1975.
18. **Huang YH.** Net effect of acute administration of desipramine on the locus coeruleus-hippocampal system. *Life Sci* 25: 739-746, 1979.
19. **Easton JD, Sherman DG.** Somatic anxiety attacks and propranolol. *Arch Neurol* 33: 689-691, 1976.
20. **Kathol RG, Noyes R, Slymen PJ, Crowe RR, Clancy J, Kerver RE.** Propranolol in chronic anxiety disorders. *Arch Gen Psychiatry* 37: 1361-1365, 1980.
21. **Redmond DE Jr.** New and old evidence for the involvement of a brain norepinephrine system in anxiety, in *The Phenomenology and Treatment of Anxiety.* Fann WE. Ed., Spectrum, New York, 1979, 153.
22. **Clark TK.** The locus coeruleus in behavior regulation: evidence for behavior-specific versus general involvement. *Behav Neural Biol* 25: 271-276, 1979.
23. **Handley SL, Mithani S.** Effects of alpha-adrenoceptor agonists and antagonists in a maze-exploration model of "fear"-motivated behaviour. *Naunyn-Schmiedeberg's Arch Pharmacol* 327: 1-5, 1984.
24. **Charney DS, Heninger GR.** Noradrenergic function and the mechanism of action of antianxiety treatment. II. The effect of long-term imipramine treatment. *Arch Gen Psychiatry* 42: 473-481, 1985.
25. **Anisman H.** Neurochemical changes elicited by stress: behavioral correlates, in *Psychopharmacology of Aversively Motivated Behavior.* Biagnami G. Ed., Plenum Press, New York, 1978, 119.

26. **File SE, Vellucci SV.** Behavioural and biochemical measures of stress in hooded rats from different sources. *Physiol Behav* 22: 31-36, 1979.

27. **Costa E, Guidotti A.** Molecular mechanisms in the receptor action of benzodiazepines. *Annu Rev Pharmacol* 19: 531-545, 1979.

28. **Guidotti A, Baraldi M, Leon A, Costa E.** Benzodiazepines: a tool to explore the biochemical and neurophysiological basis of anxiety. *Fed Proc Fed Am Soc Exp Biol* 39: 3039-3045, 1980.

29. **Gallager DW, Mallorga P, Thomas JW, Tallman JF.** GABA-benzodiazepine interactions: physiological, pharmacological and developmental aspects. *Fed Proc Fed Am Soc Exp Biol* 39: 3043-3048, 1980.

30. **Grant SJ, Galloway MP, Mayor R, Fenerty JP, Finckelstein MF, Roth RR, Redmond DE Jr.** Precipitated diazepam withdrawal elevates noradrenergic metabolism in primate brain. *Eur J Pharmacol* 107: 127-132, 1985.

31. **Sepinwall J, Cook I.** Relationships of gamma-amino-butyric acid (GABA) to anti-anxiety effects of benzodiazepines. *Brain Res Bull* 5: 839-848, 1980.

32. **Mason ST, Fibiger HC.** 6-OHDA lesion of the dorsal noradrenergic bundle alters extinction of passive avoidance. *Brain Res* 152: 209-214, 1978.

33. **Lucki I, Frazer A.** Performance and extinction of lever press behavior following chronic administration of desipramine to rats. *Psychopharmacology* 85: 253-259, 1985.

34. **Kostowski W, Valzelli L, Kozak W.** Chlordiazepoxide antagonizes locus coeruleus-mediated suppression of muricidal aggression. *Eur J Pharmacol* 91: 329-330, 1983.

35. **Oishi R, Ueki S.** Facilitation of muricide by dorsal norepinephrine bundle lesions in olfactory bulbectomized rats. *Pharmacol Biochem Behav* 8: 133-139, 1978.

36. **Yamamoto T, Watanabe T, Shibata S, Ueki S.** The effect of locus coeruleus and midbrain raphe stimulation on muricide in rats. *Jpn J Pharmacol* 29 (Suppl 41P), 1979.

37. **Kozak W, Valzelli L, Garattini S.** Anxiolytic activity on locus coeruleus-mediated suppression of muricidal aggresion. *Eur J Pharmacol* 105: 323-326, 1984.

38. **Lechin F, Van der Dijs B, Jakubowicz D, Camero RE, Villa S, Arocha L, Lechin AE.** Effects of clonidine on blood pressure, noradrenaline, cortisol, growth hormone, and prolactin plasma levels in high and low intestinal tone depressed patients. *Neuroendocrinology* 41: 156-162, 1985.

39. **Siever LJ, Pickar D, Lake CR, Cohen RM, Uhde TW, Murphy DL.** Extreme elevations in plasma norepinephrine associated with decreased alpha-adrenergic responsivity in major depressive disorder: two case reports. *J Clin Psychopharmacol* 3: 39-41, 1983.

40. **Segal DS, Mandell AJ.** Behavioral activation of rats during intraventricular infusion of norepinephrine. *Proc Natl Acad Sci* 66: 289-293, 1970.

41. **Schildkraut JJ, Orsulak PJ, Gudeman JE.** Recent studies of the role of catecholamines in the pathophysiology and classification of depressive disorders, in *Neuroregulators and Psychiatric Disorders*. Oxford Press, New York, 1977, 122.

42. **Lake CR, Pickar D, Ziegler MG, Lipper S, Slater S, Murphy DL.** High plasma norepinephrine levels in patients with major affective disorder. *Am J Psychiatry* 139: 1315-1318, 1982.

43. **Lake CR, Pickar D, Ziegler, MG.** Plasma norepinephrine and affective disorders. Abstr. American Psychiatric Association, New Orleans, May 1981.

44. **Ko GN, Elxworth JD, Roth RH, Rifkin BG, Leigh H, Redmond DE Jr.** Panic-induced elevation of plasma MHPG in phobic-anxious patients:effects of clonidine or imipramine. *Arch Gen Psychiatry* 40: 425-430, 1983.

45. **Robinson DS, Johnson GA, Nies A, Corcella J, Cooper TB, Albright D, Howard D.** Plasma levels of catecholamines and dihydroxyphenylglycol during antidepressant drug treatment. *J Clin Psychopharmacol* 3: 282-287, 1983.

46. **Nesse RM, Cameron OG, Curtis GC, McCann DS, Huber-Smith MJ.** Adrenergic functions in patients with panic anxiety. *Arch Gen Psychiatry* 41: 771-776, 1984.

47. **Charney DS, Heninger GR.** Noradrenergic function and the mechanism of action of antianxiety treatment. I. The effect of long-term alprazolam treatment. *Arch Gen Psychiatry* 42: 458-467, 1985.

48. **Sulser F, Mobley PL.** Regulation of central noradrenergic receptor function: new vistas on the mode of action of antidepressant treatments, in *Neuroreceptors: Basic Clinical Aspects*. Usdin E, Bunney WB, Davis JM. Eds., John Wiley & Sons, New York, 1981, 55.

49. **Mignot E, Laude D, Elghozi J, LeQuan-Bui KH, Meyer P.** Central administration of yohimbine increases free 3-methoxy-4-hydroxyphenylglycol in the cerebrospinal fluid of the rat. *Eur J Pharmacol* 83: 135-138, 1982.

50. **Peyrin L, Pequignot JM, Chauplannaz G, Laurent B, Aimard G.** Sulfate and glucuronide conjugates of 3-metoxy-4-hydroxy-phenylglycol (MHPG) in urine of depressed patients: central and peripheral influences. *J Neur Transm* 63: 255-269, 1985.

51. **Sweeney DR, Maas JW, Heninger GR.** State anxiety and urinary MHPG. *Arch Gen Psychiatry* 35: 1418-1423, 1978.

52. **Scatton B.** Brain 3,4-dihydroxyphenylethyleneglycol levels are dependent on central noradrenergic neuron activity. *Life Sci* 31: 495-504, 1982.

53. **Charney DS, Galloway MP, Heninger GR.** The effects of caffeine on plasma MHPG subjective anxiety, autonomic symptoms and blood pressure in healthy humans. *Life Sci* 35: 135-144, 1984.

54. **Lechin F, van der Dijs B, Gomez F, Arocha L, Acosta E, Lechin E.** Distal colon motility as a predictor of antidepressant response to fenfluramine, imipramine and clomipramine. *J Affect Dis* 5: 27-35, 1983.

55. **Lechin F, van der Dijs B, Jakubowicz D, Camero RE, Villa S, Lechin E, Gomez F.** Effects of clonidine on blood pressure, noradrenaline, cortisol, growth hormone, and prolactin plasma levels in high and low intestinal tone subjects. *Neuroendocrinology* 40: 253-261, 1985.

56. **Lechin F, van der Dijs B, Gomez F, Acosta E, Arocha L.** Comparison between the effects of D-amphetamine and fenfluramine on distal colon motility in non-psychotic patients. *Res Commun Psychol Psychiatr Behav* 7: 411-430, 1982.

57. **Lake CR, Ziegler MG, Kopin IJ.** Use of plasma norepinephrine for evaluation of sympathetic neuronal function in man. *Life Sci* 18: 1315-1326, 1976.

58. **Swann AC, Maas JW, Hattox SE.** Catecholamine metabolites in human plasma as indices of brain function: effects of debrisoquin. *Life Sci* 27: 1857-1862, 1980.

59. **Crawley JN, Hattox SE, Maas JW, Roth RH.** 3-Methoxy-4-hydroxyphenylethylglycol increase in plasma after stimulation of the nucleus locus coeruleus. *Brain Res* 141: 380-384, 1978.

60. **Mason ST, Fibiger HC.** Anxiety: the locus coeruleus disconnection. *Life Sci* 25: 2141-2147, 1979.

61. **File SE, Deakin JFW, Longden A, Crow TJ.** An investigation of the role of the locus coeruleus in anxiety and agonistic behaviour. *Brain Res* 169: 411-420,1979.

62. **Reiner PB.** Clonidine inhibits central noradrenergic neurons in unanesthetized cats. *Eur J Pharmacol* 112: 105-110, 1985.

63. **Jones BE, Harper ST, Halaris AE.** Effects of locus coeruleus lesions upon cerebral monoamine content, sleep-wakefulness states and the response to amphetamine in the cat. *Brain Res* 124: 473-496, 1977.

64. **Kostowski W.** Noradrenergic interactions among central neurotransmitters, in *Neurotransmitters, Receptors and Drug Action*. Essman W. Ed., Spectrum, New York, 1980, 47.

65. Granat **AR, Kumada M, Reis DJ.** Sympathoinhibition by A1-noradrenergic neurons is mediated by neurons in the C1 area of the rostral medulla. *J Auton Nerv System* 14: 387-395, 1985.

66. **Kostowski W.** Two noradrenergic systems in the brain and their interactions with other monoaminergic neurons. *Pol J Pharmacol Pharm* 31: 425-436, 1979.

67. **Kostowski W, Jerlicz M, Bidzinski A, Hauptmann M.** Evidence for existence of two opposite noradrenergic brain systems controlling behavior. *Psychopharmacology* 59: 311-312, 1978.

68. **Guyenet PG, Cabot JB.** Inhibition of sympathetic preganglionic neurons by catecholamines and clonidine: mediation by an alpha-adrenergic receptor. *J Neurosci* 1: 908-917, 1981.

69. **Lindvall O, Bjorklund A.** Organization of catecholamine neurons in the rat central nervous system, in *Chemical Pathways in the Brain, Handbook of Psychopharmacology*. Iversen LL, Iversen SD, Snyder SH. Eds., Plenum Press, New York, 1978, 139.

70. **Mason ST, Iversen SD.** Effects of selective forebrain noradrenaline loss on behavioral inhibition in the rat. *J Comp Physiol Psychol* 91: 165-173, 1977.

71. **Crow TJ, Deakins JFW, File SE, Longden A, Wendlant S.** The locus coeruleus noradrenergic system evidence against a role in attenuation, habituation, anxiety and motor activity. *Brain Res* 155: 249-261, 1978.

72. **Van Dongen PAM.** Locus coeruleus region: effects on behavior of cholinergic, noradrenergic, and opiate drugs injected intracerebrally into freely moving cats. *Exp Neurology* 67: 52-78, 1980.

73. **Bobillier P, Sequin S, Petitjean F, Salvert D, Touret M, Jouvet M.** The raphe nuclei of the cat brain stem: a topographical atlas of their efferent projections as revealed by autoradiography. *Brain Res* 113: 449-486, 1976.

74. **Moore RY, Halaris AE, Jones BE.** Serotonin neurons of the midbrain raphe: ascending projections. *J Comp Neurol* 180: 417-438, 1978.

75. **Slopsema JS, van der Gugten J, Bruin JPC.** Regional concentrations of noradrenaline and dopamine in the frontal cortex of the rat: dopaminergic innervation of the prefrontal subareas and lateralization of prefrontal dopamine. *Brain Res* 250: 197- 200, 1982.

76. **Tassin JP, Bockaert J, Blanc G, Stinus L, Thierry AM, Lavielle S, Premont J, Glowinski J.** Topographical distribution of dopaminergic innervation and dopaminergic receptors of the anterior cerebral cortex of the rat. *Brain Res* 154: 241-251, 1978.

77. **Herve D, Blanc G, Glowinski J, Tassin JP.** Reduction of dopamine utilization in the prefrontal cortex but not in the nucleus accumbens after selective destruction of noradrenergic fibers innervating the ventral tegmental area in the rat. *Brain Res* 237: 510-516, 1982.

78. **Herve D, Simon H, Blanc G, LeMoal M, Glowinski J, Tassin JP.** Opposite changes in dopamine utilization in the nucleus accumbens and the frontal cortex after electrolytic lesion of the median raphe in the rat. *Brain Res* 216: 422-428, 1981.

79. **Lavielle S, Tassin JP, Thierry AM, Blanc G, Herve D, Barthelemy C, Glowinski J.** Blockade by benzodiazepines of the selective high increase in dopamine turnover induced by stress in mesocortical dopaminergic neurons of the rat. *Brain Res* 168: 585-594, 1979.

80. **Bannon MJ, Roth RH.** Pharmacology of mesocortical dopamine neurons. *Pharmacol Rev* 35: 53-68, 1983.
81. **Bannon MJ, Wolf ME, Roth RH.** Pharmacology of dopamine neurons in innervating the prefrontal, cingulate and piriform cortices. *Eur J Pharmacol* 92: 119-125, 1983.
82. **Hadfield MG.** Mesocortical vs. nigrostriatal dopamine uptake in isolated fighting mice. *Brain Res* 222: 172-176, 1981.
83. **Pycock CJ, Kerwin RW, Carter CJ.** Effect of lesion of cortical dopamine terminals on subcortical dopamine receptors in rats. *Nature* 286: 74-77, 1980.
84. **Simon H, Scatton B, LeMoal M.** Dopaminergic A10 neurons are involved in cognitive functions. *Nature* 286: 150-151, 1980.
85. **Conrad ICA, Leonard CM, Pfaff DW.** Connections of the median and dorsal raphe nuclei in the rat: an autoradiographic and degeneration study. *J Comp Neurol* 156: 179-206, 1974.
86. **Plaznik A, Danysz W, Kostowski W, Bidzinski A, Hauptmann M.** Interaction between noradrenergic and serotonergic brain systems as evidenced by behavioral and biochemical effects of microinjections of adrenergic agonists and antagonists into the median raphe nucleus. *Pharmacol Biochem Behav* 19: 27-32, 1983.
87. **Bunney BS, Aghajanian GK.** Mesolimbic and mesocortical dopaminergic systems: physiology and pharmacology, in *Psychopharmacology: A Generation of Progress*. Lipton MA, DiMascio A, Killam KF. Eds. ,Raven Press, New York, 1978, 221.
88. **Moore KE, Kelly PH.** Biochemical pharmacology of mesolimbic and mesocortical dopaminergic neurons, in *Psychopharmacology: A Generation of Progress*. Lipton MA, DiMascio A, Killam KF. Eds., Raven Press, New York, 1978, 221.
89. **Nicolaou NM, Garcia-Munoz M. Arbuthnott G, Eccleston D.** Interactions between serotonergic and dopaminergic systems in rat brain demonstrated by small unilateral lesions of the raphe nuclei. *Eur J Pharmacol* 57: 295-305, 1979.
90. **Samanin R, Quattrone A, Consolo S, Ladinsky H, Algeri S.** Biochemical and pharmacological evidence of the interaction of serotonin with other aminergic systems in the brain, in *Interactions Between Putative Neurotransmitters*. Garattini S, Pujol JF, Samanin R. Eds., Raven Press, New York, 1978, 355.
91. **Phillipson OT.** Afferent projections to the ventral tegmental area of Tsay and interfascicular nucleus: a horseradish peroxidase study in the rat. *J Comp Neurol* 187: 117-144, 1979.
92. **Simon H, LeMoal M, Calas A.** Efferents and afferents of the ventral tegmental A10 region studies after local injection of (3H) leucine and horseradish peroxidase. *Brain Res* 178: 17-40, 1979.
93. **File SE, Vellucci SV.** Studies on the role of stress hormones and of 5-HT in anxiety using an animal model. *J Pharm Pharmacol* 30: 105-110, 1978.
94. **Wise CD, Berger BD, Stein L.** Benzodiazepines: anxiety reducing activity by reduction of serotonin turnover in brain. *Science* 177: 180-182, 1977.
95. **Geller I, Hartmann RJ, Croy DJ.** Attenuation of conflict behavior with cinanserin, a serotonin antagonist. Reversal of the effect with 5-hydroxytryptophan and alpha-methyltryptamine. *Res Commun Chem Pathol Pharmacol* 7: 165-174, 1974.
96. **Graeff FG.** Tryptamine antagonists and punished behavior. *J Pharmacol Exp Ther* 189: 344-350, 1974.
97. **Graeff FG.** Effect of cyproheptadine and amphetamine on intermittently reinforced lever-pressing in rats. *Psychopharmacology* 50: 65-71, 1976.
98. **Hole K, Fuxe K, Jonsson G.** Behavioral effects of 5,7-DHT lesions of ascending serotonin pathways. *Brain Res* 107: 385-399, 1976.
99. **Jacobs BL, Cohen A.** Differential behavioral effects of lesions of the median or dorsal raphe nuclei in rats: open field and pain elicited aggression. *J Comp Physiol Psychol* 46: 102-112, 1976.
100. **Deakin JFW, File SE, Hyde JRG, MacLeod NK.** Ascending 5-HT pathways and behavioural habituation. *Pharmacol Biochem Behav* 10: 687-694, 1979.
101. **Hole K, Johnson GE, Berge O-G.** 5,7-Dihydroxtryptamine lesions of the ascending 5-hydroxytryptaminergic pathways: habituation motor activity and agonistic behaviour. *Pharmacol Biochem Behav* 7: 205-210, 1977.
102. **Matte AC, Tornow H.** Parachlorophenylalanine produces dissociated effects on aggression "emotionality" and motor activity. *Neuropharmacology* 17: 555-558, 1978.
103. **Miczek KA, Altman JL, Appel JB, Boggam WO.** Parachlorophenylalanine, serotonin and behavior. *Pharmacol Biochem Behav* 3: 355-361, 1975.
104. **Morato de Carvalho S, de Aguiar JC, Graeff F.** Effect of minor tranquilizers, tryptamine antagonists and amphetamine on behavior punished by brain stimulation. *Pharmacol Biochem Behav* 15: 351-356, 1981.
105. **Segal M, Weinstock M.** Differential effects of 5-hydroxytryptamine antagonists on behaviors resulting from activation of different pathways arising from the raphe nuclei. *Psychopahrmacology* 79: 72-78, 1983.
106. **Thiebot MH, Hamon M, Soubrie P.** Attenuation of induced-anxiety in rats by chlordiazepoxide: role of raphe dorsalis benzodiazepine binding sites and serotonergic neurones. *Neuroscience* 7: 2287-2294, 1982.
107. **File SE, Hyde JRC, MacLeod NK.** 5,7-Dihydroxytryptamine lesions of dorsal and median raphe nuclei and performance in the social interaction test of anxiety and in a home cage aggression test. *J Affect Dis* 1: 115-122, 1979.

108. **Nauta WJH, Smith GP, Faull RLM, Domesick VB.** Efferent connections and nigral afferents of the nucleus accumbens septi in the rat. *Neuroscience* 3: 385-401, 1978.

109. **Wiklund L.** Studies on Anatomical, Functional, and Plastic Properties of Central Serotonergic Neurons. Doctoral dissertation, University of Lund, Sweden, 1980.

110. **Thiebot MH, Hamon M, Soubrie P.** The involvement of nigral serotonin innervation in the control of punishment-induced behavioral inhibition in rats. *Pharmacol Biochem Behav* 19: 225-229, 1983.

111. **Thiebot M-H, Soubrie P, Hamon M, Simon P.** Evidence against the involvement of serotonergic neurons in the antipunishment activity of diazepam in the rat. *Psychopharmacology* 82: 355-359, 1984.

112. **Van der Kooy D, Hattori T.** Dorsal raphe cells with collateral projections to the caudate-putamen and substantia nigra: a fluorescent retrograde double labeling study in the rat. *Brain Res* 186: 1-7, 1980.

113. **Bentivoglio M, van der Kooy D, Kuypers HGJM.** The organization of the efferent projections of the substantia nigra in the rat. A retrograde fluorescent double labeling study. *Brain Res* 174: 1-17, 1979.

114. **Dray A, Davies J, Oakley NR, Tongroach P, Vellucci S.** The dorsal and medial raphe projections to the substantia nigra in the rat: electrophysiological, biochemical and behavioural observations. *Brain Res* 151: 431-442, 1978.

115. **Pasquier DA, Kemper TL, Forbes WB, Morgane PJ.** Dorsal raphe, substantia nigra and locus coeruleus: interconnections with each other and the neostriatum. *Brain Res Bull* 2: 323-329, 1977.

116. **Pycock CJ, Carter CJ, Kerwin RW.** Effect of 6-hydroxydopamine lesions of the medial prefrontal cortex on neurotransmitters systems in subcortical sites in the rat. *J Neurochem* 34: 91-99, 1980.

117. **Blanc G, Herve D, Simon H, Lisoprawski A, Glowinski J, Tassin JP.** Response to stress of mesocortico-frontal dopaminergic neurons in rats after long-term isolation. *Nature* 284: 265-267, 1980.

118. **Fadda F, Argiolas A, Melin ME, Tissari AM, Onali PL, Gessa GL.** Stress induced increase in 3,4-dihydroxyphenylacetic acid (DOPAC) levels in the cerebral cortex and in *N. accumbens*: reversal by diazepam. *Life Sci* 23: 2219-2224, 1978.

119. **Wang RY.** Dopaminergic neurons in the rat ventral tegmental area. I. Identification and characterization. *Brain Res Rev* 3: 123-140, 1981.

120. **Wang RY.** Dopaminergic neurons in the rat ventral tegmental area. II. Evidence for autoregulation. *Brain Res* 3: 141-151, 1981.

121. **Wang RY.** Dopaminergic neurons in the rat ventral tegmental area. III. Effects of D- and L-amphetamine. *Brain Res Rev* 3: 152-165, 1981.

122. **Carter CJ, Pycock CJ.** Behavioural and biochemical effects of dopamine and noradrenaline depletion within the medial prefrontal cortex of the rat. *Brain Res* 192: 163-176, 1980.

123. **Didier M, Belin MF, Aguera M, Buda M, Pujol JF.** Pharmacological effects of GABA on serotonin metabolism in the rat brain. *Neurochem Int* 7: 481-489, 1985.

124. **Jackson DM, Anden NE, Dahlstrom A.** A functional effect of dopamine-rich parts of rat brain. *Psychopharmacologia* 45: 139-150, 1975.

125. **Tassin JP, Stinus J, Simon H, Blanc G, Thierry AM, LeMoal H, Cardo B, Glowinski J.** Relationship between the locomotor activity induced by A10 lesions and the destruction of the fronto-cortico dopaminergic innervation in the rat. *Brain Res* 141: 267-281, 1978.

126. **Beart PM, McDonald D.** 5-Hydroxytryptamine and 5-hydroxytryptaminergic-dopaminergic interactions in the ventral tegmental area of rat brain. *J Pharm Pharmacol* 34: 591-593, 1982.

127. **Fuxe K, Hökfelt T, Agnati L, Johansson D, Ljangdahl A, Perez de la Mora M.** Regulation of the mesocortical dopamine neurons, in *Advances in Biochemical Psychopharmacology. Nonstriatal Dopaminergic Neurons.* Costa E, Gessa GL. Eds. ,Raven Press, New York, 1977, 47.

128. **Warbritton JD III, Stewart RM, Baldessarini RJ.** Decreased locomotor activity and attenuation of amphetamine hyperactivity with intraventricular infusion of serotonin in the rat. *Brain Res* 143: 373-382, 1978.

129. **Lyness WH, Moore KE.** Destruction of 5-hydroxytryptaminergic neurons and the dynamics of dopamine in nucleus accumbens septi and other forebrain regions of the rat. *Neuropharmacology* 20: 327-334, 1981.

130. **Carter CJ, Pycock CJ.** The effects of 5,7-dihydroxytryptamine lesions of extra-pyramidal and mesolimbic sites on spontaneous motor behavior and amphetamine-induced stereotype. *Naunyn-Schmiedeberg's Arch Pharmacol* 208: 51-54, 1979.

131. **Herve D, Simon H, Blanc G, Lisoprawski A, LeMoal M, Glowinski J, Tassin JP.** Increased utilization of dopamine in the nucleus accumbens but not in the cerebral cortex after dorsal raphe lesion in the rat. *Neurosci Lett* 15: 127-134, 1979.

132. **Lee EHY, Geyer MA.** Indirect effects of apomorphine on serotonergic neurons in rats. *Neuroscience* 11: 437-442, 1984.

133. **Geyer MA, Puerto A, Menkes DB, Segal DS, Mandell AJ.** Behavioral studies following lesions of the mesolimbic and mesostriatal serotonergic pathways. *Brain Res* 106: 257-270, 1976.

134. **Plaznik A, Kostowski W, Bidzinski A, Hauptmann M.** Effects of lesions of the midbrain raphe nuclei on avoidance learning in rats. *Physiol Behav* 24: 257-262, 1980.

135. **Azmitia EC, Segal M.** An autoradiographic analysis of the differential ascending projections of the dorsal and medial raphe nuclei of the rat. *J Comp Neurol* 179: 641-668, 1978.

136. **Przewlocka B, Kukulka L, Tarczynska E.** The effect of lesions of dorsal or median raphe nucleus on rat behavior. *Pol J Pharmacol Pharm* 29: 573-579, 1977.

137. **Waldbilig RJ.** The role of the dorsal raphe and median raphe in the inhibition of muricide. *Brain Res* 160: 341-346, 1979.

138. **Graeff FG, Quintero S, Gray JA.** Median raphe stimulation, hippocampal theta rhythm and threat-induced behavioral inhibition. *Physiol Behav* 25: 253-261, 1980.

139. **Steranka LR, Barret RJ.** Facilitation of avoidance acquisition by lesion of the median raphe nucleus: evidence for serotonin as a mediator of shock-induced suppression. *Behav Biol* 11: 205-213, 1974.

140. **Mosko SS, Haubrich D, Jacobs BL.** Serotonergic afferents to the dorsal raphe nucleus: evidence from HRP and synaptosomal uptake studies. *Brain Res* 119: 269-290, 1977.

141. **Spealman RD.** Comparison of drug effects on responding punished by pressurized air or electric shock delivery in squirrel monkeys: pentobarbital, chlordiazepoxide, D-amphetamine and cocaine. *J Pharmacol Exp Ther* 209: 309-315, 1979.

142. **Thierry AM, Tassin JP, Blanc G, Glowinski J.** Selective activation of the mesocortical dopaminergic system by stress. *Nature* 263: 242-244, 1976.

143. **Anisman H, Ritch M, Sklar LS.** Noradrenergic and dopaminergic interactions in escape behavior: analysis of uncontrollable stress effects. *Psychopharmacology* 74: 263- 268, 1981.

144. **Herman JP, Guillonneau D, Dantzer R, Scatton B, Semerdjian-Rouquier L, LeMoal M.** Differential effects of inescapable footshocks and of stimuli previously paired with inescapable footshocks on dopamine turnover in cortical and limbic areas of the rat. *Life Sci* 30: 2207-2214, 1982.

145. **Reinhard JF Jr, Bannon MJ, Roth RH.** Acceleration by stress of dopamine synthesis and metabolism in prefrontal cortex: antagonism by diazepam. *Naunyn-Schmiedeberg's Arch Pharmacol* 318: 374-377, 1982.

146. **Lorens SA, Guldberg HC, Hole K, Kohler C, Srebro B.** Activity, avoidance learning and regional 5-hydroxytryptamine following intra-brain stem 5,7-dihydroxytryptamine and electrolytic midbrain raphe lesions in the rat. *Brain Res* 108: 97-113, 1976.

147. **Dyr W, Kostowski W, Zacharski B, Bidzinski A.** Differential clonidine effects on EEG following lesions of the dorsal and median raphe nuclei in rats. *Pharmacol Biochem Behav* 19: 177-185, 1983.

148. **Forchetti CM, Meek JL.** Evidence for a tonic GABAergic control of serotonin neurons in the median raphe nucleus. *Brain Res* 206: 208-212, 1981.

149. **Balfour DJK.** Effects of GABA and diazepam on (3H) serotonin release from hippocampal synaptosomes. *Eur J Pharmacol* 68: 11-16, 1980.

150. **Scatton B, Serrano A, Nishikawa T.** GABAmimetics decrease extracellular concentrations of 5-HIAA (as measured by in vivo voltammetry) in the dorsal raphe of the rat. *Brain Res* 341: 372-376, 1985.

151. **Nishikawa T, Scatton B.** Inhibitory influence of GABA on central serotonergic transmission. Raphe nuclei as the neuro-anatomical site of the GABAergic inhibition of cerebral serotonergic neurons. *Brain Res* 331: 91-103, 1985.

152. **Brady LS, Barret JE.** Effects of serotonin receptor antagonists on punished responding maintained by stimulus-shock termination or food presentation in squirrel monkeys. *J Pharmacol Exp Ther* 234: 106-234, 1985.

153. **Graeff FG, Rawlins N.** Dorsal periaqueductal gray punishment, septal lesions and the mode of action of minor tranquilizers. *Pharmacol Biochem Behav* 12: 41-45, 1980.

154. **Tye NC, Everitt BJ, Iversen SD.** 5-Hydroxytryptamine and punishment. *Nature* 268: 741-742, 1977.

155. **Tye NC, Iversen SD, Green AR.** The effect of benzodiazepines and serotonergic manipulations on punished responding. *Neuropharmacology* 18: 689-696, 1979.

156. **Kilts CD, Commissaris RL, Cordon JJ, Rech RH.** Lack of central 5-hydroxy- tryptamine influence on the anticonflict activity of diazepam. *Psychopharmacology* 78: 156-164, 1982.

157. **Engel JA, Hjorth S, Svensson K, Carlsson A, Liljequist S.** Anticonflict effect of the putative serotonin receptor agonist 8-hydroxy-2-(di-*n*-propylamino) tetralin (8-OH-DPAT). *Eur J Pharmacol* 105: 365-368, 1984.

158. **Soubrie P, Blas C, Ferron A, Glowinski J.** Chlordiazepoxide reduces in vivo serotonin release in the basal ganglia of encephale isole but not anesthetized cats: evidence for a dorsal raphe site of action. *J Pharmacol Exp Ther* 226: 526-532, 1983.

159. **Schlicker E, Classen K, Gothert M.** GABA receptor-mediated inhibition of serotonin release in the rat brain. *Naunyn-Schmiedeberg's Arch Pharmacol* 326: 99-105, 1984.

160. **Sainati SM, Lorens SA.** Intra-raphe benzodiazepines enhance rat locomotor activity: Interactions with GABA. *Pharmacol Biochem Behav* 18: 407-414, 1983.

161. **Przewocka B, Stala L, Scheel-Kruger J.** Evidence that GABA in the nucleus dorsalis raphe induces stimulation of locomotor activity and eating behavior. *Life Sci* 25: 937-946, 1979.

162. **Sainatl SM.** Midbrain Benzodiazepine-GABA-Serotonin Interactions: Effects on Locomotor Activity in the Rat. Doctoral dissertation, Loyola University, Chicago, 1982.

163. **Sainati SM, Lorens SA.** Intra-raphe muscimol-induced hyperactivity depends on ascending serotonin projections. *Pharmacol Biochem Behav* 17: 973-986, 1982.

164. **File SE.** Clinical lesions of both dorsal and median raphe nuclei and changes in social and aggressive behaviour in rats. *Pharmacol Biochem Behav* 12: 855-859, 1980.

165. **Gallager DW.** Benzodiazepines: potentiation of a GABA inhibitory response in the dorsal raphe nucleus. *Eur J Pharmacol* 49: 133-143, 1978.

166. **Kostowski W, Plaznik A, Pucilowski AO, Bidzinski A, Hauptmann M.** Lesion of serotonergic neurons antagonizes clonidine-induced suppression of avoidance behavior and locomotor activity in rats. *Psychopharmacology* 73: 261-264, 1981.

167. **Nishikawa T, Scatton B.** Inhibitory influence of GABA on central serotonergic transmission. Involvement of the habenulo-raphe pathways in the GABAergic inhibition of ascending cerebral serotonergic neurons. *Brain Res* 331: 81-90, 1985.

168. **Baraban JM, Aghajanian GK.** Suppression of serotonergic neuronal firing by alpha-adrenoceptor antagonists: evidence against GABA mediation. *Eur J Pharmacol* 66: 287-294, 1980.

169. **Vandermaelen CP, Aghajanian GK.** Noradrenergic activation of serotonergic dorsal raphe neurons recorded in vitro. *Soc Neurosci Abstr* 8: 482, 1982.

170. **Anden NE, Grabowska M.** Pharmacological evidence for a stimulation of dopamine neurons by noradrenaline neurons in the brain. *Eur J Pharmacol* 39: 275-282, 1976.

171. **Anden NE, Atack CV, Svensson TH.** Release of dopamine from central noradrenaline and dopamine nerves induced by a dopamine-beta-hydroxylase inhibitor. *J Neural Transm* 34: 93-100, 1973.

172. **Antelman SM, Black CA.** Dopamine-beta-hydroxylase inhibitors (DBHI) reverse the effects of neuroleptics under activating conditions: possible evidence for a norepinephrine (NE)-Dopamine (DA) interaction. *Soc Neurosci* (Abstr), Anaheim, 1977.

173. **Heym J, Trulson ME, Jacobs BL.** Effects of adrenergic drugs on raphe unit activity in freely moving cats. *Eur J Pharmacol* 74: 117-125, 1981.

174. **Simon H, LeMoal M, Stinus L, Calas A.** Anatomical relationships between the ventral mesencephalic tegmentum-A10 region and the locus coeruleus as demonstrated by anterograde and retrograde tracing techniques. *J Neural Transm* 44: 77-86, 1979.

175. **Berger B, Tassin JP, Blanc G, Moyne MA, Thierry AM.** Histochemical confirmation for dopaminergic innervation of the rat cerebral cortex after destruction of the noradrenergic ascending pathways. *Brain Res* 81: 332-337, 1974.

176. **Trulson MW, Crisp T.** Role of norepinephrine in regulating the activity of serotonin- containing dorsal raphe neurons. *Life Sci* 35: 511-515, 1984.

177. **Anderson C, Pasquier D, Forbes W, Morgane P.** Locus coeruleus-to-dorsal raphe input examined by electrophysiological and morphological methods. *Brain Res Bull* 2: 209-221, 1977.

178. **Svensson TH, Bunney BS, Aghajanian GK.** Inhibition of both noradrenergic and serotonergic neurons in brain by the alpha-adrenergic agonist clonidine. *Brain Res* 92: 291-300, 1975.

179. **Baraban JM, Aghajanian GK.** Suppression of firing activity of 5-HT neurons in the dorsal raphe by alpha-adrenoceptor antagonists. *Neuropharmacology* 19: 355-363, 1980.

180. **Waldmeier PC.** Stimulation of central serotonin turnover by beta-adrenoceptor agonists. *Naunyn-Schmiedeberg's Arch Pharmacol* 317: 115-119, 1981.

181. **Baraban J, Aghajanian GK.** Noradrenergic innervation of serotonergic neurons in the dorsal raphe: demonstration by electron microscopic autoradiography. *Brain Res* 204: 1-11, 1981.

182. **Drugan RC, Maier SF, Skolnick P, Paul SM, Crawley JN.** An anxiogenic benzodiazepine receptor ligand induces learned helplessness. *Eur J Pharmacol* 113: 453-457, 1985.

183. **Haefely W, Pieri L, Polc P, Schaffner R.** General pharmacology and neuropharmacology of benzodiazepine derivatives, in *Handbook of Experimental Pharmacology*. Hoffmeister F, Stille G. Eds., Springer-Verlag, Berlin, 1981, 136.

184. **Lippa AS, Meyerson LR, Beer B.** Molecular substrates of anxiety: clues from the heterogeneity of benzodiazepine receptors. *Life Sci* 31: 1409-1416, 1982.

185. **Seidel WF, Cohen SA, Bliwise NG, Dement WC.** Buspirone. an anxiolytic without sedative effect. *Psychopharmacology* 81: 371-373, 1985.

186. **Sanger DJ, Joly D.** Anxiolytic drugs and the acquisition of conditioned fear in mice. *Psychopharmacology* 85: 284-288, 1985.

187. **McMillen BA, Matthews RT, Sanghera MK, Shepard PD, German DC.** Dopamine receptor antagonism by the novel antianxiety drug, buspirone. *J Neurosci* 3: 733-738, 1983.

188. **Goldberg HL, Finnerty RJ.** The comparative efficacy of buspirone and diazepam in the treatment of anxiety. *Am J Psychiatry* 136: 1184-1187, 1979.

189. **Stanton HC, Taylor DP, Riblet LA.** Buspirone: an anxioselective drug with dopaminergic action, in *The Neurobiology of the Nucleus Accumbens*. Chronister RW, DeFrance JF. Eds., Haer Institute, Brunswick ME, 1981, 316.

190. Kolasa K, Fusi R, Garattini S, Consolo S, Ladinsky H. Neurochemical effects of buspirone, a novel psychotropic drug, on the central cholinergic system. *J Pharm Pharmacol* 34: 314-317, 1982.

191. Riblet L, Allen L, Hyslop D, Taylor DP, Wilderman R. Pharmacological activity of buspirone, a novel non-benzodiazepine antianxiety agent. *Fed Proc Fed Am Soc Exp Biol* 39: 752-758, 1980.

192. Cimino M, Ponzio F, Achilli G, Vantini G, Perego C, Algeri S, Garattini S. Dopaminergic effects of buspirone, a novel anxiolytic agent. *Biochem Pharmacol* 32: 1069-1074, 1983.

193. Wood PL, Nair NPV, Lal S, Etienne P. Buspirone: a potential atypical neuroleptic. *Life Sci* 33: 269-273, 1983.

Chapter 3

CENTRAL NEURONAL PATHWAYS INVOLVED IN DEPRESSIVE SYNDROME: EXPERIMENTAL FINDINGS

Fuad Lechin, Bertha van der Dijs, José Amat, and Marcel Lechin

TABLE OF CONTENTS

I. INTRODUCTION

A dictionary will describe depression and mania with opposing meanings. Yet from a clinical point of view the terms are poorly defined. The manic syndrome, which we have discussed in earlier papers, groups a series of symptoms arising from psychic and motor hyperactivity without exploring the significance of such symptoms. For example, symptoms which express the search for pleasure and reward, like others expressing anger and aggressiveness, are considered part of the manic syndrome although they spring from different emotional situations.

Animal models of the manic syndromes are easily quantified and qualified because the symptoms are visible. However, this is not true for human subjects who have the capacity to hide, repress, and disguise their emotions or states of feeling and can thus modify their behavior.

The useful animal models in studies of manic syndrome include "septal rage" and "hyperkinetic" syndromes. It will be remembered that the neurophysiological and neurochemical states corresponding to septal rage register a disappearance of dopamine (DA) in subcortical mesolimbic structures, in particular of the septum. The supersensitivity of dopaminergic (DA) receptors registered in these cases was secondary to DA denervation, experimentally induced at septal level. It is observed, furthermore, that noradrenergic (NE) activity reaching the septal region by way of ventral NE bundle (VNB), preferentially, is apparently unaffected by experimental DA denervation.[1-38] Consequently, in experimentally induced septal syndrome an increase of NE/DA ratio is produced at this level. Conversely, in the hyperkinetic syndrome experimentally induced through suppression of DA innervation to prefrontal cortex, dopamine turnover is increased at subcortical mesolimbic level, leading to reduction here of NE/DA ratio.

It becomes apparent that in both septal rage and hyperkinetic syndromes an increase in catecholaminergic activity is produced at subcortical mesolimbic level. This increase is made at the expense of NE in the septal rage syndrome and DA in the hyperkinetic syndrome.

The serotonergic (5HT) system exercises an antagonic and bridling action on both catecholaminergic systems, at all levels of the CNS.[3-6,9-14,17,18,21,39-68] From a theoretical model standpoint, it is possible to postulate the existence of two syndromes, one presenting an increased 5HT/NE ratio and the other an increased 5HT/DA ratio. These two kinds of 5HT predominance over catecholaminergic systems would constitute two types of depressive syndromes.

Leaving aside the theoretical model, numerous depressive syndromes have been described according to clinical, biochemical, and physiological characteristics. One hears different specialists mention anxious depression, anergic depression, unipolar and bipolar depression, endogenous and nonendogenous depression, psychotic and nonpsychotic depression, and NE, 5HT, and cholinergic (ACh) depressions. Lastly, an effort has been made to classify depressive syndromes according to biological markers: dexamethasone suppression test, clonidine test, TRH test, sleep disorders, etc., leading to a "Tower of Babel" of depressions.[55,56, 69-98]

It is our intention to seek the points of convergence which will make possible an examination of the common elements of depression lying beneath the clinical and biochemical features which vary with the individual's personality. In other words, we seek to verify whether the various neurochemical findings correspond to evolutionary stages of the same syndrome or whether one depressive syndrome may be expressed in different clinical ways. The common and invariable denominators, if found, may provide the key of deciphering the disguises masking the essence of the depressive syndrome.

We chose to take a new path, starting from theoretical models and testing them against clinical and biochemical pictures through trial-and-error therapy using low doses of antidepressant drugs whose neuropharmacological mechanisms are well, although not completely, known. Over 10 years of study we have developed an approach which appears to us to be valid.

II. ANIMAL MODELS OF DEPRESSION

The extrapolation of a supposed depressive state in animals to depression in humans is not possible because animal depression models are based on behavior deficits, motor hypoactivity, and decreased response to reward stimuli, etc., while human depression may exist in the absence of these outward signs and may be revealed only through psychodiagnosis. Nevertheless, although human depression has psychological components of a high intellectual order, lacking in the supposed animal models, we presume there exist enough common components between human and nonhuman depression on which to base a certain degree of extrapolation.

To date, several animal models have been evaluated through the pharmacology of depression. First, reserpine-induced depression in animals has received considerable attention following the observation that patients treated with reserpine for hypertension developed depressive states similar in many ways to clinically observed endogenous depression. Among nonhuman subjects, reserpine induces certain symptoms that are characteristic of depression.[70,71,99-104] Neurochemically, it appears that drug-induced depletion of monoamines may be responsible for observed behavioral depression. In support of this, tricyclic antidepressants, which are efficacious in treating endogenous depressions and which among other attributes block the reuptake of monoamines, alleviate reserpine-induced depression in animals. Although tricyclic antidepressants are also potent anti-ACh drugs, there is a relative paucity of data linking their anticholinergic effects to their antidepressant properties.

A second model of depression that has received considerable attention is the development of "learned helplessness" in animals following exposure to inescapable shock.[50-53,105-113] In particular, after animals are exposed to stress such as footshock with little or no possibility of controlling response to shock, in subsequent escape tasks the animals display pronounced deficit in escape performance. Such deficits following inescapable shock appear to be related to a stress-induced decrease in catecholamine activity and are mitigated by pretreatment with tricyclic antidepressants. Somewhat related is the behavioral despair model of depression seen in rats or mice which assume an immobile, floating posture when put to swim in a cylinder with little chance of escape. As is the case with learned helplessness, treatment with tricyclic antidepressants produces a marked relief of stress-induced immobility.

Finally, a third model involves the effects of chronic amphetamine treatment on self-stimulation behavior.[70,71,99,101-104] It has been demonstrated that chronic exposure to amphetamine results in a pronounced depression in rates of self-stimulation responses to intracranial brain stimulation. If animals receiving long-term amphetamine treatment are subsequently tested for intracranial self-stimulation following a saline injection, observed performance is substantially depressed relative to control animals. Moreover, the postamphetamine depression of self-stimulation response is evident through electrode placed in the lateral hypothalamus or the substantia nigra. Although not well understood, the neurochemical effects of chronic amphetamine administration include depletion of norepinephrine, decrease in DA synthesis, and neurotoxic effects on DA terminals. Thus, it is likely that hypoactive catecholamine activity induced by repeated amphetamine treatment may be involved in the observed postamphetamine depression of self-stimulation responses.

A further animal model of depression provides links between an excessive serotonin synaptic activity and depressive syndrome, since the administration of D,L-5-hydroxy-tryptophan (5HTP) induces behavioral depression in rats working on a food-reinforced operant schedule. This induced depression is suppressed by postsynaptic 5HT blockade with several antidepressant drugs (trazodone, amitriptyline, and mianserin). The potency of these drugs parallels their ability to displace radioligand binding to 5HT receptors.[54-56,79,114-122]

At the outset of this paper we postulated that depression might be due to the increase of 5HT/NE and/or 5HT/DA ratios at mesolimbic level, and that an increase of NE/DA ratio would mark schizophrenic depressive syndrome and an increase of DA/NE ratio would be found in schizoaffective depressive syndrome.

Such increments of 5HT/NE and 5HT/DA ratios would be the neurochemical patterns of the depressive stages of schizophrenics and schizoaffective-disorder patients. With respect to this, it has been observed that personality disorder patients show an increased MHPG/5HIAA ratio in their cerebral spinal fluid (CSF) during manic periods (increased aggressiveness-impulsiveness). A similar pattern of low 5HIAA emerged in studies of patients with borderline personality and schizophrenia. Additionally, aggression in animals has been linked to decreased 5HT and/or increased NE functioning rather than simply to the malfunctioning of one independent system.[1-4,6,24,38,44-47,54,66,77,123-138]

Other studies using endocrine and biochemical parameters give support to the role played by 5HT hyperactivity in depressive mechanisms. In effect, the raised plasma levels of cortisol and β-endorphins found in most depressed patients may be associated with the known stimulatory effect of central serotonin on the secretion of both substances.[85,94,95,97,139-145]

Clinical and biochemical findings demonstrate that, far from representing a homogeneous entity, depression presents several faces. According to the action of antidepressant drugs, depressive syndromes have been correlated with central NE deficiency and with central 5HT deficiency.[24,69,70,73-78,83,86-89,99,101-103,118-122,126,146-166] Other studies favor the hypothesis of a hyperactive cholinergic system underlying depression whereas still others offer strong evidence supporting DA deficiency.[73,87,89-91,167]

Although all these theories possess solid support they are unable to meet the challenge of numerous contradictory findings. In our opinion, this failure arises because most authors do not integrate their findings with those obtained by other investigators. Furthermore, most hypotheses attempt to attribute depression to one discrete, specific factor, i.e., "depression is due to supersensitivity of β-adrenergic receptors in the brain." Although this statement arises from many meticulous studies it is only a partial truth and, hence, is unable to answer questions such as: Which are the crucial brain structures? If cortex, what part of the cortex? Another statement: "hyperactivity of the ACh system underlies depression." Yet there exist many central circuits among whose links the ACh system is just one component. Moreover, hyperactivity of the ACh system in prefrontal cortex is associated with behavioral arousal, whereas ACh hyperactivity at brain stem level is associated with both arousal and sedation.

Further theories refer to NE system as a homogeneous central system. This is not true. There are at least two brain NE systems which display opposite effects. Hence, when a hypothesis postulates that depression is due to hyperactivity of the central NE system the question emerges: Which NE system — the locus coeruleus (LC) NE system, or the non-LC NE system?

Similar considerations pertain to the 5HT and DA systems. For instance, accumulated data demonstrate that 5HT-dorsal raphe (DR) and 5HT-median raphe (MR) systems are two distinct and frequently antagonist 5HT systems. Furthermore, evidence not only shows that mesolimbic and mesocortical DA systems differ, but that they exert inhibitory control over each other.

Such observations lead us to believe that any hypothesis attempting to explain depressive mechanisms must take into account all findings made by the many interdisciplinary studies of depression. We seek to apply this procedure in this paper.

III. ANTIDEPRESSANT MECHANISMS

It is accepted that antidepressant mechanisms cannot unequivocally be attributed to the specific pharmacological action of the drug, since agents having many of these pharmacological effects lack antidepressant activity. Antidepressant therapy is accompanied by a lag of 1 to 3 weeks before the onset of beneficial effects and it is known that chronic antidepressant therapy is associated with a number of adaptive changes in central monoaminergic functioning. Drug-induced adaptive modifications can occur both pre- and postsynaptically. However, when presynaptic adaptations occur in transmitter synthesis, storage, and release following chronic administration of a wide range of antidepressant therapies, no common pattern of change is observed.

Thus the changes cannot be attributed primarily to the antidepressant therapeutic action. Instead, the adaptive changes occurring postsynaptically during chronic antidepressant therapies are probably linked to the action mechanisms of the drug.

Much effort has been directed at finding a common mechanism of action for all forms of antidepressant therapies, resulting in not a little disagreement.

A. Changes in Sensitivity of Central β-Adrenoceptors

Radioligand binding studies reveal that chronic administration of tricyclic antidepresants (desipramine, imipramine, nortriptyline, and amitriptyline), MAO-inhibitors (pargyline, clorgyline, and tranylcypromine), and repeated electroconvulsive shock therapy is associated with a reduction in the number of β-adrenoceptor recognition sites present in rat cortex. However, this property does not extend to all atypical antidepressants. For example, it is generally agreed that chronic administration of mianserin fails to alter β-adrenoceptor binding in rat cortex. Moreover, antidepressants fail to change the sensitivity of β-receptors of rat hippocampal pyramidal cells to iontophoretically applied NE.[118-120,122,158-165,168-178]

B. Changes in Sensitivity of Central α-Adrenoceptors

Electrophysiological and behavioral studies indicate that sensitivity of central α_1 adrenoceptors is enhanced by chronic administration of antidepressants. Long-term desipramine, imipramine, amitryptiline, and iprindole therapies are associated with an augmented response to iontophoretically applied NE in the rat facial motor nucleus. Furthermore, the response of single dorsal lateral geniculate neurons in the rat to iontopheretic NE is enhanced by chronically administered desipramine, imipramine, amitryptiline, and iprindole.[175,177-180]

Clonidine-induced aggressiveness in mice is augmented by chronic administration of a number of antidepressants. However, in contrast to these observations and indicating an up-regulation in central α_1-adrenoceptor functioning, the chronic administration of a wide variety of antidepressant therapies is essentially devoid of an effect on rat brain α_1-adrenoceptor binding.[175-178,181-185]

The view has been expressed that a drug-induced subsensitivity of central presynaptic α_2-adrenoceptors may play a critical role in antidepressant mechanisms.[94,176,178,181,182,186,187] The presence of such subsensitivity would contribute to an increased availability of NE in the synaptic cleft. This in turn could possibly account for the reduction in β-adrenoceptor functioning cited earlier, since NE hyperinnervation is associated with a decreased density of β-adrenoceptors in rat brain.[188,189]

The significance of the α_2-adrenoceptor role in antidepressant mechanisms stems from the fact that mianserin is an α_2-receptor antagonist and from the observation that chronically administered desipramine and imipramine induce subsensitive α_2-adrenoceptors in rat heart and brain, respectively.[119,161-163]

Neurochemical experiments confirm the ability of both chronic desipramine and imipramine to down-regulate the sensitivity of rat brain α_2-adrenoceptors, as reflected by an attenuation of the ability of low doses of clonidine to decrease brain NE turnover.[190-195] Antagonism of the behavioral effects of clonidine also appears following long-term administration of either desipramine or imipramine. The pooling of these observations strongly suggests that chronic imipramine and its demethylated congener desipramine do in fact induce subsensitive α_2-adrenoceptors in rat brain, although a somewhat different conclusion has been reached by others using a more indirect approach.

Chronic mianserin has been observed in two studies to induce supersensitive α_2 adrenoceptors. If, in fact, the induction of subsensitive central presynaptic adrenoceptors represents the fundamental mode of action of antidepressants, then it is logical to assume that the effect would be observed after the chronic administration of a wide spectrum of antidepressants possessing markedly different acute pharmacological profiles.[119,161,162]

C. Changes in Sensitivity of Central Dopaminergic Receptors

The chronic administration of desipramine in rats is associated with supersensitive postsynaptic DA receptors in mesolimbic but not in striatal system. Repeated ECT produces postsynaptic DA receptor supersensitivity, but this is dependent on the presence of intact NE neurons. The induction of subsensitive presynaptic DA receptors by chronic antidepressant therapies, observed in some but not all studies, would be expected to modify central DA turnover. This, however, has shown to be unaltered by chronic administration of a variety of antidepressants. However, repeated treatment with imipramine, amitryptiline, iprindole, and ECT induces a progressive subsensitivity of somatodendritic DA autoreceptors. The induction of receptor subsensitivity is dependent upon the passage of time rather than on repeated daily treatments.[78,88,121,122,152,153,162,163,175]

D. Changes in Sensitivity of Central Serotonergic Receptors

As in the case of α_1-adrenoceptors, both behavioral and electrophysiological studies point to up-regulation in sensitivity of central 5HT receptors following chronic antidepressant therapies. Chronic amitryptiline enhances the behavioral responses of rats to 5HTP and mice to 5HT agonists 5-methoxy-*N,N*-dimethyltryptamine. In contrast, chronically administered MAO type-A inhibitors are associated with a reduction in the number of rat cortical 5HT1 binding sites, an effect which is dependent on the presence of unaltered brain 5HT levels.

In contrast to their lack of effect on central 5HT1 binding, chronically administered antidepressants down-regulate the number of 5HT2 recognition sites present in rat brain. This is a property which is common to tricyclic antidepressants, atypical antidepressants, and MAO inhibitors. However, repeated ECT is associated with an increased number of recognition sites.[55,56,79,81,82,115-117,122,152,160,178,196]

Although the above findings are insufficient in themselves to explain either the action mechanisms of antidepressants or the working of depressive syndromes, the data provide valuable information on which to base an approach to mechanisms involved in depression.

IV. TWO TYPES OF DEPRESSIVE SYNDROMES

In previously published papers we have presented physiological, biochemical, hormonal, clinical, and therapeutical data supporting the existence of at least two types of depressive syndromes: (1) depressed subjects showing high distal colon tone (high-IT) plus low norepinephrine (NE) plasma level and (2) low-IT + high NE plasma level.[92-95] Type 1 depressed patients respond to clonidine administration with an increase of intestinal tone, nonsignificant decrease of plasma NE level and nonsignificant decrease of diastolic blood pressure (DBP). Type 2 depressed patients respond to clonidine administration with no increase of intestinal tone, a great decrease of plasma NE level and a significant reduction of DBP. Type 1 depressed patients improve after a few days (5 to 10) administration of low doses of desipramine (25 mg) and maprotyline (25 mg), all antidepressants that potentiate NE activity. Moreover, Type 1 patients are vastly improved with fenfluramine (20 mg b.i.d.), a drug that depletes serotonin from both central and peripheral stores. Conversely, Type 2 depressed patients improve after a few days (5 to 10) of low doses of clorimipramine (25 mg daily), a tricyclic antidepressant that inhibits the reuptake of 5HT, preferentially. Type 1 depressed patients are worsened by clorimipramine and Type 2 depressed patients are worsened by desipramine, maprotyline, imipramine, and fenfluramine.

Improvement of Type 1 patients is accompanied by a reduction of distal colon tone and an increase of plasma NE level, whereas improvement of Type 2 patients is accompanied by an increase in IT and a decrease of plasma NE level. Finally, the clonidine-induced DBP reduction observed in Type 2 patients disappears after clorimipramine treatment. To the contrary, clonidine reduces DBP in Type 1 patients following their improvement with desipramine, maprotyline, imipramine, or fenfluramine.

Although the mean cortisol plasma level is significantly higher in Type 2 = low-IT + high NE than in Type 1 = high-IT + low NE patients, clonidine does not modify cortisol plasma levels significantly in either group of depressed patients. Likewise, growth hormone plasma levels are not modified by clonidine in either type of depressed patient. However, after improvement, a normalization of hormonal response to clonidine was observed in both types of patients. Additionally, hypercortisolemia registered in Type 2 patients is reduced following improvement.

According to the above parameter and based on accumulated data cited in this and preceding papers, we propose the existence of two types of depressive syndromes possessing different autonomic disorders. The central mechanisms underlying these autonomic disorders will be discussed.

A. Model Hypothesis for Two Brain Monoaminergic Circuits: Normal Functioning

The activity of NE neurons in the LC produces norepinephrine release at its projection levels: DR, MR, VTA, ventral hippocampus, brain cortex, some mesolimbic structures, hypothalamus, brain stem nuclei (sensory and association, preferentially), ventral spinal horn, etc.[11,16,20,22,24,31,57,58,60,61,68,165,173,174,197-224] Whereas 5HT-DR neurons are activated by NE released by LC-axons, 5HT-MR neurons are inhibited by LC-axons.[57,58,60-64,136,193,204,205,213,216,218,225-235] Thus, during LC activation, 5HT would be released at dorsal hippocampus and temporoparietal cortex, both regions receiving 5HT-DR axons.[17,18,39-41,46,47,49,61,62,65,66,136,137,194,206,221,223,224,228,229,232-234,236-252] However, the release of serotonin would be reduced at mesolimbic, ventral hippocampus, and prefrontal cortex levels (projection areas of 5HT-MR axons). During NE-LC + 5HT-DR activation = active waking periods,[17,18,39-41,46,47,49,61,62,65,66,134-137,193,194,206,221,223,224,226,228,229,232-234,236-252] DA-mesocortical neurons, located in the anterior part of VTA, are stimulated by NE released in this region by LC axons.[11,15-17,20-23,25,35,36,38,61,201-203,206,215,217,220,236,237,253-265] Conversely, DA-mesolimbic neurons located in posterolateral VTA are inhibited by serotonin released at this level by 5HT-DR axons.[11-18,22,26,39-41,49,62,65,66,134,136,193,202,219,221,228,231-234,236,239-244,246,247,266-275] During these active waking periods, NE neurons located in non-LC nuclei would be inhibited by NE released from LC axons ending in these non-LC nuclei. The A1 ventromedullary cell group receives the most dense LC innervation of all the non-LC NE nuclei.[26,190,199,234,276-279]

According to our hypothesis, during active waking the following monoaminergic activity-predominances would be found in the forebrain structures:

1. Temporoparietal cortex = NE + 5HT
2. Prefrontal cortex = NE + DA
3. Dorsal hippocampus = 5HT
4. Ventral hippocampus = NE
5. Mesolimbic structures = NE

It has been determined that 5HT exerts its inhibitory influence at brain cortex level.[18,59,114,115,136,172,205,223,224,228,280] When it is taken into account that ACh activity appears to be the basic activity of neurons at brain cortex level, it is to be expected that parietotemporal cortex would be inhibited while prefrontal cortex would be disinhibited during active waking periods.[10,12,17,21,39,41,44,46,47,50,51,53,54,60,62,212,229,281-285]

Dorsal hippocampus would receive only inhibitory 5HT influence during active waking periods, since destruction of 5HT-DR neurons induces an increase of ACh at this level.[95,173,174,207,210,212,229,286-288] The actions of NE and 5HT in hippocampal pyramidal cells are complex and not well understood. It seems that NE can stimulate β-adrenoceptors at this level without 5HT cooperation. However, the lack of serotonin in ventral hippocampus is followed by α₁-adrenoceptor up-regulation, despite NE presence in this forebrain structure.[173,174,210,286-288] Thus, β-adrenergic activity would predominate over α₁-adrenergic activity at ventral hippocampal level.

Another circuit would predominate during quiet waking and resting periods, when the NE-LC + 5HT-DR pair reduces its activity. At this time, disinhibition of 5HT-MR, DA-mesolimbic, and NE-A1 neurons would occur. Hence, the pattern of monoaminergic predominances would change at forebrain structure levels. Serotonin would predominate at prefrontal cortex (MR projection area), while a reduction of NE (LC projection area) and DA-mesocortical would occur at this cortical level. Serotonin would also predominate over NE in dorsal and ventral hippocampus, since 5HT-MR axons innervate both regions and LC is inhibited during quiet waking and resting periods.[5,10,11,14,17,18,20-22,38-42,44,57,60-64,66,134-137,190,193,194,199,202-206,213,217-219,221,225-229,231-235,237,239,241,244,246,247,249,268-270,278,287,289-296]

Although VTA mesolimbic neurons are liberated from 5HT-DR inhibitory influence, 5HT released by 5HT-MR axons in mesolimbic structures is proven to antagonize DA activity at this postsynaptic level.[64-67,297-299] Experiments at accumbens and amygdala levels clearly confirmed this.[7,10,12-14,16,20,21,26,29,30,40,45,255,300-302] Finally, NE-A1 axons which reach septal regions, but not hippocampus and cortex, would guarantee the presence of NE activity at this level, during quiet waking and resting periods.[173,174,204,207,210,214,303]

According to this second pattern of monoaminergic predominances at forebrain level, the reduction of psychic and physical activities (alertness, attention, learning, etc.) would depend on reduction of NE and DA in prefrontal cortex, NE in hippocampus, and DA in mesolimbic structures.

The reduction of peripheral sympathetic activity registered during quiet waking and resting periods fits well with the reduction of NE-LC activity (central sympatho-excitatory system) and with the disinhibition of medullary sympathoinhibitory nucleus (NE-A1). The well-established correlation between waking and resting states with high and low plasma catecholamine levels, respectively, provides data supporting our hypothesis.[94,95,179,187,194,220,233,234,248,262,270,276-279,304-313]

B. Pathophysiological Functioning

Two types of depressive syndromes would arise from the two postulated brain monoaminergic circuits: (1) a depressive syndrome based on hyperactivity of NE-LC + 5HT-DR circuit and (2) a depressive syndrome based on hyperactivity of NE-A1 + 5HT-MR circuit.

1. Type 1 Depressive Syndrome (Figure 1)

Hyperactivity of NE-LC would provoke an excessive release of NE in brain cortex, ventral hippocampus, VTA, and NE-A1 nuclei. It would also produce an excessive release of 5HT in temporoparietal cortex, dorsal hippocampus, and VTA DA neurons (posterolateral VTA region). As a result, an increase of DA at prefrontal cortex and a decrease of this neurotransmitter at mesolimbic structures would also be produced. The well-known fact that cortical dopaminergic activity exerts inhibitory influence on mesolimbic DA turnover is consistent with pathophysiological mechanisms.[11,14-16,19-21,25,27,36-38,201,236,237,254-260,267,314] Moreover, the enhancement of DA activity at prefrontal cortex level, along with the reduction of 5HT activity at this same level, is compatible with anxiety observed in some types of depressive patients (anxious depression).[3,5,10,11,14,16-18,39,44,59,62,64,66-68,134,137,172,230,239,241,244-247,249,251,252,258,259,281,295,315-317] Furthermore, the raised NE and its metabolite MHPG in plasma registered in some kinds of depressive patients is also compatible with the hyperactivity of NE-LC (sympathoexcitatory) system that we are proposing.[55,95,179,187,194,220,233,234,262,276,279,304-309,311-313]

This pathophysiological model of depression is compatible with normal or raised CSF levels

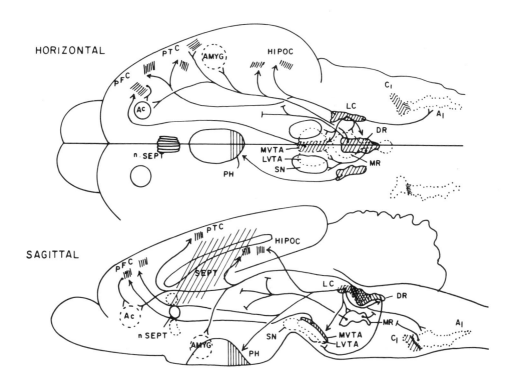

FIGURE 1. Depressive type I syndrome. According to a great bulk of experimental and clinical evidence we postulate two types of depressive syndromes. In type I depressive syndrome, NE activity of the nucleus LC or A6 cell group would predominate over the lateral ventromedullary NE-A1 cell group: A6 > A1. In other words, the sympathoexcitatory system would predominate over the sympathoinhibitory system. In addition, 5HT-DR neurons would predominate over 5HT-MR neurons: DR > MR. Finally, DA neurons located in the medial-anterior part of ventral tegmental area (MVTA) = mesocortical system, would predominate over DA neurons located in lateral parts of VTA (LVTA) = mesolimbic system; DA-mesocortical system > DA-mesolimbic system. Adrenaline neurons (A) located in C1 cell group (rostral to the lateral ventromedullary NE-A1 cell group) are released from A1 bridle and become activated. Predominance of NE-A6 plus A-C1 cell groups results in peripheral sympathetic hyperactivity. This profile of central autonomic activity is consistent with the depressive syndrome known as anxious depression. Type I depression would be accompanied by high NE + MHPG plasma levels, low-IT, raised cortisol plasma level and normal or raised 5HIAA CSF levels. See text for explanations. PFC = prefrontal cortex; PTC = parieto-temporal cortex; PH = posterior hypothalamus; n SEPT= nuclei septum; Ac = nucleus accumbens; SN = substantia nigra; and HIPOC = hippocampus. (////////) = activated; (—→) = excitatory input; (—<) = inhibitory input; and (—|) = annulled output. (Reproduced with kind permission from PJD Publications Limited, Westbury, N.Y., 11590, U.S.A., from *Res Commun Psychol Psychiatr Behav*, 11: 145-192, 1986. Copyright © by PJD Publications Ltd.)

of 5HIAA and with low CSF level of HVA. These disorders of serotonin and DA metabolites, respectively, have been reported frequently in depressed patients. Further, supersensitivity of postsynaptic DA receptors at mesolimbic level, registered by several authors, is compatible with inhibition of DA release by mesolimbic DA axons.

In published papers we report that those depressed patients showing low intestinal tone (low-IT) + high NE plasma level also showed more psychic anxiety[92] when rated by means of the Hamilton Depression Rating Scale (HDS). Furthermore, these patients showed a more pronounced and significant reduction of NE diastolic blood pressure (DBP) when challenged with clonidine. This abrupt suppression of peripheral sympathetic activity registered after clonidine administration is consistent[93] with the centrally induced sympathoinhibitory effect of the drug. Finally, the demonstration that those depressed patients were rapidly improved by very low doses of clorimipramine but not by low doses of imipramine, desipramine, maprotyline,[94] and other NE drugs, reinforces the presumption that hyperactive NE-LC neurons can be bridled by drugs which potentiate 5HT activity but not by drugs which potentiate NE activity. In effect, the firing rate of NE-LC neurons is reduced by serotonin released at this level by 5HT axons arising from 5HT-DR and 5HT-MR neurons. Potentiation of 5HT activity at prefrontal cortex and hippocampus, provoked by clorimipramine but not by the other mentioned antidepressants, can also account for the antagonism of NE hyperactivity postulated by us in the Type 1 depressed patients.

We also reported that these hypersympathetic depressed patients showed cortisol[95] plasma levels higher than hyposympathetic depressed patients, phenomenon for which we have found as yet no satisfactory explanation. The central mechanisms contributing to cortisol secretion are complex. Although direct serotonin release at mediobasal hypothalamus triggers a CRF-ACTH-cortisol cascade of events, the 5HT axons responsible for this effect seem not to arise from either DR or MR neurons.[86,140,143-145,175,318-331] Recent studies associate this cascade of events with 5HT axons provided by the medullary raphe magnus nucleus.[332] However, there exist other central monoaminergic pathways involved in CRF secretion which include septum and hippocampus structures. These pathways would integrate complex postsynaptic mechanisms in which NE-LC, 5HT-DR, and 5HT-MR nuclei would be involved[1,34,140,197,198,208,209,211,333,343] (see Figure 1).

Some evidence shows that a percentage of depressed patients are nonsuppressors when tested by the dexamethasone suppression test (DST).[96,97,139,349] Adrenalectomy has been shown to reduce serotonin turnover in hippocampus, an effect which is restored to normal by the addition of corticosterone (physiologic corticosteroid in the rat). Other findings show that natural, but not synthetic, steroids bind to hippocampus, allowing us to postulate that only natural glucocorticoids are able to antagonize serotonin receptors at hippocampal level.[98,250,350-360] Perhaps this 5HT activity in hippocampus is associated with secretion of cortisol and with an alternative feedback mechanism for its inhibition, other than the well-known hypophyseal and hypothalamic mechanisms.

2. Type 2 Depressive Syndrome (Figure 2)

Reduction of DA and NE in prefrontal cortex, along with increased 5HT release here, appear to be responsible for sedation, drowsiness, EEG synchronization, and poor alert response[9-18,21-24,26,38,44,45,57,59,60,62,64,66,67,95,136-138,194,204,205,213,220,228-230,232,235,244-246,248,249,256,270,292-295,302,303,308,309,317,361-370] typical of patients affected by Type 2 depressive syndrome. All these behavioral effects are similar but accentuated in slow wave sleep, learned helplessness, and some depressive situations.[50-53,60,71,72,105-113,136,228-230,267,303,364-370]

In this physiological disorder the terminals arising from the medullary NE-A1 nuclei would release excessive norepinephrine in septum, other mesolimbic structures, and, in addition, at the level of NE-LC nuclei. The NE released in LC would inhibit NE neurons in this nucleus; hence, would be inhibited for two main reasons: (1) a deficit of tonic excitatory influence exerted by NE-LC axons and (2) an increase of inhibitory influence by the 5HT-MR axons. This phenomenon would be secondary to disinhibition of 5HT-MR nucleus and would result in an increase of serotonin released by 5HT-MR axons at LC nuclei,[28,136,138,204,205,219,221,235,251,363,371-373] an additional factor favoring NE-LC inhibition. Taking into account that 5HT-MR axons also end in 5HT-DR nucleus, the latter 5HT system would be inhibited.[221,241,374-376] The consequent disinhibition of DA mesolimbic system liberated from DR inhibitory influence, would be potentiated through inhibition of its antagonistic DA mesocortical system, secondary to hyperactivity of 5HT-MR system. This 5HT system proves to have inhibitory influence on DA mesocortical system.[17,18,26,41,66,67,136,204,206,230,236,237,252,302,362,371,377-380] In these circumstances, DA would be reduced at prefrontal cortex while serotonin would be increased at this cortical level. On the other hand, although DA mesolimbic neurons are released from inhibitory influences of 5HT-DR and DA-mesocortical systems, the release of DA in mesolimbic structures would be presynaptically and postsynaptically antagonized by serotonin arising from 5HT-MR axons, sharing the same projection areas.[17,18,26,41,66,67,136,206,230,236, 252, 302,362,371,378-380]

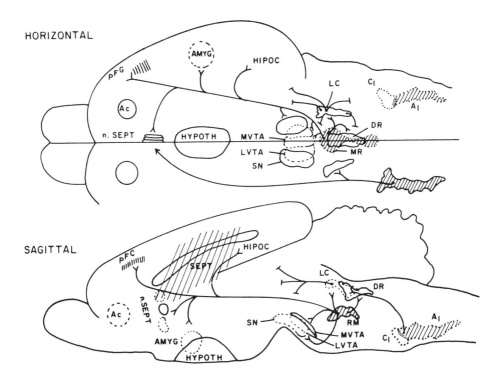

FIGURE 2. Depressive type II syndrome. In type II depressive syndrome, NE activity of the lateral ventromedullary cell group (A1) would predominate over the dorsomedial mesencephalic-pontis (or pontal) NE-A6 cell group (locus coeruleus): NE-A1 > NE-A6. In other words, sympathoinhibitory system predominates over sympathoexcitatory system. In these circumstances, adrenaline neurons (A) located in C1 cell group (rostral to NE-A1 cell group) are bridled by NE-A1 system. In addition, 5HT-MR would predominate over 5HT-DR neurons; further, it is known that 5HT-MR neurons send inhibitory axons to NE-LC and 5HT-DR neurons. Finally, DA neurons located in lateral parts of ventral tegmental area (LVTA) = DA mesolimbic system would predominate over DA-mesocortical system whose neurons are located in the medial-anterior part of ventral tegmental area (MVTA). Normally, DA-MVTA neurons are activated by NE-A6 neurons and are inhibited by 5HT-MR neurons. DA-LVTA (mesolimbic) neurons, which normally are inhibited by 5HT-DR neurons, would be freed from this bridle because the 5HT-DR system is under the inhibitory influence of 5HT-MR system. This profile of central autonomic activity is consistent with the depressive syndrome known as psychotic depression (anergic depression). PFC = prefrontal cortex; Ac = nucleus accumbens; n. SEPT= nuclei septum; AMYG = amygdala; SN = substantia nigra; and HIPOC = hippocampus. (/////////) = activated; (——>) = excitatory input; (—<) = inhibitory input; and (|——) = annulled input. Depressive type II syndrome would be accompanied by low NE + MHPG plasma levels, high-IT, normal cortisol plasma level, and low 5-HIAA CSF levels. Reproduced with kind permission from PJD Publications Limited, Westbury, N.Y., 11590, U.S.A., from *Res Commun Psychol Psychiatr Behav,* 11: 145-192, 1986. Copyright © by PJD Publications Ltd.)

REFERENCES

1. **Clemente CD, Chase MH.** Neurological substrates of aggressive behaviour, *Annu Rev Physiol* 35: 329-356, 1973.
2. **Hodge GK, Butcher LL.** Catecholamine correlates of isolation-induced aggression in mice. *Eur J Pharmacol* 31: 81-93, 1975.
3. **Matte AC, Tornow H.** Parachlorophenylalanine produces dissociated effects on aggression "emotionality" and motor activity. *Neuropharmacology* 17: 555-558, 1978.
4. **Dichiara G, Camb R, Spano PF.** Evidence for inhibition by brain serotonin of mouse killing behavior in rats. *Nature* 233: 272-273, 1971.
5. **File SE, Hyde JRG, MacLeod NK.** 5,7-Dihydroxytryptamine lesions of dorsal and median raphe nuclei and performance in the social interaction test of anxiety and in a home cage aggression test. *J Affect Dis* 1: 115-122, 1979.

6. **Randrup A, Munkvad L.** Pharmacological studies on the brain mechanisms underlying two forms of behavioral excitation: stereotyped hyperactivity and "rage". *Ann NY Acad Sci* 159: 928-938, 1969.

7. **Jones DL, Wu M, Mogenson GJ.** Cholinergic, dopaminergic and GABAergic interactions in the nucleus accumbens and globus pallidus affecting ambulatory activity. *Neurosci Abstr* 4: 855, 1978.

8. **Mogenson GJ, Jones DL, Yim CY.** From motivation to action: functional interface between the limbic and motor system. *Prog Neurobiol* 14: 69-97, 1980.

9. **Warbritton JD III, Stewart RM, Baldessarini RJ.** Decreased locomotor activity and attenuation of amphetamine hyper-activity with intraventricular infusion of serotonin in the rat. *Brain Res* 143: 373-382, 1978.

10. **Lyness WH, Moore KE.** Destruction of 5-hydroxy-trypta-minergic neurons and the dynamics of dopamine in nucleus accumbens septi and other forebrain regions of the rat. *Neuropharmacology* 20: 327-334, 1981.

11. **Bunney BS, Aghajanian GK.** Mesolimbic and mesocortical dopaminergic systems: physiology and pharmacology, in *Psychopharmacology: A Generation of Progress*. Lipton MA, DiMascio A, Killam KF. Eds., Raven Press, New York, 1978, 221.

12. **Carter CJ, Pycock CJ.** The effects of 5,7-dihydroxy-tryptamine lesions of extrapyramidal and mesolimbic sites on spontaneous motor behavior and amphetamine-induced stereotype. *Naunyn-Schmiedeberg's Arch Pharmacol* 208: 51-54, 1979.

13. **Costall B, Hui SCG, Naylor RJ.** The importance of serotonergic mechanisms for the induction of hyperactivity by amphetamine and its antagonism by intra-accumbens (3,4- dihydroxy-phenylamino)-2-imidazoline (DPI). *Neuropharmacology* 18: 605-609, 1979.

14. **Herve D, Simon H, Blanc G, Lisoprawski A, LeMoal M, Glowinski J, Tassin JP.** Increased utilization of dopamine in the nucleus accumbens but not in the cerebral cortex after dorsal raphe lesion in the rat. *Neurosci Lett* 15: 127-134, 1979.

15. **Jackson DM, Anden NE, Dahlstrom A.** A functional effect of dopamine-rich parts of rat brain. *Psychopharmacologia* 45: 139-150, 1975.

16. **Moore KE, Kelly PH.** Biochemical pharmacology of mesolimbic and mesocortical dopaminergic neurons, in *Psychopharmacology: A Generation of Progress*. Lipton MA, DiMascio A, Killam KF. Eds., Raven Press, New York, 1978, 221.

17. **Nicolaou NM, García-Muñoz M, Arbuthnott G, Eccleston D.** Interactions between serotonergic and dopaminergic systems in rat brain demonstrated by small unilateral lesions of the raphe nuclei. *Eur J Pharmacol* 57: 295-305, 1979.

18. **Samanin R, Quattrone A, Consolo S, Ladinsky H, Algeri S.** Biochemical and pharmacological evidence of the interaction of serotonin with other aminergic systems in the brain, in *Interactions Between Putative Neurotransmitters*. Garattini S, Pujol JF, Samanin R. Eds., Raven Press, New York, 1978, 355.

19. **Slopsema JS, van der Gugten J, Bruin JPC.** Regional concentrations of noradrenaline and dopamine in the frontal cortex of the rat: dopaminergic innervation of the prefrontal subareas and lateralization of prefrontal dopamine. *Brain Res* 250: 197-200, 1982.

20. **Herve D, Blanc G, Glowinski J, Tassin JP.** Reduction of dopamine utilization in the prefrontal cortex but not in the nucleus accumbens after selective destruction of nor-adrenergic fibers innervating the ventral tegmental area in the rat. *Brain Res* 237: 510-516, 1982.

21. **Herve D, Simon H, Blanc G, LeMoal M, Glowinski J, Tassin JP.** Opposite changes in dopamine utilization in the nucleus accumbens and the frontal cortex after electrolytic lesion of the median raphe in the rat. *Brain Res* 216: 422-428, 1981.

22. **Wang RY.** Dopaminergic neurons in the rat ventral tegmental area. III. Effects of D- and L-amphetamine. *Brain Res Rev* 3: 152-165, 1981.

23. **Mogenson GJ, Wu M, Manchanda SK.** Locomotor activity initiated by microinfusions of picrotoxin into the ventral tegmental area. *Brain Res* 161: 311-319, 1979.

24. **Carter CJ, Pycock CJ.** Behavioural and biochemical effects of dopamine and noradrenaline depletion within the medial prefrontal cortex of the rat. *Brain Res* 192: 163-176, 1980.

25. **Pycock CJ, Carter CJ, Kerwin RW.** Effect of 6-hydroxy-dopamine lesions of the medial prefrontal cortex on neuro-transmitters systems in subcortical sites in the rat. *J Neurochem* 34: 91-99, 1980.

26. **Jones DL, Mogenson GJ, Wu M.** Injections of dopaminergic, cholinergic, serotoninergic and GABAergic drugs into the nucleus accumbens: effects on locomotor activity in the rat. *Neuropharmacology* 20: 29-37, 1981.

27. **Tassin JP, Stinus J, Simon H, Blanc G, Thierry AM, LeMoal H, Cardo B, Glowinski J.** Relationship between the locomotor activity induced by A10 lesions and the destruction of the fronto-cortico dopaminergic innervation in the rat. *Brain Res* 141: 267-281, 1978.

28. **Scheel-Kruger J, Randrup A.** Stereotyped hyperactive behaviour produced by dopamine in the absence of noradrenaline. *Life Sci* 6: 1389-1398, 1967.

29. **Robinson SE, Malthe-Sorenssen D, Wood PL, Commisioniong J.** Dopaminergic control of the septal-hippocampal cholinergic pathway. *J Pharmacol Exp Ther* 208: 476-479, 1979.

30. **Assaf SY, Miller JJ.** Excitatory action of the mesolimbic dopamine system on septal neurons. *Brain Res* 129: 353-360, 1977.

31. **Brownstein M, Saavedra JM, Palkovits M.** Norepinephrine and dopamine in the limbic system of the rat. *Brain Re*s 79: 431-436, 1974.

32. **Lindvall, O.** Mesencephalic dopaminergic afferents to the lateral septal nucleus of the rat. *Brain Res* 87: 89-95, 1975.

33. **Moroni F, Malthe-Sorenssen D, Cheney DL, Costa E.** Modulation of acetylcholine turnover in the septal hippocampal pathway by electrical stimulation and lesioning. *Brain Res* 150: 333-341, 1978.

34. **Ohta M, Omura Y.** Inhibitory pathway from the frontal cortex to the hypothalamic ventromedial nucleus in the rat. *Brain Res Bull* 4: 231-238, 1979.

35. **Anisman H, Ritch M, Sklar LS.** Noradrenergic and dopaminergic interactions in escape behavior: analysis of uncontrollable stress effects. *Psychopharmacology* 74: 263-268, 1981.

36. **Antelman SM, Black CA.** Dopamine-beta-hydroxylase inhibitors (DBHI) reverse the effects of neuroleptics under activating conditions: possible evidence for a norepinephrine (NE)-dopamine (DA) interaction. *Soc Neurosci Abstr* 1977; cited by **Anisman H, Ritch M, Sklar LS.** Noradrenergic and dopaminergic interactions in escape behavior: analysis of uncontrollable stress effects. *Psychopharmacology* 74: 263-268, 1981.

37. **Antelman SM, Caggiula AR.** Norepinephrine-dopamine inter-actions and behavior. *Science* 195: 646-652, 1977.

38. **Galey S, LeMoal M.** Behavioural effects of lesions in the A10 dopaminergic area of the rat. *Brain Res* 124: 83-97, 1977.

39. **Hole K, Johnson GE, Berge O-G.** 5,7-Dihydroxytryptamine lesions of the ascending 5-hydroxytryptaminergic pathways: habituation motor activity and agonistic behaviour. *Pharmacol Biochem Behav* 7: 205-210, 1977.

40. **Costall B, Naylor RJ, Marsden CB, Pycock CJ.** Serotonergic modulation of the dopamine response from the nucleus accumbens. *J Pharm Pharmacol* 28: 523-526, 1976.

41. **Trimbach C.** Hippocampal Modulation of Behavioral Arousal: Mediation by Serotonin. Doctoral dissertation, Princeton University, Princeton, NJ, 1972.

42. **Aghajanian GK, Haigler HJ.** Studies on the physiological activity of 5-HT neurons, in *Pharmacology and the Future of Man*. Acheson G. Ed., S. Karger, Basel, 1973, 269.

43. **Breese GR, Cooper BR, Mueller RA.** Evidence for involvement of 5-hydroxytryptamine in the actions of amphetamine. *Br J Pharmacol* 52: 307-314, 1974.

44. **Stein L, Wise DC.** Serotonin and behavioral inhibition, in *Serotonin — New Vistas*, Vol. 2. Costa E, Gessa GL, Sandler M. Eds., Raven Press, New York, 1974, 281.

45. **Persip GL, Hamilton LW.** Behavioral effects of serotonin or a blocking agent applied to the septum of the rat. *Pharmacol Biochem Behav* 1: 139-147, 1973.

46. **Giambalvo CT, Snodgrass SR.** Biochemical and behavioral effects of serotonin neurotoxins on the nigrostriatal dopamine system: comparison of injection sites. *Brain Res* 152: 555-566, 1978.

47. **Dray A, Davies J, Oakley NR, Tongroach P, Vellucci S.** The dorsal and medial raphe projections to the substantia nigra in the rat: electrophysiological, biochemical and behavioural observations. *Brain Res* 151: 431-442, 1978.

48. **James TA, Starr MS.** Rotational behaviour elicited by 5-HT in the rat: evidence for an inhibitory role of 5-HT in the substantia nigra and corpus striatum. *J Pharm Pharmacol* 32: 196-200, 1980

49. **Nojyo Y, Sano Y.** Ultrastructure of the serotonergic nerve terminals in the suprachiasmatic and interpeduncular nuclei of rat brains. *Brain Res* 149: 482-488, 1978.

50. **Brown L, Rosellini RA, Samuels OB, Riley EP.** Evidence for a serotonergic mechanism of the learned helplessness phenomenon. *Pharmacol Biochem Behav* 17: 877-883, 1982.

51. **Petty F, Sherman AD.** Learned helplessness induction decreases in vivo cortical serotonin release. *Pharmacol Biochem Behav* 18: 649-650, 1983.

52. **Hellhammer DH.** Learned helplessness — an animal model revisited, in *The Origins of Depression*. Agnst J. Ed., Springer-Verlag, Berlin, 1983, 147.

53. **Weiss JM, Goodman PA, Losito PG, Corrigan S, Charry JM, Bailey WH.** Behavioral depression produced by an uncontrollable stressor: relationship to norepinephrine, dopamine and serotonin levels in various regions of rat brain. *Brain Res Rev* 3: 167 205, 1981.

54. **Maas JW, Redmond DE, Gauen R.** Effects of serotonin depletion on behavior in monkeys, in *Serotonin and Behavior*. Barchas J, Usdin E. Eds., Academic Press, New York, 1973, 351.

55. **Aprison MH, Hingtgen JN, Nagayama M.** Testing a new theory of depression with an animal model: neurochemical-behavioural evidence for postsynaptic serotonergic receptor involvement, in *New Vistas in Depression*. Langer S, Takahashi R, Segawa T, Briley M. Eds., Pergamon Press, Elmsford, NY, 1982, 171.

56. **Aprison MH, Hingtgen JN.** Hypersensitive serotonergic receptors: a new hypothesis for one subgroup of unipolar depression derived from an animal model, in *Serotonin: Current Aspects of Neurochemistry and Function*. Haber B, Gabay S, Issidorides MR, Alivisatos SGA. Eds., Plenum Press, New York, 1981, 627.

57. **Geyer MA, Lee EHY.** Effects of clonidine, piperoxane and locus coeruleus lesion on the serotonergic and dopaminergic systems in raphe and caudate nucleus. *Psychopharmacology* 33: 3399-3404, 1984.

58. **Yamamoto T, Watanabe T, Shibata S, Ueki S.** The effect of locus coeruleus and midbrain raphe stimulation on muricide in rats. *Jpn J Pharmacol* 29(Suppl.): 41, 1979.

59. **Kostowski W.** Brain serotonergic and catecholaminergic system: facts and hypothesis, in *Current Developments in Psychopharmacology*. Essman WB, Valzelli L. Eds., Spectrum, New York, 1975, 39.

60. **Plaznik A, Danysz W, Kostowski W, Bidzinski A, Hauptmann M.** Interaction between noradrenergic and serotonergic brain systems as evidenced by behavioral and biochemical effects of microinjections of adrenergic agonists and antagonists into the median raphe nucleus. *Pharmacol Biochem Behav* 19: 27-32, 1983.

61. **Pasquier DA, Kemper TL, Forbes WB, Morgane PJ.** Dorsal raphe, substantia nigra and locus coeruleus: interconnections with each other and the neostriatum. *Brain Res Bull* 2: 323-329, 1977.

62. **Przewlocka B, Kukulka L, Tarczynska E.** The effect of lesions of dorsal or median raphe nucleus on rat behavior. *Pol J Pharmacol Pharm* 29: 573-579, 1977.

63. **Waldbillig RJ.** The role of the dorsal raphe and median raphe in the inhibition of muricide. *Brain Res* 160: 341-346, 1979.

64. **Graeff FG, Quintero S, Gray JA.** Median raphe stimulation, hippocampal theta rhythm and threat-induced behavioral inhibition. *Physiol Behav* 25: 253-261, 1980.

65. **Fibiger HC, Miller JJ.** An anatomical and electro-physiological investigation of the serotonergic projection from the dorsal raphe nucleus to the substantia nigra in the rat. *Neuroscience* 2: 975-987, 1977.

66. **Waldmeier PC.** Serotonergic modulation of mesolimbic and frontal cortical dopamine neurons. *Experientia* 36: 1092-1094, 1980.

67. **Pycock CJ, Horton RW, Carter CJ.** Interactions of 5-hydroxytryptamine and gamma-aminobutyric acid with dopamine. *Adv Biochem Psychopharmacol* 19: 323-341, 1978.

68. **Fuxe K, Hökfelt T, Agnati L, Johansson D, Ljangdahl A, Perez de la Mora M.** Regulation of the mesocortical dopamine neurons, in *Advances in Biochemical Psychopharmacology. Nonstriatal Dopaminergic Neurons*. Costa E, Gessa GL. Eds., Raven Press, New York, 1977, 47.

69. **Schildkraut JJ.** Norepinephrine metabolites as biochemical criteria for classifying depressive disorders and predicting responses to treatment: preliminary findings. *Am J Psychiatry* 130: 695-698, 1973.

70. **Maas JW.** Biogenic amines and depression. Biochemical and pharmacological separation of two types of depression. *Arch Gen Psychiatry* 32: 1357-1361,1975.

71. **Garver DL, Davis JM.** Biogenic amine hypotheses of affective disorders. *Life Sci* 24: 383-394, 1979.

72. **Roy A, Pickar D, Linnoila M, Doran AR, Ninan P, Paul SM.** Cerebrospinal fluid monoamine and monoamine metabolite concentrations in melancholia. *Psychiatr Res* 15: 281-292, 1985.

73. **Van Praag HM.** Significance of biochemical parameters in the diagnosis, treatment and prevention of depressive disorders. *Biol Psychiatry* 12: 101-131, 1977.

74. **Van Praag HM.** Central monoamine metabolism in depression. *Comprehen Psychiatry* 21: 30-43, 1980.

75. **Goodwin FK, Post RM.** 5-Hydroxytryptamine and depression: a model for the interaction of normal variance with pathology. *Br J Clin Pharmacol* 15 : 3935-4055, 1983.

76. **Agren H.** Symptom patterns in unipolar and bipolar depression correlating with monoamine metabolites in the cerebrospinal fluid. I. General patterns. *Psychiatr Res* 3: 211-223, 1980.

77. **Agren H.** Symptom patterns in unipolar and bipolar depression correlating with monoamine metabolites in the cerebrospinal fluid. II. Suicide. *Psychiatr Res* 3: 225-236, 1980.

78. **Asberg M, Ringberger VA, Sjoqvist F, Thoren P, Transman L, Tuck JR.** Monoamine metabolites in cerebrospinal fluid and serotonin uptake inhibition during treatment with clorimipramine. *Clin Pharmacol Ther* 21: 201-207, 1977.

79. **Ogren S-O, Fuxe K, Agnati LF, Gustafsson J-A, Johansson G, Holm AC.** Reevaluation of the indoleamine hypothesis of depression. Evidence for a reduction of functional activitiy of central 5HT systems by antidepressant drugs. *J Neural Transm* 46: 85-103, 1979.

80. **Brown WA, Qualls CB.** Pituitary adrenal disinhibition in depression: marker of a subtype with characteristic clinical features and response to treatment? *Psychiatr Res* 4: 115-128, 1981.

81. **Asberg M, Thoren P, Traskman L, Bertilsson L, Ringberger V.** Serotonin depression: a biochemical subgroup within the affective disorders? *Science* 191: 478-483, 1976.

82. **Brown WA, Haier RJ, Qualls CB.** Dexamethasone suppression test identifies subtypes of depression which respond to different antidepressants. *Lancet* 1: 928-933, 1980.

83. **Goodwin FK, Rubovits R, Jimerson D, Post RM.** Serotonin and norepinephrine "subgroups" in depression. *Sci Proc Am Psychiatr Assoc* 130: 108, 1977.

84. **Checkley SA, Slade PA, Shur E.** Growth hormone and other responses to clonidine in patients with endogenous depression. *Br J Psychiatry* 138: 51-55, 1981.

85. **Berger M, Doerr P, Lund R, Bronisch T, von Zerssen D.** Neuroendocrinological and neurophysiological studies in major depressive disorders: are there biological markers for the endogenous subtype? *Biol Psychiatry* 17: 1217-1242, 1982.

86. **Rosenbaum AH, Schatzberg AF, Maruta T, Orsualak PJ, Cole JO, Grab EL, Schildkraut JJ.** MHPG as a predictor of anti-depressant response to imipramine and maprotiline. *Am J Psychiatry* 137: 1090-1097, 1980.

87. **Schatzberg AF.** Classification of depressive disorders, in *Depression, Biology, Psychodynamics and Treatment*. Cole JO, Schatzberg AF, Frazier SH. Eds., Plenum Press, New York, 1978, 13.

88. **Goodwin FK, Rubovits R, Jimerson DC, Post RM.** Serotonin and norepinephrine "subgroups" in depression: metabolite findings and clinical-pharmacological correlations. *Sci Proc Am Psychiatr Assoc* 130: 108-115, 1977.

89. **Zimmerman M, Coryell W, Pfohl B.** The categorical and dimensional models of endogenous depression. *J Affect Dis* 9: 181-186, 1985.

90. **Siever LJ, Risch SC, Murphy DL.** Central cholinergic-adrenergic imbalance in the regulation of affective state. *Psychiatr Res* 4: 108-114, 1981.

91. **Dube S, Kumar N, Ettedgui E, Pohl R, Jones D, Sitaram N.** Cholinergic REM induction response: separation of anxiety and depression. *Biol Psychiatry* 20: 408-418, 1985.

92. **Lechin F, van der Dijs B, Acosta E, Gomez F, Lechin E, Arocha L.** Distal colon motility and clinical parameters in depression. *J Affect Dis* 5: 19-26, 1983.

93. **Lechin F, van der Dijs B, Gomez F, Arocha L, Acosta E, Lechin E.** Distal colon motility as a predictor of anti-depressant response to fenfluramine, imipramine and clomipramine. *J Affect Dis* 5: 27-35, 1983.

94. **Lechin F, van der Dijs B, Jakubowicz D, Camero RE, Villa S, Arocha L, Lechin AE.** Effects of clonidine on blood pressure, noradrenaline, cortisol, growth hormone, and prolactin plasma levels in high and low intestinal tone depressed patients. *Neuroendocrinology* 41: 156-162, 1985.

95. **Lechin F, van der Dijs B.** Slow wave sleep (SWS), REM sleep (REMS), and depression. *Res Commun Psychol Psychiatr Behav* 9: 227-262, 1984.

96. **Coryell W, Gaffrey G, Burkhardt PE.** The dexamethasone suppression test and familial subtypes of depression — a naturalistic replication. *Biol Psychiatry* 17: 33-40, 1982.

97. **Carroll BJ.** Neuroendocrine diagnosis of depression: the dexamethasone suppression test, in *Treatment of Depression: Old Controversies and New Approaches*. Clayton PJ, Barret J. Eds., Raven Press, New York, 1982.

98. **Schatzberg AF, Rothschild AJ, Stahl JB, Bond RA, Rosembaun AH, Lofgren SB, MacLaughlin RA, Sullivan MA, Cole JO.** The dexamethasone suppression test: identification of subtypes of depression. *Am J Psychiatry* 140: 88-91, 1983.

99. **Zis AP, Goodwin FK.** The amine hypothesis, in *Handbook of Affective Disorders*. Paykel ES. Ed., Churchill Livingstone, Edinburgh, 1982, 175.

100. **Kraemer GW, Ebert MH, Lake CR, McKinney WT.** Cerebrospinal fluid measures of neurotransmitter changes associated with pharmacological alteration of the despair response to social separation in Rhesus monkeys. *Psychiatr Res* 11: 303-315, 1984.

101. **Schildkraut J.** The catecholamine hypothesis of affective disorders: a review of supporting evidence. *Am J Psychiatry* 122: 508-522, 1965.

102. **Bunney WE Jr, Davis JM.** Norepinephrine in depressive reactions. *Arch Gen Psychiatry* 13: 483-494, 1965.

103. **Schildkraut JJ, Kety SS.** Biogenic amines and emotion. *Science* 156: 21-30, 1967.

104. **Bloom FE.** Central monoaminergic transmission, in *Golgi Centennial Symposium: Perspectives in Neurobiology*. Santini M. Ed., Raven Press, New York, 1975, 489.

105. **Sherman AD, Petty F.** Neurochemical basis of the action of antidepressants on learned helplessness. *Behav Neural Biol* 30: 119-134, 1980.

106. **Hellhammer DH, Rea MA, Bell M, Belkien L, Ludwig M.** Learned helplessness: effects on brain monoamines and the pituitary-gonadal axis. *Pharmacol Biochem Behav* 21: 481-485, 1984.

107. **Anisman H, Irwin J, Sklar LS.** Deficits of escape performance following catecholamine depletion: implications for behavioral deficits induced by uncontrollable stress. *Psychopharmacology* 64: 163-170, 1979.

108. **Anisman H, Sklar LS.** Catecholamine depletion in mice upon reexposure to stress: mediation of the escape deficits produced by inescapable shock. *J Comp Physiol* 93: 610-625, 1979.

109. **Anisman H, Suissa A, Sklar LS.** Escape deficits produced by uncontrollable stress: antagonism by dopamine and norepinephrine agonists. *Behav Neurol Biol* 28: 34-47, 1980.

110. **Anisman H, Zacharko RM.** Depression: the predisposing influence of stress. *Behav Brain Sci* 5: 89-137, 1982.

111. **Collu R, Gibb W, Ducharne JR.** Role of catecholamines in the inhibitory effect of immobilization stress on testosterone secretion in rats. *Biol Reprod* 30: 416-422, 1984.

112. **Petty F, Sherman A.** A neurochemical differentiation between exposure to stress and the development of learned helplessness. *Drug Dev Res* 2: 43-45, 1982.

113. **Drugan RC, Maier SF, Skolnick P, Paul SM, Crawley JN.** An anxiogenic benzodiazepine receptor ligand induces learned helplessness. *Eur J Pharmacol* 113: 453-457, 1985.

114. **Fuller RW, Snoddy HD, Cohen ML.** Interactions of trazodone with serotonin neurons and receptors. *Neuropharmacology* 23: 539-544, 1984.

115. **Fuxe K, Ogren S-O, Agnati LF, Calza L.** Evidence for stabilization of cortical 5HT neurotransmission by chronic treatment with antidepressant drugs: induction of a high and low affinity component in 3H-5HT blinding sites. *Acta Physiol Scand* 114: 477-480, 1982.

116. **Fuxe K, Ogren S-O, Agnati LF, Eneroth P, Holm AC, Andersson K.** Long-term treatment with zimelidine leads to a reduction in 5-hydroxytryptamine neurotransmission within the central nervous system of the mouse and rat. *Neurosci Lett* 21: 57-62, 1981.

117. **Ogren S-O, Fuxe K, Archer T, Johansson G, Holm AC.** Behavioural and biochemical studies on the effects of acute and chronic administration of antidepressant drugs on central serotonergic receptor mechanisms, in *New Vistas in Depression.* Langer S, Takahashi R, Segawa T, Briley M. Eds., Pergamon Press, Elmsford, NY, 1982, 171.

118. **Roffman M, Kling MA, Cassens G, Orsulak PJ, Reigle TG, Schildkraut JJ.** The effects of acute and chronic administration of tricyclic antidepressants of MHPG-SO4 in rat brain. *Commun Psychopharmacol* 1: 195-206, 1977.

119. **Sugrue MF.** Effects of acutely and chronically administered desipramine and mianserine on the clonidine-induced decrease in rat brain 3-methoxy-4-hydroxyphenyl-ethyleneglycol sulphate content. *Br J Pharmacol* 69: 299P, 1980.

120. **Tang SW, Helmeste DM, Stancer HC.** The effect of acute and chronic desipramine and amitriptyline treatment on rat brain total 3-methoxy-4-hydroxyphenylglycol. *Naunyn-Schmiedeberg's Arch Pharmacol* 305: 207-211, 1978.

121. **Kraemer GW, McKinney WT.** Interactions of pharmacological agents which alter biogenic amine metabolism and depression: an analysis of contributing factor within a primate model of depression. *J Affect Dis* 1: 33-39, 1979.

122. **Nagayama H, Hingtgen JN, Aprison MH.** Postsynaptic action by four antidepressive drugs in an animal model of depression. *Pharmacol Biochem Behav* 15: 650-655, 1981.

123. **Ost RM, Ballenger JC, Goodwin FK.** Cerebrospinal fluid studies on neurotransmitter function in manic and depressive illness, in *The Neurobiology of Cerebrospinal Fluid.* Wood JH. Ed., Plenum Press, New York, 1980, 685.

124. **Vestergaard P, Sorensen T, Hoppe E, Rafaelsen OJ, Yates CM, Nicolaou N.** Biogenic amine metabolites in cerebrospinal fluid of patients with affective disorders. *Acta Psychiatr Scand* 58: 88-96, 1978.

125. **Traskman L, Asberg M, Bertilsson L, Sjostrand L.** Monoamine metabolites in CSF and suicidal behaviour. *Arch Gen Psychiatr* 38: 631-636, 1981.

126. **Berger PA, Faull KF, Kilkowski J, Anderson PJ, Kraemer H, Davis KL, Barchas JD.** CSF monoamine metabolites in depression and schizophrenia. *Am J Psychiatry* 137: 174- 180, 1980.

127. **Brown GL, Ballenger JC, Minichiello MD, Goodwin FK.** Human aggression and its relationship to cerebrospinal fluid 5-hydroxy-indoleacetic acid, 3-methoxy-4-hydroxyphenyl-glycol, and homovanillic acid, in *Psychopharmacology of Aggression.* Sandler M. Ed., Raven Press, New York, 1979.

128. **Brown GL, Ebert MH, Goyer PF, Jimerson DC, Klein WJ, Bunney WE, Goodwin FK.** Aggression, suicide and serotonin: relationship to CSF amine metabolites. *Am J Psychiatry* 139: 741-746, 1982.

129. **Brown GL, Goodwin FK, Ballenger JC, Goyer PF, Major LF.** Aggression in humans correlates with cerebrospinal fluid amine metabolites. *Psychiatr Res* 1: 131-140, 1979.

130. **Brown GL, Goodwin FK, Bunney WJ Jr.** Human aggression and suicide: their relationship to neuropsychiatric diagnoses and serotonin metabolism, in *Serotonin in Biological Psychiatry.* Ho BT. Ed., Raven Press, New York, 1982.

131. **Lloyd KG, Farley IJ, Deck JHN, Hornykiewicz O.** Serotonin and 5-hydroxyindoleacetic acid in discrete areas of the brainstem of suicide victims and control patients, in *Advances in Biochemical Pharmacology.* Costa E, Gessa GL, Sandler M. Eds., Raven Press, New York, 1974, 387.

132. **Pare CM, Yeung DP, Price K, Stacey RS.** 5-Hydroxytryptamine, noradrenaline, and dopamine in brain-stem, hypothalamus and caudate nucleus of controls and of patients commiting suicide by coal gas poisoning. *Lancet* ii: 133-135, 1969.

133. **Ziegler MG, Lake CR, Wood JH, Ebert MH.** Norepinephrine in cerebro-spinal fluid: basic studies, effects of drugs and diseases, in *Neurobiology of Cerebrospinal Fluid.* Wood JH. Ed., Plenum Press, New York, 1979, 141.

134. **Miczek KA, Altman JL, Appel JB, Boggam W.** Para-chlorophenylalanine, serotonin and behaviour. *Pharmacol Biochem Behav* 3: 355-361, 1975.

135. **Sheard MH.** The effect of *p*-chlorophenylalanine on behaviour in rats: relation to 5-hydroxytryptamine and 5-hydroxyindoleacetic acid. *Brain Res* 15: 524-528, 1969.

136. **Kostowski W.** Interactions between serotonergic and catecholaminergic systems in the brain. *Pol J Pharmacol Pharm* Suppl 27: 15-24, 1975.

137. **Jacobs BL, Cohen A.** Differential behavioral effects of lesions of the median or dorsal raphe nuclei in rats: open field and pain elicited aggression. *J Comp Physiol Psychol* 46: 102-112, 1976.

138. **Ellison G.** Behavior and the balance norepinephrine and serotonin. *Acta Neurobiol Exp* 35: 499-515, 1975.

139. **Carroll BJ, Greden JF, Freinberg M, et al.** Neuroendocrine dysfunction in genetic subtypes on primary unipolar depression. *Psychiatr Res* 2: 251-258, 1980.

140. **Maran JW, Carlson DE, Grizzle WE, Ward DG, Gann DS.** Organization of the medial hypothalamus for control of adrenocorticotropin in the cat. *Endocrinology* 103: 957-970, 1978.

141. **Gann DS, Ward DE, Carlson DE.** Neural pathways controlling release of corticotropin (ACTH), in *Interaction Within the Brain-Pituitary-Adrenocortical System.* Jones MT, Gillham B, Dallman MF, Chattopadhyay S. Eds., Academic Press, London, 1979, 75.

142. **Asnis GM, Halbreich U, Sachar EJ, Natham RS, Ostrow LC, Novacenko H, Davis M, Endicott J, Puig-Antich J.** Plasma cortisol secretion and REM period latency in adult endogenous depression. *Am J Psychiatry* 140: 750-753, 1983.

143. **Steiner JA, Grahame Smith DG.** Central pharmacological control of corticosterone secretion in the intact rat. Demonstration of cholinergic and serotoninergic facilitatory and alpha-adrenergic inhibitory mechanisms. *Psychopharmacology* 71: 213-217, 1980.

144. **Kennett GA, Joseph MH.** Stress induced increases in 5HT release, measured in vivo, depend upon increased tryptophan availability. *Neurosci Lett Suppl* 7: 56, 1981.

145. **Joseph MH, Kennett GA.** Corticosteroid response to stress depends upon increased tryptophan availability. *Psychopharmacology* 79: 79-81, 1983.

146. **Shaw DM, O'Keefe R, MacSweeney DA, et al.** 3-Methoxy-4-hydroxyphenylglycol in depression. *Psychol Med* 3: 333-336, 1973.

147. **Maas JW, Dekirmenjian H, DeLeon-Jones F.** The identification of depressed patients who have a disorder of norepinephrine metabolism and/or disposition, in *Frontiers in Catecholamine Research.* Usdin E, Snyder SH. Eds., Pergamon Press, Elmsford, NY, 1974, 1091.

148. **Schildkraut JJ.** Catecholamine metabolism and affective disorders, in *Frontiers in Catecholamine Research.* Usdin E, Snyder SH. Eds., Pergamon Press, Elmsford, NY, 1974, 1165.

149. **Post RM, Gordon EK, Goodwin FK, et al.** Central norepinephrine metabolism in affective illness: MHPG in the cerebrospinal fluid. *Science* 179: 1002-1003, 1973.

150. **Shopsin B, Wilk S, Gershon S, et al.** Cerebrospinal fluid MHPG: an assessment of norepinephrine metabolism in affective disorders. *Arch Gen Psychiatry* 28: 230-233, 1973.

151. **Asberg M, Thoren P, Traskman L, Bertilsson L, Ringberger V.** Serotonin depression: a biochemical subgroup within the affective disorders? *Science* 191: 478-480, 1976.

152. **Bower MB.** Cerebrospinal fluid 5-hydroxyindoleacetic acid (5-HIAA) and homovanillic acid (HVA) following probenecid in unipolar depressives treated with amitriptyline. *Psychopharmacology* 23: 26-33, 1972.

153. **Sugrue MF.** Changes in rat brain monoamine turnover following chronic antidepressant administration. *Life Sci* 26: 423-429, 1980.

154. **Schildkraut JJ, Keeler BA, Papousek M, et al.** MHPG excretion in depressive disorders: relation to clinical subtypes and desynchronized sleep. *Science* 181: 762-764, 1973.

155. **Maas JW, Fawcett JA, Dekirmenjian H.** Catecholamine metabolism and the depressive states. read before the Annu Meet Am Psychiatr Assoc, Boston, 1968.

156. **Goodwin FK, Murphy DL, Brodie HKH, et al.** L-Dopa, catecholamines and behavior: a clinical and biochemical study in depressed patients. *Biol Psychiatry* 2: 341-366, 1970.

157. **Bunney WE Jr, Davis JM.** Norepinephrine in depressive reactions. *Arch Gen Psychiatry* 13: 483-487, 1965.

158. **Przegalinski E, Kordecka-Magiera A, Mogilnicka E, Maj J.** Chronic treatment with some atypical antidepressants increases the brain level of 3-methoxy-4-hydroxy-phenylglycol (MHPG) in rats. *Psychopharmacology* 74: 187-190, 1981.

159. **Clements-Jewery S.** The development of cortical beta-adrenoceptor subsensitivity in the rat by chronic treatment with trazodone, doxepin and mianserin. *Neuropharmacology* 17: 779-781, 1978.

160. **Fuxe K, Ogre S-O, Agnati L, Gustafsson JA, Jonsson G.** On the mechanisms of action of the antidepressant drugs amitriptyline and nortriptyline. Evidence for 5-hydroxy-tryptamine receptor blocking activity. *Neurosci Lett* 6: 339-343, 1977.

161. **Klimek V, Mogilnicka E.** The influence of mianserin and danitracen on the disappearance of noradrenaline in the rat brain. *Pol J Pharmacol Pharm* 30: 255-261, 1978.

162. **Leonard BE, Kafoe W.** A comparison of acute and chronic effects of four antidepressant drugs on the turnover of serotonin, dopamine and noradrenaline in the rat brain. *Biochem Pharmacol* 25: 1939-1942, 1976.

163. **Maj J.** Pharmacological spectrum of some new anti-depressants, in *Advances in Pharmacology and Therapeutics*, Vol. 5. Dumond C. Ed., Pergamon Press, Elmsford, NY, 1978, 161.

164. **Nielsen M, Braestrup C.** Chronic treatment with desipramine caused a sustained decrease of 3,4-dihydroxy-phenylglycol-sulfate and total 3-methoxy-4-hydro-hyphenylglycol in rat brain. *Naunyn-Schmiedeberg's Arch Pharmacol* 300: 87-92, 1977.

165. **Pujol JF, Stein D, Blondaux Ch, Petitjean F, Frament JL, Jouvet M.** Biochemical evidence for interaction phenomena between noradrenergic and serotonergic system in the cat brain, in *Frontiers in Catecholamine Research.* Usdin E, Snyder SH. Eds., Pergamon Press, Elmsford, NY, 1973, 771.

166. **Maas JW, Fawcett JA, Dekirmenjian H.** 3-Methoxy-4-hydroxy-phenylglycol (MHPG) excretion in depressive states. *Arch Gen Psychiatry* 19: 129-134, 1968.

167. **Janowsky DS, El-Yousef MK, Davis JM, et al.** A cholinergic-adrenergic hypothesis of mania and depression. *Lancet* 11: 632-635, 1972.

168. **Banerjee SP, Kung LS, Riggi SJ, Chanda SK.** Development of beta-adrenergic receptor subsensitivity by antidepressants. *Nature* 268: 455-456, 1977.

169. **Bergstrom DA, Kellar KJ.** Adrenergic and serotonergic receptor binding in rat brain after chronic desmethyl-imipramine. *J Pharmacol Exp Ther* 209: 256-261, 1979.

170. **Bergstrom DA, Kellar KJ.** Effect of electroconvulsive shock on monoaminergic binding sites in rat brain. *Nature* 278: 363-466, 1979.

171. **Bloom FE, Hoffer BJ, Siggins GR.** Norepinephrine mediated synapses: a model system for neuropharmacology. *Biol Psychiatry* 4: 157-177, 1972.

172. **Foote SL, Freedman R, Oliver AP.** Effects of putative transmitters on neuronal activity in monkey cortex. *Brain Res* 86: 229-242, 1975.

173. **Segal M, Bloom FE.** The action of norepinephrine in the rat hippocampus.I. Iontophoretic studies. *Brain Res* 107: 513-525, 1976.

174. **Segal M, Bloom FE.** The action of norepinephrine in the rat hippocampus. II. Activation of the input pathway. *Brain Res* 72: 99-114, 1974.

175. **Schwartz JC, Costentin J, Martres MP, Protais P, Baudry M.** Modulation of receptor mechanisms in the CNS: hyper- and hypo-sensitivity in catecholamine. *Neuropharmacology* 17: 665-672, 1978.

176. **Tang SW, Helmeste DM, Stancer HC.** Interaction of anti-depressants with clonidine on rat brain total 3-methoxy-4-hydroxyphenylglycol. *Can J Physiol Pharmacol* 57: 435-437, 1979.

177. **Vetulani J, Sulser F.** Action of various antidepressant treatments reduces reactivity of noradrenergic cyclic AMP-generating system in limbic forebrain. *Nature* 257: 495-496, 1975.

178. **Sulser F, Mobley PL.** Regulation of central noradrenergic receptor function: new vistas on the mode of action of antidepressant treatments, in *Neuroreceptors: Basic Clinical Aspects.* Usdin E, Bunney WB, Davis JM. Eds., John Wiley & Sons, New York, 1981, 55.

179. **Lake CR, Pickar D, Ziegler MG, Lipper S, Slater S, Murphy DL.** High plasma norepinephrine levels in patients with major affective disorders. *Am J Psychiatry* 139: 1315-1319, 1982.

180. **Siever LJ, Insel T, Uhde T.** Noradrenergic challenges in the affective disorders. *J Clin Psychopharmacol* 1: 193-198, 1981.

181. **Siever LJ, Uhde TW.** New studies and perspectives on the noradrenergic receptor system in depression: effects of the alpha2-adrenergic agonist clonidine. *Biol Psychiatry* 19: 131-156, 1984.

182. **Eriksson E, Eden S, Modigh K.** Up- and down-regulation of central postsynaptic alpha2-receptors reflected in the growth hormone response to clonidine in reserpine-pretreated rats. *Psychopharmacology* 77: 327-335, 1982.

183. **Vetulani J, Stawarz RJ, Dingell JV, Sulser F.** A possible common mechanism of action of antidepressant treatments. Reduction in the sensitivity of the noradrenergic cyclic AMP generating system in the rat limbic forebrain. *Naunyn-Schmiedeberg's Arch Pharmacol* 293: 109-114, 1976.

184. **Vetulani J, Antkiewicz-Michaluk L, Rokosz-Pelc A, Pilc A.** Chronic electroconvulsive treatment enhances the density of [^{3}H] prazosin binding sites in the central nervous system of the rat. *Brain Res* 275: 392-395, 1983.

185. **Vetulani J, Antkiewicz-Michaluk L, Rokosz-Pelc A.** Chronic administration of anti-depressant drugs increased the density of cortical [^{3}H] prazosin binding site in the rat. *Brain Res* 310: 360-362, 1984.

186. **Jimerson DC, Post RM, Stoddard FJ, Gillin JC, Bunney WE Jr.** Preliminary trial of the noradrenergic agonist clonidine in psychiatric patients. *Biol Psychiatry* 139: 1315-1319, 1980.

187. **Kohno Y, Tanaka M, Glavin GB, Hoaki Y, Tsuda A, Nagasaki N.** Time course of brain MHPG-SO4 level following stimulation of pre- and post-synaptic alpha-adrenoceptors by clonidine. *Jpn J Pharmacol* 34: 125-127, 1984.

188. **Wang CH, U'Prichard DC.** Reciprocal alterations in rat brain beta- and alpha2-adrenergic receptor sites after chronic intracerebroventricular infusion of isoproterenol. *Soc Neurosci Abstr* 5: 3, 1980.

189. **Wolfe BB, Harden TK, Sporn JR, Molinoff PB.** Presynaptic modulation of beta- adrenergic receptors in rat cerebral cortex after treatment with antidepressant. *J Pharmacol Exp Ther* 207: 446-457, 1978.

190. **Van Dongen PAM.** Locus coeruleus region: effects on behavior of cholinergic, noradrenergic, and opiate drugs injected intracerebrally into freely moving cats. *Exp Neurol* 67: 52-78, 1980.

191. **Skolnick P, Daily JW, Segal DS.** Neurochemical and behavioral effects of clonidine and related imidazolines: interaction with alpha-adrenoreceptors. *Eur J Pharmacol* 47: 451-455, 1978.

192. **Huang YH.** Net effect of acute administration of desipramine on the locus coeruleus-hippocampal system. *Life Sci* 25: 739-746, 1979.

193. **Svensson TH, Bunney BS, Aghajanian GK.** Inhibition of both noradrenergic and serotonergic neurons in brain by the alpha-adrenergic agonist clonidine. *Brain Res* 92: 291-300, 1975.

194. **Gaillard J-M.** Brain noradrenergic activity in wakefulness and paradoxical sleep: the effect of clonidine. *Neuropsychobiology* 13: 23-25, 1985.

195. **Scuvee-Moreau JJ, Dresse AL.** Effect of various anti-depressant drugs on the spontaneous firing rate of locus coeruleus and raphe dorsalis neurons of the rat. *Eur J Pharmacol* 57: 219-225, 1979.

196. **Peroutka SJ, Snyder SH.** Long-term antidepressant treatment decreases spiroperidol-labeled serotonin receptor binding. *Science* 210: 88-90, 1980.

197. **Chi CC.** Afferent connections to the ventromedial nucleus of the hypothalamus. *Brain Res* 17: 439-445, 1970.

198. **Saper CB, Swanson LW, Cowan WM.** An autoradiographic study of the afferent connections of the lateral hypothalamic area in the rat. *J Comp Neurol* 183: 689-706, 1979.

199. **Kostowski W, Jerlicz M, Bidzinski A, Hauptmann M.** Evidence for existence of two opposite noradrenergic brain systems controlling behavior. *Psychopharmacology* 59: 311-312, 1978.

200. **Powell EW, Leman RB.** Connections of the nucleus accumbens. *Brain Res* 105: 389-403, 1976.

201. **Blanc G, Herve D, Simon H, Lisoprawski A, Glowinski J, Tassin JP.** Response to stress of mesocortico-frontal dopaminergic neurons in rats after long-term isolation. *Nature* 284: 265-267, 1980.

202. **Wang RY.** Dopaminergic neurons in the rat ventral tegmental area. I. Identification and characterization. *Brain Res Rev* 3: 123-140, 1981.

203. **Wang RY.** Dopaminergic neurons in the rat ventral tegmental area. II. Evidence for autoregulation. *Brain Res Rev* 3: 141-151, 1981.

204. **Kostowski W.** Two noradrenergic systems in the brain and their interactions with other monoaminergic neurons. *Pol J Pharmacol Pharm* 31: 425-436, 1979.

205. **Kostowski W.** Noradrenergic interactions among central neurotransmitters, in *Neurotransmitters, Receptors and Drug Action*. Essman W. Ed., Spectrum, New York, 1980, 47.

206. **Simon H, LeMoal M, Calas A.** Efferents and afferents of the ventral tegmental A10 region studies after local injection of (3H) leucine and horseradish peroxidase. *Brain Res* 178: 17-40, 1979.

207. **Storm-Mathisen J, Goldberg HC.** 5-Hydroxytryptamine and noradrenaline in the hippocampal region: effect of transection of afferent pathways on endogenous levels, high affinity uptake and some transmitter-related enzymes. *J Neurochem* 22: 793-803, 1974.

208. **Krieger MS, Conrad LCA, Pfaff DW.** An autoradiographic study of the efferent connections of the ventromedial nucleus of the hypothalamus. *J Comp Neurol* 183: 785-816, 1979.

209. **Saper CB, Swanson LW, Cowan WM.** The efferent connections of the ventromedial nucleus of the hypothalamus of the rat. *J Comp Neurol* 169: 409-442, 1976.

210. **Segal M, Bloom FE.** The action of norepinephrine in the rat hippocampus. II. Activation of the input pathway. *Brain Res* 72: 99-114, 1974.

211. **Marchand JE, DeFrance JF, Stanley JC.** Ventromedial nucleus of the hypothalamus: convergent excitatory and inhibitory responses to fimbria and stria terminals stimulation. *J Neurosci Res* 8: 613-629, 1982.

212. **Mason ST, Fibiger HC.** 6-OHDA lesion of the dorsal noradrenergic bundle alters extinction of passive avoidance. *Brain Res* 152: 209-214, 1978.

213. **Lindbrink P.** The effect of lesions of ascending noradrenaline pathways on sleep and waking in the rat. *Brain Res* 74: 19-40, 1974.

214. **Lindvall O, Bjorklund A.** Organization of catecholamine neurons in the rat central nervous system, in *Chemical Pathways in the Brain, Handbook of Psychopharmacology*. Iversen LL, Iversen SD, Snyder SH. Eds., Plenum Press, New York, 1978, 139.

215. **Thierry AM, Tassin JP, Blanc G, Glowinski J.** Selective activation of the mesocortical dopaminergic system by stress. *Nature* 263: 242-244, 1976.

216. **Vandermaelen CP, Aghajanian GK.** Noradrenergic activation of serotonergic dorsal raphe neurons recorded in vitro. *Soc Neurosci Abstr* 8: 482, 1982.

217. **Simon H, LeMoal M, Stinus L, Calas A.** Anatomical relationships between the ventral mesencephalic tegmentum-A10 region and the locus coeruleus as demonstrated by anterograde and retrograde tracing techniques. *J Neural Transm* 44: 77-86, 1979.

218. **Anderson C, Pasquier D, Forbes W, Morgane P.** Locus coeruleus-to-dorsal raphe input examined by electro-physiological and morphological methods. *Brain Res Bull* 2: 209-221, 1977.

219. **Kostowski W, Samanin R, Bareggi SR, Mark V, Garattini S, Valzelli L.** Biochemical aspects of the interaction between midbrain raphe and locus coeruleus in the rat. *Brain Res* 82: 178-182, 1974.

220. **Jones BE, Harper ST, Halaris AE.** Effects of locus coeruleus lesions upon cerebral monoamine content, sleep-wakefulness states and the response to amphetamine in the cat. *Brain Res* 124: 473-496, 1977.

221. **Conrad ICA, Leonard CM, Pfaff DW.** Connections of the median and dorsal raphe nuclei in the rat: an autoradiographic and degeneration study. *J Comp Neurol* 156: 179-206, 1974.

222. **Segal M, Bloom FE.** The action of norepinephrine in the rat hippocampus. IV. The effects of locus coeruleus stimulation on evoked hippocampal unit activity. *Brain Res* 107: 513-525, 1976.

223. **Waterhouse BD, Moises HC, Woodward DJ.** Noradrenergic modulation of somatosensory cortical neuronal responses to iontophoretically applied putative neurotransmitters. *Soc Neurosci Abstr* 4: 286, 1978.

224. **Woodward DJ, Waterhouse BD.** Interaction of norepinephrine with cerebrocortical activity evoked by stimulation of somatosensory afferent pathways in the rat. *Soc Neurosci Abstr* 4: 287, 1978.

225. **Baraban JM, Aghajanian GK.** Suppresion of serotonergic neuronal firing by alpha-adrenoceptor antagonists: evidence against GABA mediation. *Eur J Pharmacol* 66: 287-294, 1980.

226. **Baraban JM, Aghajanian GK.** Suppression of firing activity of 5HT neurons in the dorsal raphe by alpha-adrenoceptor antagonists. *Neuropharmacology* 19: 335-341, 1980.

227. **Trulson ME, Crisp T.** Role of norepinephrine in regulating the activity of serotonin-containing dorsal raphe neurons. *Life Sci* 35: 511-515, 1984.

228. **Key B, Krzywosinski L.** Electrocortical changes induced by the perfusion of noradrenaline, acetylcholine and their antagonists directly into the dorsal raphe nucleus of the cat. *Br J Pharmacol* 61: 297-305, 1977.
229. **Dyr W, Kostowski W, Zacharski B, Bidzinski A.** Differential clonidine effects on EEG following lesions of the dorsal and median raphe nuclei in rats. *Pharmacol Biochem Behav* 19: 177-185, 1983.
230. **Aghajanian GK, Wang RY.** Physiology and pharmacology of central serotonergic neurons, in *Psychopharmacology: A Generation of Progress*. Lipton MA, DiMascio A, Killam KF. Eds., Raven Press, New York, 1978, 171.
231. **Baraban JM, Aghajanian GK.** Suppression of firing activity of 5-HT neurons in the dorsal raphe by alpha-adrenoceptor antagonists. *Neuropharmacology* 19: 355-363, 1980.
232. **Kiianmaa K, Fuxe K.** The effects of 5,7-DHT-induced lesions of the ascending 5-HT pathways on the sleep wake-fulness cycle. *Brain Res* 131: 287-301, 1977.
233. **Gumulka W, Samanin R, Valzelli L, Consolo S.** Behavioural and biochemical effects following the stimulation of the nucleus raphis dorsalis in rats. *J Neurochem* 18: 533-535, 1971.
234. **Lorens SA, Sorensen JP, Yunger IM.** Behavioral and neurochemical effects of lesions in the raphe system of the rat. *J Comp Physiol Psychol* 77: 48-52, 1971.
235. **Steranka LR, Barrett RJ.** Facilitation of avoidance acquisition by lesion of the median raphe nucleus: evidence for serotonin as a mediatior of shock-induced suppression. *Behav Biol* 11: 205-213, 1974.
236. **Maeda H, Mogenson GJ.** An electrophysiological study of inputs to neurons of the ventral tegmental area from the nucleus accumbens and medial preoptic-anterior hypothalamic areas. *Brain Res* 197: 365-377, 1980.
237. **Phillipson OT.** Afferent projections to the ventral tegmental area of Tsai and interfascicular nucleus: a horseradish peroxidase study in the rat. *J Comp Neurol* 187: 117-144, 1979.
238. **Van Der Kooy D, Hattori T.** Dorsal raphe cells with collateral projections to the caudate-putamen and substantia nigra: a fluorescent retrograde double labeling study in the rat. *Brain Res* 186: 1-7, 1980.
239. **Bobillier P, Sequin S, Petitjean F, Salvert D, Touret M, Jouvet M.** The raphe nuclei of the cat brain stem: a topographical atlas of their efferent projections as revealed by autoradiography. *Brain Res* 113: 449-486, 1976.
240. **Dray A, Gonye TJ, Oakley NR, Tanner T.** Evidence for the existence of a raphe projection to the substantia nigra in rat. *Brain Res* 113: 45-57, 1976.
241. **Moore RY, Halaris AE, Jones BE.** Serotonin neurons of the midbrain raphe: ascending projections. *J Comp Neurol* 180: 417-438, 1978.
242. **Lorez HP, Richards JG.** 5-HT nerve terminals in the fourth ventricle of the rat brain: their identification and distribution studied by fluorescence histochemistry and electron microscopy. *Cell Tissue Res* 165: 37-48, 1975.
243. **Grabowska M.** Influence of midbrain raphe lesions on some pharmacological and biochemical effects of apomorphine in rats. *Psychopharmacologia* 39: 315-322, 1974.
244. **Lorens SA, Guldberg HC, Hole K, Kohler C, Srebro B.** Activity, avoidance learning and regional 5-hydroxy-tryptamine following intra-brain stem 5,7-dihydroxy-tryptamine and electrolytic midbrain raphe lesions in the rat. *Brain Res* 108: 97-113, 1976.
245. **Hole K, Fuxe K, Jonsson G.** Behavioral effects of 5,7-DHT lesions of ascending serotonin pathways. *Brain Res* 107: 385-399, 1976.
246. **Deakin JFW, File SE, Hyde JR, MacLod NK.** Ascending 5-HT pathways and behavioural habituation. *Pharmacol Biochem Behav* 10: 687-694, 1979.
247. **Azmitia EC, Segal M.** An autoradiographic analysis of the differential ascending projections of the dorsal and medial raphe nuclei of the rat. *J Comp Neurol* 179: 641-668, 1978.
248. **Jacobs BL, Asher R, Dement WC.** Electrophysiological and behavioral effect of electrical stimulation of raphe nuclei in cats. *Physiol Behav* 11: 489-495, 1973.
249. **Wiklund L.** Studies on Anatomical, Functional, and Plastic Properties of Central Serotonergic Neurons. Doctoral dissertation, University of Lund, Sweden, 1980.
250. **Van Loon GR, Shum A, Sole MJ.** Decreased brain serotonin turnover after short-term (two-hour) adrenalectomy in rats: a comparison of four hour turnover methods. *Endocrinology* 108: 1392-1402, 1981.
251. **Diaz J, Ellison G, Masuoka D.** Opposed behavioral syndromes in rats with partial and more complete central serotonergic lesions made with 5,6-dihydroxytryptamine. *Psychopharmacologia* 37: 67-69, 1974.
252. **Reader TA.** Distribution of catecholamines and serotonin in the rat cerebral cortex: absolute levels and relative proportions. *J Neural Transm* 50: 13-27, 1981.
253. **Anden NE, Grabowska M.** Pharmacological evidence for a stimulation of dopamine neurons by noradrenaline neurons in the brain. *Eur J Pharmacol* 39: 275-282, 1976.
254. **Anden NE, Atack CV, Svensson TH.** Release of dopamine from central noradrenaline and dopamine nerves induced by a dopamine-beta-hydroxylase inhibitor. *J Neural Transm* 34: 93-100, 1973.
255. **Fadda F, Argiolas A, Melin ME, Tissari AM, Onali PL, Gessa GL.** Stress induced increase in 3,4-dihydroxyphenylacetic acid (DOPAC) levels in the cerebral cortex and in *N. accumbens*: reversal by diazepam. *Life Sci* 23: 2219-2224, 1978.
256. **LaVielle S, Tassin JP, Thierry AM, Blanc G, Herve D, Barthelemy C, Glowinski J.** Blockade by benzodiazepines of the selective high increase in dopamine turnover induced by stress in mesocortical dopaminergic neurons of the rat. *Brain Res* 168: 585-594, 1979.

257. **Leonard CM.** The prefrontal cortex of the rat. I. Cortical projection of the mediodorsal nucleus. II. Efferent connections. *Brain Res* 12: 321-343, 1969.

258. **Bannon MJ, Roth RH.** Pharmacology of mesocortical dopamine neurons. *Pharmacol Rev* 35: 53-68, 1983.

259. **Bannon MJ, Wolf ME, Roth RH.** Pharmacology of dopamine neurons innervating the prefrontal, cingulate and piriform cortices. *Eur J Pharmacol* 92: 119-125, 1983.

260. **Berger B, Tassin JP, Blanc G, Moyne MA, Thierry AM.** Histochemical confirmation for dopaminergic innervation of the rat cerebral cortex after destruction of the noradrenergic ascending pathways. *Brain Res* 81: 332-337, 1974.

261. **Simon H, Scatton B, LeMoal M.** Dopaminergic A10 neurons are involved in cognitive functions. *Nature* 286: 150-151, 1980.

262. **Mason ST.** Noradrenaline and selective attention: a review of the model and the evidence. *Life Sci* 27: 617-631, 1980.

263. **Versteeg DHG, van der Gugten J, DeJonc W, Palkovits M.** Regional concentrations of noradrenaline and dopamine in rat brain. *Brain Res* 113: 563-574, 1976.

264. **Berger B, Thierry AM, Tassin JP, Moyne MA.** Dopaminergic innervation of the rat prefrontal cortex: a fluorescence histochemical study. *Brain Res* 106: 133-145, 1976.

265. **Reinhard JF Jr, Bannon MJ, Roth RH.** Acceleration by stress of dopamine synthesis and metabolism in prefrontal cortex: antagonism by diazepam. *Naunyn-Schmiedeberg's Arch Pharmacol* 318: 374-377, 1982.

266. **LeMoal M, Stinus L, Galay D.** Radiofrequency lesion of the ventral mesencephalic tegmentum: neurological and behavioural considerations. *Exp Neurol* 50: 521-535, 1976.

267. **Herman JP, Guillonneau D, Dantzer R, Scatton B, Semerdjian-Rouquier L, LeMoal M.** Differential effects of inescapable foot-shocks and of stimuli previously paired with inescapable foot-shocks on dopamine turnover in cortical and limbic areas of the rat. *Life Sci* 30: 2207-2214, 1982.

268. **Yim CY, Mogenson GJ.** Electrophysiological studies of neurons in the ventral tegmental area of Tsai. *Brain Res* 181: 301-313, 1980.

269. **Jacobs BL, Foote SL, Bloom FE.** Differential projections of neurons within the dorsal raphe nucleus of the rat: a horseradish peroxidase (HRP) study. *Brain Res* 147: 149-153, 1978.

270. **Trulson ME, Jacobs BL, Morrison AR.** Raphe unit activity across the sleep-waking cycle in normal cats and in pontine lesioned cats diplaying REM sleep without atonia. *Brain Res* 226: 75-91, 1981.

271. **Neill DB, Grant LD, Grossman SP.** Selective potentiation of locomotor effects of amphetamine by midbrain raphe lesions. *Physiol Behav* 9: 655-657, 1972.

272. **Segal DS, Mandell AJ.** Long-term administration of D-amphetamine progressive augmentation of motor activity and stereotype. *Pharmacol Biochem Behav* 2: 249-255, 1974.

273. **Tissari AH, Argiolas A, Fadda F, Serra G, Gessa GL.** Foot-shock stress accelerates non-striatal dopamine synthesis without activating tyrosine hydroxylase. *Naunyn-Schmiedeberg's Arch Pharmacol* 308: 155-157, 1979.

274. **Hartmann RJ, Geller I.** P-CPA effects on a conditioned emotional response in rats. *Life Sci* 10: 927-933, 1971.

275. **Seligman MEP, Maier SF, Solomon RL.** Unpredictable and uncontrollable aversive events, in *Aversive Conditioning and Learning.* Brush FR. Ed., Academic Press, New York, 1971.

276. **Elam M, Svensson TH, Thoren P.** Differentiated cardiovascular afferent regulation of locus coeruleus neurons and sympathetic nerves. *Brain Res* 358: 77-84,1985.

277. **Howe PRC.** Blood pressure control by neurotransmitters in the medulla oblongata and spinal cord. *J Auton Nerv System* 12: 95-115, 1985.

278. **Granat AR, Kumada M, Reis DJ.** Sympathoinhibition by A1-noradrenergic neurons is mediated by neurons in the C1 area of the rostral medulla. *J Auton Nerv Syst* 14: 387-395, 1985.

279. **Sourkes TL.** Neurotransmitters and central regulation of adrenal functions. *Biol Psychiatry* 20: 182-191, 1985.

280. **Woodward DJ, Moises HC, Waterhouse BD, Hoffer BJ, Freedman R.** Modulatory actions of norepinephrine in the central nervous system. *Fed Proc Fed Am Soc Exp Biol* 38: 2109-2116, 1979.

281. **File SE.** Clinical lesions of both median and median raphe nuclei and changes in social and aggressive behaviour in rats. *Pharmacol Biochem Behav* 12: 855-859, 1980.

282. **Hanin L, Masarelli R, Costa E.** Acetylcholine concentrations in rat brain: diurnal oscillation. *Science* 170: 341-342, 1970.

283. **Lewis PR, Shute CCD.** The cholinergic limbic system: projections to hippocampal formation, medial cortex, nuclei of the ascending cholinergic reticular system and the subfornical organ and supraoptic crest. *Brain* 90: 521-539, 1967.

284. **Racagni G, Cheney DL, Trabucchi M, Wang C, Costa E.** Measurement of acetylcholine turnover rate in discrete areas of rat brain. *Life Sci* 15: 1961-1975, 1974.

285. **Shute CCD, Lewis PR.** Cholinergic nervous pathways in the forebrain. *Nature* 189: 332-333, 1961.

286. **Frankhuyzen AL, Mulder AH.** Pharmacological characterization of presynaptic alpha-adrenoceptors modulating (3H) noradrenaline and (3H) 5-hydroxytryptamine release from slices of the hippocampus of the rat. *Eur J Pharmacol* 81: 97-106, 1982.

287. **Frankhuyzen AL, Mulder AH.** Noradrenaline inhibits 3H-serotonin release from slices of rat hippocampus. *Eur J Pharmacol* 63: 179-187, 1980.

288. **Green JD, Arduini AA.** Hippocampal electrical activity in arousal. *J Neurophysiol* 17: 533-557, 1954.

289. **Redmond DE Jr, Huang YH, Snyder DR, Maas JW.** Behavioral effects of stimulation of the locus coeruleus in the stumptail monkey (*Macaca arctoides*). *Brain Res* 116: 502-510, 1976.

290. **Redmond DE Jr, Huang YH, Snyder DR, Maas JW, Baulu J.** Behavioral changes following lesions of the locus coeruleus in Macaca arctoides. *Neurosci Abstr* 1: 472, 1976.

291. **German DC, Dalsass M, Kiser RS.** Electrophysiological examination of the ventral tegmental (A10) area in the rat. *Brain Res* 181: 191-197, 1980.

292. **Stein L, Wise CD, Belluzi JD.** Effects of benzodiazepines on central serotonergic mechanisms, in *Mechanism of Action of Benzodiazepines*. Costa E, Greengard P. Eds., Raven Press, New York, 1975, 29.

293. **McGinty DJ, Harper RM.** Dorsal raphe neurons: depression of firing during sleep in cats. *Brain Res* 101: 569-575, 1976.

294. **Trulson ME, Jacobs BL.** Raphe unit activity in freely moving cats: correlation with level of behavioral arousal. *Brain Res* 163: 135-142, 1979.

295. **Trulson ME, Jacobs BL.** Activity of serotonin-containing neurons in freely moving cats, in *Serotonin Neurotransmission and Behavior*. Jacobs BL, Gelperin A. Eds., MIT Press, Cambridge, 1981, 339.

296. **Trulson ME, Preussler DW, Howell GA, Frederickson CJ.** Raphe unit activity in freely moving cats: effects of benzodiazepines. *Neuropharmacology* 21: 1045-1050, 1982.

297. **Gray JA.** *The Neuropsychology of Anxiety: an Inquiry into the Functions of the Septo-Hippocampal System*. Oxford University Press, New York, 1982.

298. **Beckstead EM, Domesick VB, Nauta WJH.** Efferent connections of substantia nigra and ventral tegmental area in the rat. *Brain Res* 175: 191-217, 1979.

299. **Carter CJ, Pycock CJ.** A study of the sites of interaction between dopamine and 5-hydroxytryptamine for the production of fluphenazine-induced catalepsy. *Naunyn-Schmiedeberg's Arch Pharmacol* 304: 135-139, 1978.

300. **Fonnum F, Walaas I, Iversen E.** Localization of GABAergic, cholinergic and aminergic structures in the limbic system. *J Neurochem* 29: 221-230, 1977.

301. **Nauta WJH, Smith GP, Faull RLM, Domesick VB.** Efferent connections and nigral afferents of the nucleus accumbens septi in the rat. *Neuroscience* 3: 385-401, 1978.

302. **Bentivoglio M, van der Kooy D, Kuypers HGJM.** The organization of the efferent projections of the substantia nigra in the rat. A retrograde fluorescent double labeling study. *Brain Res* 174: 1-17, 1979.

303. **Sinha AK, Henricksen S, Dement WC, Barchas JD.** Cat brain amine content during sleep. *Am J Physiol* 224: 381-383, 1973.

304. **Sweeney DR, Maas JW, Heninger GR.** State anxiety and urinary MHPG. *Arch Gen Psychiatry* 35: 1418-1423, 1978.

305. **Lake CR, Ziegler MG, Kopin IJ.** Use of plasma norepinephrine for evaluation of sympathetic neuronal function in man. *Life Sci* 18: 1315-1321, 1976.

306. **Maura G, Bonanno G, Raiteri M.** Chronic clonidine induces functional down-regulation of presynaptic alpha2-adrenoceptors regulating (3H) noradrenaline and (3H) 5-hydroxytryptamine release in the rat brain. *Eur J Pharmacol* 112: 105-110, 1985.

307. **Elam M, Svensson TH, Thoren P.** Differentiated cardio-vascular afferent regulation of locus coeruleus neurons and sympathetic nerves. *Brain Res* 358: 77-84, 1985.

308. **Smee ML, Weston PF, Kinner DS, Day T.** Dose-related effects of central noradrenaline stimulation of behavioural arousal in rats. *Psychopharmacol Commun* 1: 123-130, 1975.

309. **Foote SL, Aston-Jones G, Bloom FE.** Impulse activity of locus coeruleus neurons in awake rats and monkeys is a function of sensory stimulation and arousal. *Proc Natl Acad Sci USA* 77: 3033-3039, 1980.

310. **Sheu Y-S, Nelson JP, Bloom FE.** Discharge patterns of cat raphe neurons during sleep and waking. *Brain Res* 73: 263-276, 1974.

311. **Ko GN, Elsworth JD, Roth RH, Rifkin BG, Leigh H, Redmond DE Jr.** Panic-induced elevation of plasma MHPG in phobic-anxious patients: effects of clonidine or imipramine. *Arch Gen Psychiatry* 40: 425-430, 1983.

312. **Mignot E, Laude D, Elghozi J, LeQuan-Bui KH, Meyer P.** Central administration of yohimbine increases free 3-methoxy-4-hydroxyphenylglycol in the cerebrospinal fluid of the rat. *Eur J Pharmacol* 83: 135-138, 1982.

313. **Charney DS, Heninger GR, Redmond DE Jr.** Yohimbine induced anxiety and increased noradrenergic functions in humans: Effects of diazepam and clonidine. *Life Sci* 33: 19-30, 1983.

314. **Dickinson SL, Slater P.** Effect of lesioning dopamine, noradrenaline and 5-hydroxytryptamine pathways on tremorine-induced tremor and rigidity. *Neuropharmacology* 21: 787-794, 1982.

315. **Gray JA.** Precis of the neuropsychology of anxiety: an enquiry into the functions of the septo-hippocampal system. *Behav Brain Sci* 5: 469-534, 1982.

316. **Geller L, Blum K.** The effects of 5-HT on *p*-chloro-phenylalanine (pCPA) attenuation of "conflict" behaviour. *Eur J Pharmacol* 9: 319-324, 1970.

317. **Collinge J, Pycock C.** Differential actions of diazepam on the release of (3H)-5-hydroxytryptamine from cortical and midbrain raphe slices in the rat. *Eur J Pharmacol* 85: 9-14, 1982.

318. **Chan LT, Schall SM, Saffran M.** Properties of the corticotrophin releasing factor of the rat median eminence. *Endocrinology* 85: 664-651, 1969.

319. **Hashimoto K, Ohno N, Yunoki S, Kageyama J, Aoki Y, Takahara J, Ofuji T.** Characterization of corticotropin-releasing factor (CRF) and arginine vasopressin in median eminence extracts on Sephadex gel-filtration. *Endocrinol Jpn* 28: 1-7, 1981.

320. **Krieger DT, Liotta A, Brownstein MJ.** Corticotropin-releasing factor distribution in normal and Brattleboro rat brain, and effect of deafferentation, hypophysectomy and steroid treatment in normal animals. *Endocrinology* 100: 227-237, 1977.

321. **Makara GB, Stark E, Karteszi M, Palkovits M, Rappy G.** Effects of paraventricular lesions on stimulated ACTH release and CRF in stalk-median eminence of the rat. *Am J Physiol* 240: E441-E446, 1981.

322. **Vale W, Rivier C.** Effects of a putative hypothalamic CRF and known substances on the secretion of radioimmunoassayable ACTH by cultures anterior pituitary cells. *Abstr 59th Annu Meet Endocrine Society*, 1977, 217.

323. **Buckingham JC, Hodges JR.** Hypothalamic receptors influencing the secretion of corticotropin releasing hormone in the rat. *J Physiol* 290: 421-431, 1979.

324. **Jones MT.** Control of corticotropin (ACTH) secretion, in *The Endocrine Hypothalamus*. Jeffcoate SL, Hutchinson JSM. Eds., Academic Press, New York, 1978, 385.

325. **Jones MT, Hillhouse EW, Burden J.** Effect of various putative neurotransmitters on the secretion of corticotrophin releasing hormone from the rat hypothalamus in vitro. A model of the neurotransmitters involved. *J Endocrinol* 69: 1-20, 1976.

326. **Kennett GA, Joseph MH.** The functional importance of increased brain tryptophan in the serotonergic response to restraint stress. *Neuropharmacology* 20: 39-43, 1981.

327. **Rose JC, Ganong WF.** Neurotransmitter regulation of pituitary secretion, in *Current Developments in Psychopharmacology*. Essman WB, Valzelli L. Eds., Spectrum, New York, 1976, 86.

328. **Van Loon R.** Brain catecholamines and ACTH serotonin, in *Frontiers in Neuroendocrinology*. Martin L, Ganong WF. Eds., Oxford University Press, New York, 1973, 209.

329. **Agren H, Terenius L.** Hallucinations in patients with major depression. Interactions between CSF monoaminergic and endorphinergic indices. *J Affect Dis* 9: 25-34, 1985.

330. **Sourkes TL.** Neurotransmitters and central regulation of adrenal functions. *Biol Psychiatry* 20: 182-191, 1985.

331. **Hashimoto K, Ohno N, Aoki Y, Kageyama J, Takahara J, Ofuji T.** Distribution and characterization of corticotropin-releasing factor and arginine vasopressin in rat hypothalamic nuclei. *Neuroendocrinology* 34: 32-37, 1982.

332. **Descarries L, Beaudet A.** The serotonin innervation of adult rat hypothalamus, in *Cell Biology of Hypothalamic Neurosecretion*. Vincent JD, Kordon C. Eds., CNRS, Paris, 1978.

333. **Brown JS, Hunsperger RW, Rosvold HE.** Interaction of defense and flight reactions produced by simultaneous stimulation at two points in the hypothalamus of the cat. *Exp Brain Res* 8: 130-149, 1969.

334. **Dreifuss JJ, Murphy JT, Gloor P.** Contrasting effects of two identified amygdaloid efferent pathways on single hypothalamic neurons. *J Neurophysiol* 31: 237-248, 1968.

335. **Cowan WM, Raisman G, Powell TPS.** The connections of the amygdala. *J Neurol Neurosurg Psychiatry* 28: 137-151, 1965.

336. **Van Atta L, Sutin J.** Relationships among amygdaloid and other limbic structures in influencing activity of lateral hypothalamic neurons, in *The Neurobiology of the Amygdala*. Eleftheriou BE. Ed., Plenum Press, New York, 1972, 343.

337. **Carlson DE, Dornhorst A, Maran JW, Gann DS.** Hypothalamic neurons responding to hemodynamic input and to stimulation in the pons may influence adrenocorticotropin release. *J Neurosci* 4: 897-907, 1984.

338. **Bobillier P, Petitjean F, Salvert D, Lighier M, Seguin S.** Differential projections of the nucleus raphe dorsalis and nucleus raphe centralis as revealed by autoradiography. *Brain Res* 85: 205-210, 1975.

339. **Carlson DE, Dornhorst A, Gann DS.** Organization of the lateral hypothalamus for control of adrenocorticotropin release in the cat. *Endocrinology* 107: 961-969, 1980.

340. **Gann DS, Ward DG, Baertschi AJ, Carlson DE, Maran JW.** Neural control of ACTH release in response to hemorrhage. *Ann NY Acad Sci* 294: 477-497, 1977.

341. **Gann DS, Ward DG, Carlson DE.** Neural control of ACTH: a homeostatic reflex. *Recent Prog Horm Res* 34: 357-400, 1978.

342. **Grizzle WE, Dalman MF, Schramm LP, Gann DS.** Inhibitory and facilitatory hypothalamic areas mediating ACTH release in the cat. *Endocrinology* 95: 1450-1461, 1974.

343. **Kawata M, Hashimoto K, Takahara J, Sano Y.** Immunohistochemical demonstration of the localization of corticotropin releasing factor-containing neurons in the hypothalamus of mammals including primates. *Anat Embryol* 165: 303-313, 1982.

344. **Saper CB, Swanson LW, Cowan WM.** The efferent connections of the anterior hypothalamic area of the rat, cat and monkey. *J Comp Neurol* 182: 575-600, 1978.

345. **Casady RL, Taylor AN.** Effect of electrical stimulation of the hippocampus upon corticosteroid levels in the freely behaving, non-stressed rat. *Neuroendocrinology* 20: 68-78, 1976.

346. **Meibach RC, Siegel A.** Efferent connections of the hippocampal formation in the rat. *Brain Res* 124: 197-224, 1977.

347. **Poletti CE, Kinnard MA, MacLean PD.** Hippocampal influence on unit activity of hypothalamus, preoptic region, and basal forebrain in awake, sitting squirrel monkeys. *J Neurophysiol* 36: 308-324, 1973.

348. **Polletti CE, Sujatanond M.** Evidence for a second hippocampal efferent pathway to hypothalamus and basal forebrain comparable to fornix system: a unit study in the awake monkey. *J Neurophysiol* 44: 514-531, 1980.

349. **Brown WA, Keitner G, Qualls B, Haier R.** The dexamethasone suppression test and pituitary-adrenocortical function. *Arch Gen Psychiatry* 42: 121-123, 1985.

350. **Glaser T, Traber J.** Binding of the putative anxiolytic TVX Q 7821 to hyppocampal 5-hydroxytryptamine (5-HT) recognition sites. *Naunyn-Schmiedeberg's Arch Pharmacol* 329: 211-215, 1985.

351. **Nishikawa T, Scatton B.** Inhibitory influence of GABA on central serotonergic transmission. Raphe nuclei as the neuroanatomical site of the GABAergic inhibition of cerebral serotonergic neurons. *Brain Res* 331: 91-103, 1985.

352. **Schutz MTB, de Aguiar JC, Graeff FG.** Anti-aversive role of serotonin in the dorsal periaqueductal gray matter. *Psychopharmacology* 85: 340-345, 1985.

353. **Schmidt RH, Bjorklund A, Lindvall O, Loren I.** Prefrontal cortex: dense dopaminergic input in the newborn rat. *Dev Brain Res* 5: 222-228, 1982.

354. **Hamilton TC, Hunt AAE, Poyser RH.** Involvement of central alpha2-adrenoceptors in the mediation of clonidine-induced hypotension in the cat. *J Pharm Pharmacol* 32: 788-789, 1980.

355. **Sakakura M, Yoshioka M, Kobayashi M, Takebe K.** The site of inhibitory action of a natural (corticosterone) and synthetic steroid (dexamethasone) in the hypothalamus-pituitary-adrenal axis. *Neuroendocrinology* 32: 174-178, 1981.

356. **Bohus B, Strashimirov D.** Localization and specificity of corticoid "feedback receptors" at the hypothalamo-hypophyseal level: comparative effects of various steroids implanted in the median eminence or the anterior pituitary of the rat. *Neuroendocrinology* 6: 197-209, 1970.

357. **Martini L, Focchi M, Gavazzi G, Pecile A.** Inhibitory action of steroids on the release of corticotrophin. *Arch Int Pharmacodyn* 140: 156-163, 1962.

358. **Micco DJ Jr, McEwen BS.** Glucocorticoids, the hippocampus and behaviour: interactive relation between task activation and steroid hormone binding specificity. *J Comp Physiol Psychol* 94: 624-633, 1980.

359. **Rousseau GG, Baxter JD, Tomkins GM.** Glucocorticoid receptors: relation between steroid binding and biologic effects. *J Mol Biol* 67: 99-107, 1972.

360. **Vermes I, Smelik PG, Mulder AH.** Effects of hypophysectomy, adrenalectomy and corticosterone treatment on uptake and release of putative central neurotransmitters by rat hypothalamic tissue in vitro. *Life Sci* 19: 1719-1726, 1976.

361. **Foote S, Bloom FE.** Activity of locus coeruleus neurons in the anesthetized squirred monkey, in *Catecholamines: Basic and Clinical Frontiers*. Usdin E. Ed., Pergamon Press, Elmsford, NY, 1979, 625.

362. **Lees AJ, Fernando JCR, Curzon G.** Serotonergic involvement in behavioral responses to amphetamine at high dosage. *Neuropharmacology* 18: 153-158, 1979.

363. **Rudorfer MV, Scheinin M, Karou F, Ross RJ, Potter WZ, Linnoila M.** Reduction of norepinephrine turnover by serotonergic drug in man. *Biol Psychiatry* 19: 179-185, 1984.

364. **Spyraki C, Fibiger HC.** Clonidine-induced sedation in rats: evidence for mediation by postsynaptic alpha2-adrenoceptors. *J Neural Transm* 54: 153-163, 1982.

365. **Drew GM, Gower AJ, Marriott AS.** Alpha2-adrenoceptors mediated clonidine-induced sedation in the rat. *Br J Pharmacol* 67: 133-141, 1979.

366. **Langer SZ, Massingham R.** Alpha-adrenoceptors and the clinical pharmacology of clonidine, in *Proc World Conference on Clinical Pharmacology and Therapeutics*. Turner P. Ed., Macmillan, New York, 1980, 158.

367. **Laverty R, Taylor KM.** Behavioral and biochemical effects of 2-(2,6-dichloro-phenylamino)-2-imidazoline hydrochloride (ST 155) on the central nervous system. *Br J Pharmacol* 35: 253-264, 1969.

368. **Strombon U, Svensson T.** Clonidine: attenuation of sedative action by facilitated central noradrenergic neurotransmission. *J Neural Transm* 47: 29-39, 1980.

369. **Palkovits M, Zaborszky L, Brownstein MJ, Fekete MIK, Herman JP, Kanyicska B.** Distribution of norepinephrine and dopamine in cerebral cortical areas of the rat. *Brain Res Bull* 4: 593-601, 1979.

370. **Bunney BS.** The electrophysiological pharmacology of mid-brain dopaminergic systems, in *The Neurobiology of Dopamine*. Horn AS, Korf J, Westerink BHC. Eds., Academic Press, New York, 1979, 417.

371. **McRae-Degueurce A, Milon H.** Serotonin and dopamine afferents to the rat locus coeruleus: a biochemical study after lesioning of the ventral mesencephalic tegmental A10 region and the raphe dorsalis. *Brain Res* 263: 344-347, 1983.

372. **Leger L, McRae-Degueurce A, Pujol JE.** Origine de l'innervation serotoninergique du locus coeruleus chez le rat. *CR Acad Sci* 290: 807-810, 1980.

373. **Sladek J, Walker P.** Serotonin-containing neuronal peri-karya in the primate locus coeruleus and subcoeruleus nuclei. *Brain Res* 134: 359-366, 1977.

374. **Mosko SS, Haubrich D, Jacobs BL.** Serotonergic afferents to the dorsal raphe nucleus: evidence from HRP and synaptosomal uptake studies. *Brain Res* 119: 269-290, 1977.
375. **Trulson ME, Crisp T, Howell GA.** Raphe unit activity in freely moving cats: effects of quipazine. *Neuropharmacology* 21: 681-686, 1982.
376. **Felten DL, Harrigan P.** Dendritic bundles in nuclei raphe dorsalis and centralis superior of the rabbit. A possible substrate for local control of serotonergic neurons. *Neurosci Lett* 16: 275-280, 1980.
377. **Heym J, Trulson ME, Jacobs BL.** Effects of adrenergic drugs on raphe unit activity in freely moving cats. *Eur J Pharmacol* 74: 117-125, 1981.
378. **Milon H, McRae-Degueurce A.** Pharmacological investigation on the role of dopamine in the rat locus coeruleus. *Neurosci Lett* 30: 297-301, 1982.
379. **Swanson LW.** The projections of the ventral tegmental area and adjacent regions: a combined fluorescent retrograde and immunofluorescence study in the rat. *Brain Res Bull* 9: 321-354, 1982.
380. **Ochi J, Shimizu K.** Occurrence of dopamine-containing neurons in the midbrain raphe nuclei of the rat. *Neurosci Lett* 8: 317-320, 1978.

Chapter 4

CENTRAL NEURONAL PATHWAYS INVOLVED IN PSYCHOTIC SYNDROMES

Fuad Lechin, Bertha van der Dijs, José Amat, and Marcel Lechin

TABLE OF CONTENTS

I. AGGRESSIVE BEHAVIOR

Aggressive behavior is determined by genetic and environmental factors. However, although this behavior has been extensively studied it is little understood.

Aggression is not a single phenomenon. There are several classes of aggression, each with its own neuronal and endocrinal basis. In laboratory animals aggression may be induced by a variety of methods which include electrical stimulation or lesion of the brain, pharmacological manipulations, foot shock or other aversive stimuli, changes in environment, or feeding schedules.[1-20] Isolation-induced aggression in mice has been studied extensively for elucidation of the neurochemical basis of aggression.[21-25]

Isolation, however, not only induces aggression in mice, but produces a variety of symptoms termed collectively the "isolation syndrome". This syndrome consists of increased spontaneous activity, increased reactivity to pain, tremors, vocalization, and an increase in muscle tone. The aggressive behavior pattern seen after isolation is quite complex and may be preceded by chasing, rearing, and tail rattling. However, violent aggressive display of attacks and biting is not manifested by all isolated rats or mice, and aggression studies utilize only those isolated animals showing the aggressive components of the syndrome, specifically biting. Other aggressive behavior in rodents includes mouse killing (muricide) and shock-induced fighting in the rat.

A. Experimental Evidence Involving Noradrenergic System in Aggressive Behavior

Ample evidence suggests that stimulation of brain noradrenergic (NE) neurons plays an inhibitory role in rat mouse-killing (muricidal) aggression. The muricidal reaction is inhibited by microinjections of NE into the amygdala and potentiated by destruction of the dorsal NE bundle (DNB). Other findings report that stimulation of the locus coeruleus (LC), main origin of ascending NE fibers, inhibits muricidal aggression induced by olfactory bulbectomy in rats.[1,2,10,21,22,24,26-42]

The role of LC in the production of anxiety and fear has been discussed in preceding papers. In fact, electrolytic lesioning of the LC reduces the fear response in primates, while electrical stimulation of LC elicits overt behavioral signs of fear. Thus, it seems that suppression of muricidal aggression by LC stimulation may depend partly on enhanced anxiety and fear. With respect to this, it has been demonstrated that LC sends axons to the olfactory tubercle (OT) where they exert an inhibitory influence. In turn, the (OT) sends fibers to the amygdala which is clearly involved in aggressive behavior. So, there is anatomical and physiological evidence of an LC-amygdala-OT circuit related to aggressive behavior.[43-48]

Lesions of the septal forebrain area in the rat have long been known to produce a dramatic increase in irritability and reactivity. The resulting syndrome, often referred to as "septal rage", is characterized by initially explosive levels of irritability and reactivity which gradually subside to prelesion levels over a fairly prolonged period of behavioral recovery. Similar aggressive symptoms are observed following lesions of the lateral hypothalamus. Septal rage is associated with destruction of dopaminergic (DA) afferents since prelesion administration of haloperidol significantly blocks the postlesion behavioral syndrome, while prelesion administration of desipramine (which prevents destruction of NE fibers following 6-OHDA microinjection) does not interfere with the appearance of septal rage syndrome.[12,15,49-68]

Various authors report that prelesion administration of pCPA mitigates the septal rage syndrome induced by destruction of DA innervation fibers. The fact that pCPA induces DA as well as NE and serotonergic (5HT) depletion suggests that septal rage is not produced when DA fiber destruction is accompanied by NE fiber destruction. In other words, it seems that the onset of septal rage depends on the absence of dopamine and the presence of norepinephrine at septal level.[52,56-72]

The role of this NE/DA imbalance in mechanisms determining aggressive behavior is supported by studies showing that the NE steady-state level was significantly lower in OT and substantia nigra (SN) and significantly higher in the septal area of aggressive mice when compared to isolated nonfighter controls. NE turnover was only higher in ventral tegmental area (VTA) regions of the aggressors. In fact, microinjection of NE into VTA induces inhibition of DA neurotransmission, suggesting that NE and DA are inversely correlated at mesolimbic level and that NE system bridles DA system at this level.[12,21,22,24,29-32,73-85]

If we assume as correct the possibility that increase of NE/DA ratio at mesolimbic level is positively correlated with aggression, then NE predominance would be derived from ventral NE bundle (VNB) innervation which provides the majority of NE axons to mesolimbic area. Furthermore, as previously mentioned, LC activation is positively correlated with fear and anxiety and negatively correlated with aggression. LC activity is able to stimulate DA mesocortical system, but not DA mesolimbic system. Finally, it is well known that DA mesolimbic and DA mesocortical activities are negatively correlated.[29,30,65,66,71,73-76,78-80,85-92]

On the basis of the above experimental findings it may be proposed that aggressive behavior is positively correlated with NE/DA ratio at septal-mesolimbic level and negatively correlated with DA mesocortical activity. This profile of catecholaminergic activity is antagonic to that proposed for anxiety.

Studies reinforcing the above models employ pharmacological manipulations with D-amphetamine which demonstrate that high doses of this drug provoke predominance of NE activity and correlate positively with aggression, whereas low doses of D-amphetamine provoke DA predominance and do not correlate with aggression.[10,12,20,24,29,30,39-41,65,66,68,71,75,76,78-80,86,92-99]

B. Experimental Evidence Involving Serotonergic System in Aggressive Behavior

Social isolation has been shown to induce a decrease of 5HT turnover in the whole brain as well as in the diencephalon, and a decrease of brain tryptophane concentration in mice, while leaving the concentration of 5HT in the brain unaffected. Also, decreased brain tryptophane hydroxylase activity in the septal area of isolated rats has been reported.[3,5,6,8,9,15,26,100-104]

Recent studies have reported that after 8 weeks of isolation "aggressive" DBA/2 mice, when compared with "nonaggressive" C57 B1/6 mice, are characterized by hyperactivity, increased motor activity, and increased excitability. In these aggressive mice, isolation produces a decrease of 5HT turnover in specific brain areas. For example, changes in lateral hypothalamus may be related to the aggressive effects.[12-14,22,23,27,28,55,80,82-84,104-106]

Other studies in rodents have shown an inverse relationship between 5HT levels in the brain and different kinds of aggressive behavior such as mouse killing and shock-induced fighting in the rat, while isolation-induced aggressive behavior in the mouse is inhibited by central 5HT depletion induced by pCPA, 5,7-DHT, raphe lesions, or dietary tryptophane manipulation.[8,9,23,24,29-31,54,58,80,81,101,103,104,107,108]

Isolated-aggressive mice show reduction of 5HT activity in amygdala and lateral hypothalamus. The changes of NE activity found in these areas have been discussed earlier.

More precise studies show that dorsal raphe (DR) and median raphe (MR) nuclei function differently in the inhibition of mouse-killing behavior. However, it should be noted that differential function among the raphe nuclei is not confined to muricide. For example, lesions of MR have been shown to produce hyperactivity while lesions of DR do not.[1,2,49,50,83,84,104,108-118]

There is evidence of a projection from DR and MR to amygdala. This is of particular interest since lesions of the amygdala provoke muricide. The possibility that muricide following destruction of DR may be due to loss of the DR 5HT input to the amygdala has now been demonstrated. Conceived in this way, the amygdala is seen as the nodal point of inhibition ascending from DR and descending from OT. It is possible then that muricide induced by pCPA is due to serotonin depletion in the DR-amygdala pathway. The fact that electrical stimulation of dorsal tegmentum in MR vicinity induces muricide reinforces the hypothesis that in this tegmental area, lateral to central gray, there exists circuitry employed in both naturally occurring and DR-lesion-released muricide. In short, it appears that the mesencephalon may contain two separate mechanisms for control of mouse-killing behavior: an inhibitory raphe system (DR) and a facilitatory raphe system (MR and nearby tegmental system). This facilitatory role of MR on muricidal behavior is supported by experimental studies showing that lesions in this 5HT nucleus or nearby tegmental region interfere with muricidal behavior.[1,2,33,49-51,55,58,83,84,101,104,110-112,119,120]

Antagonism between DR and MR 5HT systems is suggested by many kinds of experimental studies (anatomical, physiological, biochemical, pharmacological, and behavioral).

5HT neurons from DR and MR innervate different brain areas. In addition, MR axons appear to tonically inhibit cholinergic neurons in cortex and stimulate cholinergic (ACh) neurons in hippocampus; whereas 5HT-DR axons appear to tonically inhibit ACh neurons in both cortex and hippocampus. Neurons from MR project to medial septum, site of cholinergic cell bodies which project to hippocampus, and to hippocampus itself. When MR is lesioned, the level of hippocampal 5HT drops greatly, but the ACh turnover level in hippocampus is unaffected. Thus, although there is substantial 5HT innervation of hippocampus from MR, it appears that these 5HT neurons do not interact with hippocampal ACh neurons, at least in a tonic action. On the other hand lesion of DR, which projects to lateral septum and to hippocampus as well, significantly increases ACh turnover in hippocampus. Moreover, DR 5HT projection to hippocampus inhibits elicitation of hippocampal θ rhythm through septal stimulation, which involves the septal-hippocampal ACh neurons receiving excitatory influence from 5HT-MR projections.[50,58,61,83,110,116,121-158]

The diverse anatomical projections of DR and MR to septal area may explain why septal-lesion induced muricide depends on destruction of DR or MR projections. In effect, numerous experimental studies in this matter suggest the existence of two complex and opposing circuits passing through distinct septal regions. DR, MR, amygdala, OT, and lateral hypothalamus are proven to be among structures related to the two circuits dealing with muricidal behavior.[50,83,110,111,116,124,125,143-145,159]

Other anatomical evidence shows that MR sends axons to DR but DR nucleus does not project to the former. This fact, along with the well-known finding that serotonin microinjection in the 5HT perykaria of DR inhibits the neuronal firing of these cells, strongly suggests that 5HT released by MR axons at DR level displays an inhibitory effect on DR neurons. This suggestion has been corroborated by neurochemical and surgical destruction of afferents to DR nucleus.[128,160-172]

Other examples of the DR-MR antagonism arise from experiments showing that NE-LC neurons exert a tonic excitatory effect on 5HT-DR neurons, while NE axons reaching MR nucleus exert an inhibitory effect on 5HT neurons at this level. Both excitatory and inhibitory effects are mediated through α_1-adrenoceptors. So, NE-LC and 5HT-DR nuclei operate synergically, while NE-LC and 5HT-MR nuclei operate antagonically. The inhibitory influence of MR nucleus on NE-LC neurons is coherent with DR-MR antagonism.[105,109-112,166-169,172-183]

The effect of pharmacological manipulations using α_1- and α_2-agonist and antagonist drugs on intact and DR or MR lesioned rats is coherent with the DR-MR antagonism hypothesis. Sedation and EEG synchronization result from admininstration of α_1-antagonists or α_2-agonists, suggesting DR inhibition and MR disinhibition. To the contrary, arousal and EEG de-synchronization result from use of α_1-agonists or α_2-antagonists, suggesting DR stimulation and MR inhibition.[134,168,169,173-176,180-186]

Other studies show that D-amphetamine and methylphenidate decrease intracellular 5HT fluorescence and increase 5HT extracellular fluorescence at DR but not at MR level. These findings suggest that catecholamines released through the action of these drugs stimulate the firing of 5HT-DR neurons but not of 5HT-MR neurons. Further evidence shows that, while activation of 5HT-DR neurons inhibits subcortical mesolimbic DA system, 5HT-MR neurons exert inhibitory influence on DA mesocortical system. These findings fit well with the fact that when mesolimbic DA turnover is elevated, mesocortical DA turnover is low, and vice versa.[12,62,64,65,91,92,113,130,131,133,156,157,177,187-206]

Behavioral evidence correlates aggression with MR activity and suppression of aggression, fear, and anxiety with DR activity. Moreover, MR correlates with sedation, EEG synchronization, and reward, whereas DR correlates with arousal, EEG desynchronization, and punishment.[1-3,6,8,9,13-15,26,27,33,38,43-45,51,101,103-105,108-112,119,120,156,173,179,184,207-209]

Summarizing, we postulate the association of DR-5HT system with arousal, learning ability, memory, intellectual, and other high-level brain functions, while MR-5HT activity would be associated with aggression, reward, sedation, and low-level (primitive) brain functions.

II. MANIC SYNDROME

The central symptoms of mania are widely accepted. These include disturbances of mood, thought, and behavior. According to RDC criteria the following symptoms are included: (1) increase in activity (socially, at work, or sexually) or physical restlessness; (2) talkativeness; (3) flight of ideas; (4) inflated self-esteem; (5) decreased need for sleep; (6) distractibility (attention too easily drawn to unimportant or irrelevant external stimuli); (7) irritability; (8) extreme gregariousness; and (9) excessive involvement in pleasurable activities.

Although mania and hypomania have been classified respectively as primary and secondary, and psychotic and nonpsychotic, it is widely accepted that manic syndrome is related to overactivity of the DA system. Supporting this hypothesis are clinical observations including the findings that direct and indirect DA agonists are able to provoke appearance of manic syndrome, and that anti-DA drugs reduce manic symptoms. Additionally, many experimental studies in animals, employing diverse experimental designs, have been able to reproduce or suppress most components of the manic syndrome (accelerated motor activity, irritability, aggressiveness, sexual hyperactivity, distractibility, extreme gregariousness, brain stimulation reward, etc.).[199,210-230]

We examine below some possible mechanisms involved in the most important symptoms of manic syndrome.

A. Increased Motor Behavior

Increase of motor activity is associated with augmented dopamine turnover in subcortical mesolimbic area, specially in nucleus accumbens. Experimentally, this raised DA activity is obtained following injection of DA agonists in nucleus accumbens, or through systemic administration of high doses of such agonists. Moreover, stimulation of DA neurons located at VTA or DA-A10 cell group, as well as destruction or inhibition of 5HT neurons located in MR nucleus or B8 cell group, are also followed by increased motor activity. In all these cases, an enhancement of DA turnover is registered at accumbens level. We know also that the hyperactive syndrome induced by MR lesion is mediated by hippocampus because destruction of hippocampus interferes with the syndrome.[73,75,86-91,93,98,99,107,189,191-194,231-237]

Paradoxically, both lesion and stimulation of VTA nucleus provoke motor hyperactivity. The fact that destruction of VTA region interferes not only with DA mesolimbic system but also DA mesocortical system explain this apparent contradiction. In effect, DA cortical activity is known to exert a bridle on DA subcortical activity. Hence, suppression of cortical DA releases subcortical DA from cortical inhibition in such a way that ablation of prefrontal cortex or lesioning of DA terminals at this level is followed by hyperactivity and raised subcortical DA turnover.[75,87,89,90,92,238-242]

Another line of evidence shows that 5HT activity at accumbens level exerts a bridle on DA-induced hyperactivity. 5HT axons to the nucleus arise from DR nucleus or B7 cell group, mainly.[189,243]

Summarizing, the hyperactive syndrome is caused first through inhibition of DR-5HT, thus releasing DA mesolimbic system from its bridle (indirect stimulation); second, through disinhibition of MR-5HT, which in turn inhibits mesocortical system.

B. Distractibility or Attention-Deficit Disorder

This behavior is experimentally obtained by reducing DA activity at cortical prefrontal level. DA axons which reach this area end in the deepest cortical layers (V and VI). Ablation of cortical prefrontal area and selective pharmacologic destruction of its DA terminals (preserving NE terminals) are followed by increase of motor activity and deficit of attention. During such experimentally induced behavior in rats, the rodents pay overattention to irrelevant external stimuli. The animals show a loss of fear and anxiety and an enhancement of exploratory activity. They are unable to learn or retain previously acquired learning. Therefore, these experimental animals have been proposed as a model for hyperkinetic or deficit attention disorder syndrome observed in humans. Lack of DA activity, produced experimentally in cortical prefrontal area, is in striking contrast to the increase of DA activity in the same area registered in animals during anxiety states and during the first stage of stress when the animal is in an alert state and suppresses motor activity.[73,87,88,91,92,206,231,232,234,244-246]

NE terminals reaching cortical areas end homogenously throughout the six cortical layers. These NE terminals arise from the LC or A6-cell group. Although both dopamine and norepinephrine are inhibitory transmitters, cortical neurons in deeper layers are more readily inhibited by DA than by NE, while the ability of these transmitters to reduce the firing rate of cortical cells is inverted regarding each other towards external layers.[35-37,46,74,92,126,202,203,20247-258]

Some kind of DA-NE antagonism seems to exist at cortical level since elimination of cortical NE by lesioning NE terminals at prefrontal cortex is followed by DA concentration increase of up to 40% at this level. Furthermore, the experiments mentioned earlier in which reduction or suppression of DA activity at cortical level is accompanied by enhancement of DA activity at subcortical mesolimbic level confirm an antagonism between DA mesolimbic and DA mesocortical systems.[200,202,203,258-263]

Shedding light on NE-DA interactions between NE-LC and DA-VTA systems are studies showing that stimulation of NE-LC neurons increases DA release in cortical prefrontal area but not in nucleus accumbens. This DA release is suppressed by GABA-mimetic drugs such as the benzodiazepines. As mentioned earlier, the anxiolytic properties of benzodiazepines are due to inhibition of NE-LC neurons. It should be noted here that other anxiolytic drugs exist which lack the ability to reduce NE-LC activity. Also, other areas of GABA input exist, e.g., the VTA region.[34,77,78,196,197,200,264-278]

Experimental studies in rats using foot shock provoked stress which selectively elevated DA metabolism in prefrontal cortex but not in nucleus accumbens or caudate nucleus. Pretreatment with low doses of naloxone, an opiate antagonist, reversed this DA elevation. These data support the hypothesis that stress-induced release of endogenous opioid causes excitation of mesocortical DA neurons. According to these findings opioid mechanisms would stimulate, while GABA mechanisms would inhibit, DA mesocortical system.[200,279,280]

C. Decreased Need for Sleep

From results obtained in experiments on cats, Jouvet postulated that slow wave sleep (SWS) may be initiated by central 5HT system. He believed the onset of SWS to be an active process involving orbitofrontal cortex, intralaminar thalamic nuclei, and MR-5HT nuclei. Reduction in SWS by pCPA-induced inhibition of 5HT synthesis and its reversal by the administration of serotonin support the 5HT hypothesis. There is a wealth of data showing the existence of an antagonism between DA and 5HT systems. In effect, functional antagonism between DA-mesolimbic and 5HT-MR systems is found not only in DA-VTA and 5HT-MR cell bodies but in their projection areas: subcortical mesolimbic structures, hippocampus, prefrontal cortex, etc.[186,237,281-286]

There is also ample evidence showing that 5HT-MR system is hyperactive during SWS. Thus, the hyperactivity of DA-mesolimbic system postulated as the base of manic syndrome might reduce the activity of 5HT-MR system, contributing to a decreased need for sleep. The fact that stimulation of 5HT-MR system induces sedation and EEG synchronization is coherent with the role of this 5HT system in sleep. Also consistent is the finding that all antimanic drugs are DA antagonists, proven in turn to raise serotonin brain levels.[86,147,287-294]

D. Irritability

Lesions of the septal forebrain area in the rat have long been known to produce a dramatic increase in irritability and reactivity. The resulting syndrome, often called "septal rage", is characterized by intially explosive levels of irritation giving way to a fairly prolonged period of behavioral recovery.

Irritability is a frequent symptom of the different types of manic syndromes observed in humans. Depending on the clinical types, irritability can be found as a permanent manifestation or only during certain periods.

The septal rage syndrome, induced by surgical lesion, intracisternal, intraventricular, or localized intraseptal injections of the catecholamine neurotoxin 6-OHDA, can be mitigated or avoided by prelesion chronic treatment with DA-blocking agents (haloperidol, pimozide). Conjunctive administration of desipramine, which preserves the integrity of NE terminals, to increase DA specificity of 6-OHDA lesions, does not significantly alter the intensity of the ensuing lesion-induced syndrome. Moreover, neurochemical assays of brain tissue from septally lesioned animals have revealed postoperative decreases in forebrain dopamine associated with lesion-induced irritability. The possible involvement of DA system in septal rage syndrome is supported by the fact that administration of L-dopa following septal lesion dramatically accelerates the time of postlesion behavioral recovery. This capability extends to other DA-agonists. Finally, the supersensitivity of DA receptors in the forebrain of septally-lesioned animals is consistent with a reduction of DA at this level.[12,15,49-68]

Irritability induced in rodents by isolation (isolation syndrome) is also accompanied by reduction of DA activity, but not NE activity, in VTA and some of its mesolimbic projection areas. Hence, both septal rage and isolation irritability syndromes seem to be associated with an increase of NE/DA ratio at mesolimbic level. With respect to this, our speculations are addressed to the possibility that the presence or absence of irritability and aggressiveness in human manic syndrome might depend on variations in NE/DA ratio at mesolimbic level.[21-25]

The NE-DA antagonism at mesolimbic level has been amply demonstrated in experimental animals. Taking into account that NE innervation of mesolimbic area arises mainly from non-LC-NE nuclei, then NE-DA antagonism would exist between these non LC-NE nuclei and DA-VTA neurons innervating subcortical mesolimbic structures. Conversely, as mentioned above, NE-LC system stimulates DA mesocortical system, but not DA subcortical (mesolimbic) system. Since a clear antagonism has been established between DA cortical and DA subcortical systems, indirect antagonism between NE-LC system and DA mesolimbic system might be postulated.[34,52,58,64,71,77,87,92,116,117,145,155,196,197,200,254,255,259,260,263,269, 270,276,278,295-297]

The VNB is the main ascending fiber system originating within ventral pontine tegmentum from several groups of neuronal cell bodies containing norepinephrine (A1 and A2 cell groups, preferentially). This VNB innervates a variety of forebrain subcortical structures including hypothalamus, septum, and amygdala. Apparently no VNB fibers end in the cortex or hippocampus, however.[298-304]

Selective bilateral radiofrequency destruction of VNB results in a pronounced enhancement of ambulation and motor activity in rats.[86-89,91,98] This induced excitation is suppressed by administration of DA blocking agents. The fact that lesions of VNB deplete NE from nucleus accumbens and microinjections of NE into VTA or accumbens produce a locomotor depression support the NE-VNB/DA-VTA antagonism.[78,274,305-308] It follows from this antagonism arising in nuclei projecting through VNB and DA mesolimbic system, that two types of behavioral expressions would theoretically occur: (1) NE-VNB predominance and (2) DA-mesolimbic predominance. In NE-VNB predominance, aggression (septal rage and isolation aggressiveness) plus motor depression would dominate, whereas DA mesolimbic predominance would be characterized by increased exploratory activity, gregariousness, decreased need for sleep, distractibility, attention deficit, etc. In the former syndrome NE-VNB predominance would be due to a deficiency of dopamine at mesolimbic level along with a supersensitivity of mesolimbic postsynaptic DA receptors.

Taking into account that aggressiveness and supersensitivity of mesolimbic DA receptors are behavioral and neuropharmacological findings typical of human schizophrenia, the septal rage syndrome and the isolation-induced aggressive psychotic syndrome might be good animal models for human schizophrenia. Furthermore, the fact that haloperidol, pimozide, and other DA blocking agents improve both animal and human syndromes is consistent with this hypothesis. In the second syndrome, DA mesolimbic predominance, the symptoms registered fit well with manic manifestations observed in schizoaffective disorder patients. The fact that carbamazepine, clonazepam, apomorphine, and other DA autoreceptor agonists improve these patients, reinforces this hypothesis.

E. Excessive Involvement in Pleasurable Activities

Sexual hyperactivity and excessive involvement in pleasurable activities are characteristic manifestations of the manic syndrome in schizoaffective disorder patients. It should be remembered here that DA innervation of septal and other mesolimbic structures primarily involves libido and endocrine physiology.[14,50,309-313] Administration of DA antagonists to both normal and manic subjects induces libido disappearance, raises prolactin plasma levels, and promotes infertility, all of which are consistent with the positive correlation between DA mesolimbic behavior and sexual activity.[213-216,218-222,273,314-315]

Electrical stimulation of many regions of the brain produces pleasurable, or reward, response. Accumulated data suggest that the anatomical substrate of brain stimulation reward is closely related to DA mesolimbic (VTA or A10 cell group) and 5HT-MR nucleus or B8 cell group.[73,187,188,200,316-334]

Brain stimulation reward is strong and has low threshold when electrodes are within the borders of VTA cell layer, although it is not obtained with reasonable currents when electrodes are dorsal, ventral, caudal, or lateral to DA cells.

DA-VTA cells have two kind of neurons: lateral DA neurons which innervate mesolimbic structures, and medial-anterior neurons innervating prefrontal cortex. Sufficient evidence exists to support a hypothesis that rewarding is related to septal and other mesolimbic but not cortical DA innervated structures. Furthermore, treatments which interfere with DA metabolism or synaptic action, clearly interfere with brain stimulation reward. Moreover, they do so by reducing the rewarding impact of stimulation and not merely by impairing the response capacity of the animal. Also, neuroleptics (DA blocking agents) have been found to cause major disruption of self-stimulation in every stimulation site thus far reported. All the above leads us to postulate a positive correlation between rewarding and DA mesolimbic but not DA mesocortical system. The latter seems to be closely related to punishment behavior and anxiogenic activity.[200,316,317,319-346]

Another line of investigation links 5HT-MR system with reward effects. In fact, strong evidence suggests that the rewarding effect of habenular stimulation is mediated by 5HT neurons of MR nucleus. Further, there is clearcut evidence showing that rewarding effects provoked by MR stimulation are suppressed by cerebral 5HT depletion. Conversely, 5HT-DR system is closely related to punishment behavior and suppression of reward.[187,188,347-358]

The fact that 5HT-DR system depresses DA-mesolimbic system, whereas 5HT-MR system depresses DA-mesocortical system, fits in well with the above findings.

Summarizing, present evidence suggests two opposite circuits: (1) reward and (2) punishment. The former would be positively correlated with DA-mesolimbic and 5HT-MR systems, whereas the second would be correlated with DA-mesocortical and 5HT-DR systems. There is evidence that posterior septum which receives projections from DA-mesolimbic and 5HT-MR nuclei is crucially involved in reward response. These considerations offer support for the hypothesis that overactivity of DA-mesolimbic and 5HT-MR systems underlie the excessive involvement in pleasurable activities which is a typical symptom of manic syndrome.

III. TWO TYPES OF PSYCHOTIC SYNDROMES

On the basis of data presented above and on the two postulated brain monoaminergic circuits, two types of psychotic syndromes may be proposed with two distinct physiopathological mechanisms: (1) schizophrenic and (2) schizoaffective. In the schizophrenic type a NE over DA predominance would prevail during acute psychotic symptoms, at mesolimbic level. During remission periods, an attenuation of catecholaminergic activity plus disinhibition of 5HT system (5HT-MR system, mainly) would be found. In other words, during acute periods NE > DA, whereas during remission periods 5HT-MR > 5HT-DR. Taking into account that NE innervation of mesolimbic structures is supplied by non-LC-NE nuclei = VNB, the NE predominance we postulate would result from NE-VNB terminals. Similarly, since DA mesocortical system is under tonic excitatory influence from NE-LC system, DA mesolimbic system would predominate over DA mesocortical system.[52,55,64,66,71,75,131,154,155,200,205,254,255,297,302,308,335,359-363]

In schizoaffective type a predominance of DA over NE system would be found during acute periods. In these patients, DA mesolimbic activity would predominate over DA mesocortical activity. However, during remission periods, an attenuation of DA mesolimbic activity would occur and DA mesocortical activity would be disinhibited. For this reason, schizoaffective patients recuperate their normal mental state completely during remission periods. Conversely, schizophrenic patients would be affected by an absolute lack of both mesolimbic and mesocortical DA activities; hence, they never attain total normalcy during remission periods. The fact that tolerance to DA blockade is produced at mesolimbic level but never at mesocortical level during prolonged antipsychotic treatment suggests that improvement in these patients is found only when an increase of DA release is obtained at mesocortical levels, which predominates over DA mesolimbic system.

In the following pages we explain the bases on which these postulations rest.

It is known that many descending A6 or LC-NE axons end at A1-NE cell group level, while an important NE-A1 fiber system ends in LC. Since NE and other α_2-agonists microinjected into NE nuclei induce inhibition of NE-neuron firing rate, it follows that A6 and A1-NE nuclei inhibit each other. Hence, from a theoretical point of view it is difficult to accept that both NE nuclei can be simultaneously active; logically, they would display alternating activity.[300,302-303,362,364-371]

Hyperactivity of LC is associated with fear and anxiety which are accompanied by inhibition of motor activity. The manifestations are in striking contrast to the aggressiveness, lack of fear, and motor hyperactivity observed during acute psychotic episodes. Hence, LC does not predominate during these periods.[38,43,209,372]

Experimental activation of NE-LC is known to be followed by 5HT-MR inhibition and by activation of DA-mesocortical system. Hence, the reduction in LC activity we proposed for schizophrenics would produce disinhibition of 5HT-MR and inhibition of DA-mesocortical systems.[183,287,372] It should be remembered that muricidal aggression has been associated with 5HT-MR system and that this system displays antagonistic activity to 5HT-DR, a 5HT nucleus which is negatively correlated with muricidal aggression and positively correlated with fear and anxiety. Dopamine hypoactivity at mesolimbic level fits well with the lack of affectivity and libido observed in schizophrenics, while the lack of dopamine at cortical prefrontal level would explain schizophrenics' observed lack of fear and anxiety.[109,110,178,184,372-375]

While LC-NE exerts sympathoexcitatory influence,[376,377] non-LC NE nuclei display sympa-

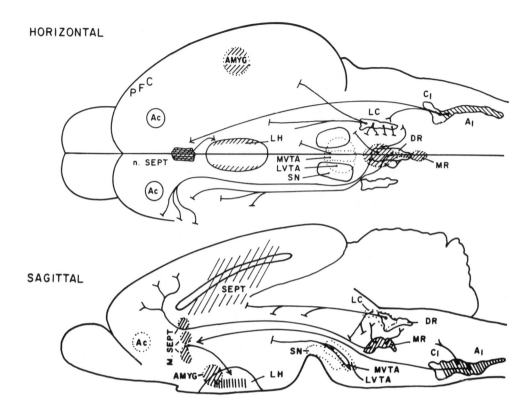

FIGURE 1. Schizophrenic type syndrome. Direct and indirect experimental evidence suggests that, during aggressive behavior, the lateral ventromedullary NE-A1 cell predominates over the dorsomedial mesencephalic-pontis NE-A6(LC) cell group. Also, 5HT neurons of the MR would predominate over 5HT-DR neurons. DA neurons of both medial and lateral ventral tegmental areas (MVTA) and (LVTA) would be inhibited. In effect, DA-MVTA neurons which are normally stimulated by NE-LC neurons and bridled by 5HT-MR neurons would receive only the latter influence. In turn, DA-LVTA neurons would receive an enhanced inhibitory influence from the hyperactive NE-A1 cell group. Finally, adrenaline (A) neurons located in the rostral part of A1 cell group (C1 cell group), would be strongly bridled. The lack of anxiety, fear, and motor activity observed during aggressive behavior is consistent with the above profile of central autonomic activity. Increased NE-A1 activity + reduced DA-LVTA activity would result in raised NE/DA ratio at mesolimbic projection areas, i.e., septum, amygdala, and some diencephalic areas such as lateral hypothalamus (LH). This is registered during experimentally induced "septal rage syndrome". The predominant 5HT-MR neurons would exert inhibitory influences on NE-LC, 5HT-DR, and PFC (prefrontal cortex). SEPT (septum), n. SEPT (nuclei of septum), Ac (nucleus accumbens), SN (substantia nigra), M. SEPT (medial septum), (||||||||||) activated structures, (⟶) excitatory inputs, (◀—) inhibitory inputs, and (——|) annulled outputs. (Reproduced with kind permission from PJD Publications Limited, Westbury, N.Y., 11590, U.S.A., from *Res Commun Psychol Psychiatr Behav*, 11: 207-260, 1986. Copyright © by PJD Publications Ltd.)

thoinhibitory activity because they project to sympathetic preganglionic cells from the intermediolateral spinal horn, ventromedullary nucleus reticularis lateralis, and anterior hypothalamus. Norepinephrine released at these levels inhibits sympathetic preganglionic neurons. These facts might explain schizophrenics' predominance of peripheral parasympathetic activity (raised colon motility) which disappears upon administration of anti-NE or anti-ACh drugs.[302,362,366,367,378-385]

A. Schizophrenic Type Syndrome (Figure 1)

The animal models for this psychotic syndrome are septal rage and isolation-induced psychosis. In both experimentally induced animal syndromes a great reduction of dopamine at mesolimbic level plus a supersensititvity of DA receptors has been demonstrated.[12,22,31,32,52,55,57,61,66,68,71,94] Although a majority of human studies dealing with quantification of central DA activity in schizophrenics shows a deficit rather than an excess of this neurotransmitter (CSF, plasma and post-mortem brains), there are as yet no conclusive findings in this matter.[186,386-398] It is possible that the difficulty arises from the existence of several DA systems (nigrostriatal, mesolimbic, mesocortical, and hypothalamic). However, a consensus does exist among investigators about supersensitivity of DA receptors (D_2 type) in the mesolimbic region of schizophrenic brain (post-mortem studies).[386,387,389,396,397,399-404] This finding, similar to that observed in experimental animals after destruction of DA mesolimbic innervation, cannot be interpreted in any way except as a "denervation supersensitivity". In our opinion, the only logical explanation for the D_2 supersensitivity found in schizophrenic brain would be a lack of DA at this level. The fact that both animal and human schizophrenic syndromes are improved by DA blocking agents strongly supports this assumption.

At mesolimbic level DA deficit would result in relative NE predominance. Although in animal septal rage, DA blockade suppresses the deficiency symptoms despite a demonstrated lack of dopamine at this level, the DA blocking agents would interfere with the action on these supersensitive DA receptors of some neurotransmitter other than dopamine. The most probable candidate would be norepinephrine. In effect, NE has been proven to act as an agonist on DA receptors. Moreover, clonidine, an inhibitor of NE release, suppresses dramatically the acute psychotic symptoms in schizophrenic patients. These facts are consistent with others showing raised CSF-NE levels in schizophrenics during acute periods.[186,391,398,405]

Although LC provides some projections to septal and other mesolimbic structures, the main NE innervation of this area arises from non-LC nuclei which project through VNB (A1, A2, A5). Thus, the NE predominance we postulated during acute schizophrenic periods would result from NE released by VNB terminals.

B. Schizoaffective-Type Syndrome (Figure 2)

Formerly considered psychotics, these patients are now included in the nonpsychotic but affective disorder group (DSM III). Schizoaffective disorder subjects show manic or hypomanic episodes alternating with normal or depressive periods.

During manic periods, schizoaffective patients are quickly improved by clonazepam,[229,230,405] apomorphine,[213,215,216,406,407] and bromocriptine,[218,220-222] all drugs which stimulate DA autoreceptors and thus suppress release of dopamine from DA terminals. In our experience, manic periods observed in schizoaffective disorder patients are greatly and sharply worsened by administration of clonidine, an α_2-agonist which inhibits NE release.[405] Such findings suggest an imbalance between DA and NE at mesolimbic level: DA > NE. This increase in the DA/NE ratio postulated by us for schizoaffective syndrome is opposed to the increased NE/DA ratio postulated for schizophrenic syndrome.

NE deficiency postulated for schizoaffectives would coexist with an α-adrenoceptor supersensitivity at mesolimbic level. Therapeutical evidence shows that schizoaffective patients are greatly improved by administration of α_1-antagonists (dihydroergotamine, levopromazine).[405] It is possible that the excess of dopamine released in these patients at mesolimbic level would act not only on DA receptors but also on α_1-supersensitive receptors.[274,408-410]

The postulation of an NE deficiency at mesolimbic level is reinforced by patients' improvement following administration of desipramine, a NE potentiating drug, during nonmanic periods (Lechin, personal communications).

During manic or hypomanic periods, schizoaffective patients show most of the well-known symptoms: decreased need for sleep, more energy than usual, inflated self-esteem, increased productivity, sharpened and unusually creative thinking, extreme gregariousness, uninhibited behavior, hypersexuality without recognition of possible painful consequences, excessive involvement in pleasurable activities, squandering, foolish business investments, physical restlessness, unusual talkativeness, overoptimism, inappropriate laughing, and joking. Associ-

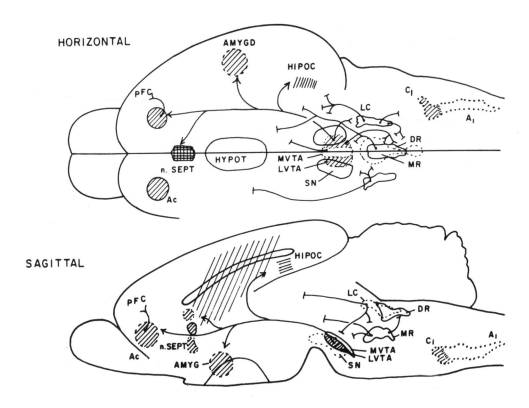

FIGURE 2. Schizoaffective type syndrome. Physiological disorders underlying at the bottom of this syndrome fit well with hyperactivity of DA mesolimbic systems, which neurons are located in the lateral parts of ventral mesolimbic system, which neurons are located in the lateral parts of ventral tegmental area (LVTA). These DA neurons send axons to mesolimbic structures: (Ac) n. accumbens, (SEPT) septum, (AMYG) amygdala, and (HIPOC) hippocampus. It is a known fact that hyperactivity of DA mesolimbic system exerts inhibitory influence over DA mesocortical activity, which neurons are locates in the medial part of ventral tegmental area (MVTA). Further, LVTA neurons send inhibitory axons to NE neurons in the LC and 5HT neurons in the DR nucleus. The lack of activity in the prefrontal cortex (PFC), LC, DR, and MVTA, fits well with the lack of anxiety observed in these patients. The hypermotility, decreased need for sleep, sexual hyperactivity, deficit of attention, and inability to be depressed, registered in these patients are consistent with the reduction of activity of both NE-LC and 5HT systems, and of DA mesocortical system: (——➤) excitatory input, (◄——) inhibitory input, (——┤) annulled output, and (||||||||) activated structure.

ated psychotic features such as delusions, hallucinations, incoherence, or loosening of associations appear occasionally.

Peripheral autonomic system manifestations registered during schizoaffective manic periods are greatly increased distal colon motility which disappears following administration of clonazepam, bromocriptine, or apomorphine, but not clonidine or anticholinergics, and low plasma levels of NE which are not reduced by clonidine. Both findings are consistent with peripheral hyposympathetic activity, correlated with a reduction of central sympathetic activity (LC posterior hypothalamus pathway). Although similar sympathetic reduction has been observed in schizophrenic syndrome, a different mechanism would be proposed for hyposympathetic activity registered in schizoaffectives. In effect, whereas an inhibition of NE-LC system secondary to predominant non-LC-NE nuclei is postulated for schizophrenics, the LC inhibition in schizoaffectives would derive from a hyperactive DA mesolimbic system. Experimental evidence in animals affords proof of the bridling influences that DA mesolimbic systems exert on NE-LC system.[37,197,205,246,270,274,277,279,280,410]

The improvement induced by DA autoreceptor agonists (clonazepam, bromocriptine, and apomorphine) during schizoaffective manic or hypomanic periods would be interpreted as a reduction of dopamine release at mesolimbic level, but not mesocortical, since the mesocortical system lacks DA autoreceptors. The sudden disappearance of manic symptoms observed in these patients, who preserved attention, ability to concentrate, and other intellectual functions, supports the postulation that dopamine is not suppressed at prefrontal cortical level.[89,90,92,196,200,205,245,411-415]

Anatomical and pharmacological evidence shows that while 5HT-DR system sends inhibitory axons to DA-VTA mesolimbic neurons, 5HT-MR system sends inhibitory axons to DA-VTA mesocortical neurons. On the other hand, DA neurons from VTA (A10) nucleus send inhibitory axons to both 5HT nuclei. Hence, it is possible to postulate a "see-saw" between dopaminergic and serotonergic systems with DA predominance during manic episodes and 5HT predominance during depressive periods.[27,87,90,92,133,193,200,246,414,416-423]

Attenuation of DA activity (remission periods) would favor the liberation of NE-LC activity from the inhibitory influences of the DA mesolimbic system which predominates during manic periods. The inhibitory influences contributed by 5HT-MR system predominate during depressive periods.[27,121-123,130,131,135,136,141,156,158,190,192,310,374,416-420,424-427]

The fact that NE plasma levels and diastolic blood pressure rise significantly in all schizoaffective patients during remission periods supports a postulated increase in central sympathoexcitatory system (NE-LC) activity (Lechin, personal communication).

Schizoaffective type syndrome can present two clinical pictures: (1) with psychotic symptoms and (2) without psychotic symptoms. In the first case, we assume dopamine would be absent or significantly reduced in cortical prefrontal level, whereas in the second case dopamine would be present at both mesocortical and mesolimbic levels. The fact that DA mesocortical neurons receive inhibitory influences from both 5HT-MR neurons and non-LC NE nuclei may link DA cortical deficiency with hyperactivity of one of these systems.[428,429] According to such a hypothesis, in psychotic manifestations of both schizophrenic and schizoaffective syndromes, dopamine would be absent or reduced in cortical prefrontal area. Further, in schizophrenics dopamine would also be absent in septal region, while in schizoaffectives dopamine would be found in excess in septal region.

According to these profile models, the monoaminergic pathways in mesolimbic and prefrontal cortex form circuits which interact and display alternating predominance. In some pathological circumstances, there would be a failure in the alternating patterns of the circuits; some circuits would maintain predominance over others, thus generating the physiological disorders underlying different psychotic syndromes. In line with the above reasoning, we postulate the existence of two main monoaminergic circuits:

1. Circuit I = NE-LC (A6) — 5HT-DR — DA-VTA or (A10) mesocortical system
2. Circuit II = NE-non-LC (A1)— 5HT-MR — DA-VTA or (A10) mesolimbic system

Predominance of circuit I generates fear, anxiety, and motor hyperactivity. During predominance of circuit I, the subject is in an alert state, awaiting other signals. He is concentrating on the appearance of external or internal signals while paying attention to a discrete phenomenon ocurring at the moment. The impact of this phenomenon on the prefrontal cortex would trigger the appearance from the memory store of a cascade of associated images. Mesocortical system would be necessary for concentration, learning, association of ideas, memory, and higher intellectual functions. circuit I, more developed in man than in other mammals, would be responsible for repression of emotional impulses and primitive behavior. In other words, the activity of circuit I protects mammals against psychotic behavior.

There would be some synergistic-antagonistic activities among the different nuclei integrating circuit I. Thus, NE activity of LC exerts a tonic excitatory influence on 5HT-DR and DA-mesocortical activity. In turn, 5HT-DR and DA-mesocortical systems are able to limit NE-LC activity. These interactions between the three nuclei are exerted through the interchange of axons ending at cell body levels, receptors located on the intermingled terminals, and postsynaptic receptors located in those structures where they share innervation and, finally, through polysynaptic mechanisms.

Predominance of circuit II would generate loss of fear and anxiety, facilitating aggressive and primitive impulses and psychotic behavior. There would be two variants in circuit II predominance: (1) NE-mesolimbic activity greater than DA-mesolimbic activity and (2) DA-mesolimbic activity greater than NE-mesolimbic activity. In case (1), aggressive impulses, lack of affectivity, libido, and pleasurable activities are observed. Additionally, motor activity is reduced (minor stereotype absent). In case (2), affectivity, libido, pleasurable activities, and motor activity are in excess. Case (1) would be represented by the schizophrenic syndrome. Case (2) would be represented by the schizoaffective syndrome. There would be two other possibilities within the schizoaffective type syndrome: (1) a psychotic modality due to deficiency of dopamine at prefrontal cortical level, and (2) a nonpsychotic modality in which dopamine is in excess at both mesolimbic and mesocortical levels.

REFERENCES

1. **Eclancher R, Schmitt P, Karli P.** Effects de lessions précoses de l'amygdale sur le developpement de l'aggressivité interspécifique du rat. *Physiol Behav* 14: 277-283, 1975.
2. **Vergnes M.** De clenchement de reactions d'aggression inter-spécifique apres lesion amygdalienne chez le rat. *Physiol Behav* 14: 271-276, 1975.
3. **Vergnes M, Mach G, Kempf E.** Lésions du raphé et réaction d'aggression interspécifique rat-souris. Effects comportementaux et biochimiques. *Brain Res* 57: 67-76, 1973.
4. **Waldbillig RJ.** Attack, eating, drinking and gnawing elicited by electrical stimulation of rat mesencephalon and pons. J *Comp Physiol Psychol* 89: 200-212, 1975.
5. **Gibbons JL, Barr GA, Bridger WH, Leibowitz SF.** Manipulations of dietary tryptophan: effects on mouse killing and brain serotonin in the rat. *Brain Res* 169: 139-153, 1979.
6. **Bocknik SE, Kulkarni AS.** Effect of a decarboxylase inhibitor (Ro 4-4602) on 5-HTP induced muricide blockade in rats. *Neuropharmacology* 13: 279-281, 1974.
7. **Bowers D.** Facilitate Effects of Electric Shock on Mouse Killing by Hooded Rats. Doctoral dissertation, Temple University, Philadelphia, 1974.
8. **Gibbons JL.** Serotonergic Mechanisms and Predatory Aggression: The Effects Produced by pCPA, Tryptophan Injections, and a Tryptophan-Free Diet on Mouse Killing Behavior by Rats. Abstr 37:1955-B, Doctoral dissertation, Carnegie Mellon University, Pittsburgh, 1976.
9. **Grant LD, Coscina DV, Grossman SP, Freedman DX.** Muricide after serotonin depleting lesions of midbrain raphe nuclei. *Pharmacol Biochem Behav* 1: 77-80, 1973.
10. **Katz RJ.** Catecholamines in predatory behavior: a review and critique. *Aggressive Behav* 4: 153-172, 1978.
11. **Malick JB.** Effects of age and food deprivation on the development of muricidal behavior in rats. *Physiol Behav* 14: 171-175, 1975.
12. **Marotta RF.** Pharmacological Manipulations of the Septal Irritability Syndrome: Role of Dopaminergic Mechanisms in Recovery of Function. Doctoral dissertation. City University of New York, 1977.
13. **Miczek KA, Altman JL, Appel JB, Boggan WO.** Para-chlorophenylalanine, serotonin and killing behavior. *Pharmacol Biochem Behav* 3: 355-361, 1975.
14. **Miczek KA, Barry H III.** Pharmacology of sex and aggression, in *Behavioral Pharmacology*, Glick SD, Goldfarb J. Eds., C.V. Mosby, St Louis, 1976, 176.
15. **Miczek KA, Grossman SP.** Effects of septal lesions on inter- and intra-species aggression in rats. J *Comp Physiol Psychol* 79: 37-45, 1972.
16. **Polsky RH.** Hunger, prey feeding, and predatory aggression. *Behav Biol* 13: 81-93, 1975.
17. **Potegal M, Marotta R, Gimino F.** Factors in the waning of muricide in the rat. I. Analysis of intra- and intersession decrement. *Aggressive Behav* 1: 277-290, 1975.
18. **Woodworth CH.** Attack elicited in rats by electrical stimulation of the lateral hypothalamus. *Physiol Behav* 6: 345-353, 1971.

19. **Miczek KA.** A new test for aggression in rats without aversive stimulation. Different effects of D-amphetamine and cocaine. *Psychopharmacology* 60: 253-259, 1979.

20. **Mizcek KA, O'Donnell JM.** Intruder-evoked aggression in isolated and monisolated mice: effects of psychomotor stimulants and L-dopa. *Psychopharmacology* 57: 47-55, 1978.

21. **Tizabi Y, Massari VJ, Jacobowitz DM.** Isolation induced aggression and catecholamine variations in discrete brain areas of the mouse. *Brain Res Bull* 5: 81-86, 1979.

22. **Garattini S, Giacolone E, Valzelli E.** Biochemical changes during isolation-induced aggressiveness in mice, in *Aggressive Behavior*. Garattini S, Sigg EB. Eds., Excerpta Medica, Amsterdam, 1969, 179.

23. **Krsiak M, Janku I.** The development of aggressive behaviour in mice by isolation, in *Aggressive Behavior*. Garattini S, Sigg EB. Eds., Excerpta Medica, Amsterdam, 1969, 101.

24. **Modigh K.** Effects of isolation and fighting in mice on the rate of synthesis of noradrenaline, dopamine and 5-hydroxytryptamine in the brain. *Psychopharmacologia* 33: 1-17, 1973.

25. **Valzelli L.** The "isolation syndrome" in mice. *Psychopharmacologia* 31: 305-320, 1973.

26. **Gibbons JJ, Barr GA, Schimmel GT, Bridger WH.** Inescapable shock alters mescaline's disruption of active avoidance acquisition. *Psychopharmacology* 74: 336-338, 1981.

27. **McLain WC III, Cole BT, Schrieber R, Powell DA.** Central catechol- and indoleamine systems and aggression. *Pharmacol Biochem Behav* 2: 123-126, 1974.

28. **Johansson G.** Relation of biogenic amines to aggressive behavior. *Med Biol* 52: 189-192, 1974.

29. **Pradhan SN.** Aggression and central neurotransmitters. *Int Rev Neurobiol* 18: 213-261, 1975.

30. **Reis DJ.** Central neurotransmitters in aggressive behavior, in *Neural Bases of Violence and Aggression*. Fields WS, Sweet WH. Eds., Warren H Green, St. Louis, 1975, 57.

31. **Thoa NB, Tizabi Y, Jacobowitz DM.** The effect of isolation on catecholamine concentration and turnover in discrete areas of the rat brain. *Brain Res* 131: 259-269, 1977.

32. **Tizabi Y, Thoa NB, Maengwyn-Davies GD, Kopin IJ, Jacobowitz DM.** Behavioral correlation of catecholamine concentration and turnover in discrete brain areas of three strains of mice. *Brain Res* 166: 199-205, 1977.

33. **Vergnes M, Bochrer A, Karli P.** Interspecific aggressiveness and reactivity in the mouse-killing and non-killing rats: compared effects of olfactory bulb removal and raphe lesions. *Aggressive Behav* 1: 1-15, 1974.

34. **Solano-Flores LP, Aguilar-Baturoni HU, Guevara-Aguilar R.** Locus coeruleus influences upon the olfactory tubercle. *Brain Res* 5: 383-389, 1980.

35. **Descarries L, Lapierre Y.** Noradrenergic axon terminals in the cerebral cortex of rat. I. Radioautographic visualization after topical application of DL-3H-norepinephrine. *Brain Res* 51: 141-160, 1973.

36. **Dillier N, Laszlo J, Muller B, Koella WP, Olpa H-R.** Activation of an inhibitory noradrenergic pathway projecting from the locus coeruleus to the cingulate cortex of the rat. *Brain Res* 154: 61-68, 1978.

37. **Olpe H-R, Glatt A, Laszlo J, Schellenberg A.** Some electrophysiological and pharmacological properties of the cortical, noradrenergic projection of the locus coeruleus in the rat. *Brain Res* 186: 9-19, 1980.

38. **Kozak W, Valzelli L, Garattini S.** Anxiolytic activity on locus coeruleus-mediated suppression of muricidal aggression. *Eur J Pharmacol* 105: 323-326, 1984 .

39. **Barr GA.** Facilitation of mouse killing behavior by decreases in catecholamine function, paper presented to the Eastern Psychological Association, Bethesda, 1976.

40. **Barr GA, Gibbons JL, Bridger WH.** Neuropharmacological regulation of mouse killing by rats. *Behav Biol* 17: 143-159, 1976.

41. **Horovitz Z, Piala J, High J, Burke J, Leaf R.** Effects of drugs on the mouse killing (muricide) test and its relationship to amygdaloid function. *Int J Neuropharmacol* 5: 405-411, 1966.

42. **Reis DJ.** The chemical coding of aggression in brain, in *Advances in Behavioral Biology*, Vol. 10. Myers RD, Drucker RR. Eds., Plenum Press, New York, 1974, 125.

43. **Clark TK.** The locus coeruleus in behavior regulation: evidence for behavior-specific versus general involvement. *Behav Neural Biol* 25: 271-273, 1979.

44. **Hoehn-Saric R.** Neurotransmitters in anxiety. *Arch Gen Psychiatry* 39: 735-740, 1982.

45. **Hoehn-Saric R, Merchant AF, Keyser ML, Smith VK.** Effects of clonidine on anxiety disorders. *Arch Gen Psychiatry* 38: 1278-1283, 1981.

46. **Sanghera MK, German DC.** The effects of benzodiazepine and non- benzodiazepine anxiolytics on locus coeruleus unit activity. *J Neural Transm* 57: 267, 1983.

47. **Shibata S, Watanabe S, Liou SY, Ueki S.** Effects of adrenergic blockers on the inhibition of muricide by desipramine and noradrenaline injected into the amygdala in olfactory bulbectomized rats. *Pharmacol Biochem Behav* 18: 203-211, 1983.

48. **Ossipov MH, Chatterjee TK, Gebhart GF.** Locus coeruleus lesions in the rat enhance the antinoceptive potency of centrally administered clonidine but not morphine. *Brain Res* 341: 1985, 320-330.

49. **Miley WM, Baenninger R.** Inhibition and facilitation of interspecies aggression in septal lesioned rats. *Physiol Behav* 9: 379-384, 1972.

50. **Paxinos G.** Interruption of septal connections: effects on drinking, irritability and copulation. *Physiol Behav* 17: 81-88, 1976.
51. **Penot C, Vergnes M.** Déclenchement de réactions d' agression interspécifique par lésion septale aprés lésion préalable de l'amigdale chez le rat. *Physiol Behav* 17: 445-450, 1976.
52. **Marotta RF.** Mitigation of the septal lesion syndrome by prelesion chronic treatment with haloperidol. *Pharmacol Biochem Behav* 16: 769-775, 1982.
53. **Albert DJ, Richmond SE.** Hyperreactivity and aggressiveness following infusion of local anesthetic into the lateral septum or surrounding structures. *Behav Biol* 18: 211-226, 1976.
54. **Balagura S, Harrell LE.** The lateral hypothalamic syndrome: its modification by obesity and leanness. *Physiol Behav* 13: 345-347, 1974.
55. **Bernard BJ, Berchek J, Yutzey D.** Alterations in brain monoaminergic functioning associated with septal lesion induced hyperreactivity. *Pharmacol Biochem Behav* 3: 121-126, 1975.
56. **Coscina DV, Seggie J, Godse DD, Stancer HC.** Induction of rage in rats by central injection of 6-hydroxydopamine. *Pharmacol Biochem Behav* 1: 1-6, 1973.
57. **Cage FH, Olton DS.** L-Dopa reduces hyperreactivity induced by septal lesions in rats. *Behav Biol* 17: 213-218, 1976.
58. **Cage FH, Thompson RG, Valdes JJ.** Endogenous norepinephrine and serotonin within the hippocampal formation during the development and recovery from septal hyperreactivity. *Pharmacol Biochem Behav* 9: 359-367, 1978.
59. **Glusman M.** The hypothalamic "savage" syndrome. *Res Publ Assoc Res Nerv Ment Dis* 52: 52-92, 1974.
60. **Gotsick J, Marshall R.** Time course of the septal rage syndrome. *Physiol Behav* 9: 685-687, 1972.
61. **Grossman SP.** An experimental dissection of the septal syndrome, in *Functions of the Septo-Hippocampal System*. Ciba Foundation Symposium, Series No. 58. Elsevier, Amsterdam, 1978, 227.
62. **Harrell LE, Balagura S.** Septal rage: mitigation by presurgical treatment with *p*-chlorophenylalanine. *Pharmacol Biochem Behav* 3: 157-159, 1975.
63. **Hynes M, Anderson C, Gianutsos G, Lal H.** Effects of haloperidol, methyltyrosine and morphine on recovery from lesions of lateral hypothalamus. *Pharmacol Biochem Behav* 3: 755-759, 1975.
64. **Lindvall O.** Mesencephalic dopaminergic afferents to the lateral septal nucleus of the rat. *Brain Res* 87: 89-95, 1975.
65. **Fried PA.** The septum and hyper reactivity: a review. *Br J Psychol* 64: 267-275, 1973.
66. **Marotta RF, Logan N, Potegal M, Glusman M, Gardner EL.** Dopamine agonists induce recovery from surgically-induced septal rage. *Nature* 269: 513-515, 1977.
67. **Marotta RF, Logan NA, Riverso SM, Gardner EL, Potegal M.** Dopaminergic mechanisms in the septal hyperirritability syndrome, paper presented at the Annu. Meet. Eastern Psychological Association, Boston, 1977.
68. **Marotta RF, Potegal M, Gardner E, Glusman M.** Abolition of the septal syndrome in the rat by L-dopa, paper presented at the Annu. Meet. Am. Psychological Association, Chicago, 1975.
69. **Muller P, Seeman P.** Dopaminergic supersensitivity after neuroleptics: time-course and specificity. *Psychopharmacology* 60: 1-11, 1978.
70. **Munoz C, Grossman SP.** Behavioral consequences of selective destruction of neuron perikarya in septal area of rats. *Physiol Behav* 24: 779-788, 1980.
71. **Olton DS, Gage FH.** Behavioral, anatomical and biochemical aspects of septal hyperreactivity, in *The Septal Nuclei*. DeFrance JF. Ed., Plenum Press, New York, 1976, 507.
72. **Stark P, Henderson J.** Central cholinergic suppression of hyperreactivity and aggression in septal-lesioned rats. *Neuropharmacology* 11: 839-847, 1972.
73. **Costall B, Naylor RJ.** The behavioural effects of dopamine applied intracerebrally to areas of the mesolimbic system. *Eur J Pharmacol* 32: 87-92, 1975.
74. **Bunney BS, Aghajanian GK.** Dopamine and norepinephrine innervated cells in the rat prefrontal cortex: pharmacological differentiation using microiontophoretic techniques. *Life Sci* 19: 1783-1792, 1976.
75. **Galey D, Simon H, LeMoal M.** Behavioral effects of lesions in the A10 dopaminergic area of the rat. *Brain Res* 124: 83-97, 1977.
76. **Barr GA, Gibbons JL, Bridger WH.** Inhibition of rat predatory aggression by acute and chronic D- and L-amphetamine. *Brain Res* 124: 565-570, 1977.
77. **German DC, Dalsass M, Kiser RS.** Electrophysiological examination of the ventral tegmental (A10) area in the rat. *Brain Res* 181: 191-197, 1980.
78. **Dalsass M, German DC, Kiser RS, Speciale S.** Effects of D-amphetamine on dopaminergic neurons in the ventral tegmental area of the rat. *Neurosci Abstr* 5: 553-556, 1979.
79. **Daruna JH.** Patterns of brain monoamine activity and aggressive behavior. *Neurosci Biobehav Rev* 2: 101-113, 1978.
80. **Karczmar AG, Scudder CL.** Aggression and neurochemical changes in different strains and genera of mice, in *Aggressive Behavior*. Garattini S, Sigg EB. Eds., Excerpta Medica, Amsterdam, 1969, 209.

81. **Slotnick BM, McMullen MF.** Intraspecific fighting in Albino mice with septal forebrain lesion. *Physiol Behav* 8: 333-337, 1972.

82. **Welch BL, Welch AS.** Isolation reactivity and aggression: evidence for an involvement of brain catecholamine and serotonin, in *Physiology of Fighting and Defeat*. Eleftheriou BE. Ed., University of Chicago Press, Chicago, 1971, 91.

83. **Vergnes M, Bandler R, Kempf E.** Muricide induced by diagonal band damage: role of 5-HT pathways. *Brain Res* 185: 203-207, 1980.

84. **Blander R, Vergnes M.** Interspecies aggression in the rat: the role of the diagonal band of Broca. *Brain Res* 175: 327-333, 1979.

85. **Broderick PA, Barr GA, Sharpless NS, Bridger WH.** Biogenic amine alterations in limbic brain regions of muricidal rats. *Res Commun Chem Pathol Pharmacol* 48: 3-15, 1985.

86. **Taylor KM, Snyder SH.** Differential effects of D- and L-amphetamine on behavior and on catecholamine disposition in dopamine and norepinephrine containing neurons of rat brain. *Brain Res* 28: 295-309, 1971.

87. **Tassin JP, Stinus L, Simon H, Blanc G, Thierry AM, LeMoal M, Cardo B, Glowinski J.** Relationship between the locomotor hyperactivity induced by A10 lesions and the destruction of the fronto-cortical dopaminergic innervation in the rat. *Brain Res* 141: 267-281, 1978.

88. **Galey D, LeMoal M.** Locomotor activity after various radiofrequency lesions of the limbic midbrain area in the rat. Evidence for a particular role of the ventral mesencephalic tegmemtum. *Life Sci* 19: 677-684, 1976.

89. **LeMoal M, Stinus L, Galey D.** Radiofrequency lesion of the ventral mesencephalic tegmentum: neurological and behavioural considerations. *Exp Neurol* 50: 521-535, 1976.

90. **LeMoal M, Galey D, Cardo B.** Behavioral effects of local injection of 6-hydroxydopamine in the medial ventral tegmentum in the rat. Possible role of the mesolimbic dopaminergic system. *Brain Res* 88: 190-194, 1975.

91. **Sloviter RS, Drust EG, Conner JD.** Evidence that serotonin mediates some behavioral effects of amphetamine. *J Pharmacol Exp Ther* 206: 348-353, 1978.

92. **Tassin JP, Stinus L, Simon H, Blanc G, Thierry AM, Cardo B, Glowinski J.** Distribution of dopaminergic terminals in rat cerebral cortex. Role of dopaminergic mesocortical system in "ventral tegmental area syndrome", in *Non-Striatal Dopaminergic Neurons, Advances in Biochemical Psychopharmacology*, Vol. 16. Costa E, Gessa GL. Eds., Raven Press, New York, 1977, 21.

93. **Rolinski Z, Scheel-Kruger J.** The effect of dopamine and noradrenaline antagonists on amphetamine induced locomotor activity in mice and rats. *Acta Pharmacol Toxicol* 33: 385-392, 1973.

94. **Avis HH.** The neuropharmacology of aggression: a critical review. *Psychol Bull* 81: 47-63, 1974.

95. **Barr GA, Moyer KE, Gibbons JL.** Effects of imipramine, D-amphetamine, and tripelennamine on mouse and frog killing by the rat. *Physiol Behav* 16: 267-269, 1976.

96. **Gay PE, Leaf RC, Arble FB.** Inhibitory effects of pre- and post test D-amphetamine on mouse killing by rats. *Pharmacol Biochem Behav* 3: 33-45, 1975.

97. **McCarty RC, Whitesides GH.** Effects of D- and L-amphetamine on the predatory behavior of southern grasshopper mice *Onychomys torridus*. *Agressive Behav* 2: 99-105, 1976.

98. **Svensson TH.** Functional and biochemical effects of D- and L-amphetamine on behaviour and catecholamine disposition in dopamine and norepinephrine containing neurons of rat brain. *Arch Pharmakol* 271: 170-180, 1971.

99. **Kelly PH, Seviour PW, Iversen SD.** Amphetamine and apomorphine responses in the rat following 6-OHDA lesion of the nucleus accumbens septi and corpus striatum. *Brain Res* 94: 507-522, 1975.

100. **Gibbons JL, Barr GA, Bridger WH, Leibowitz SF.** Effects of parachlorophenylalanine and 5-hydroxytryptophan on mouse killing behavior in killer rats. *Pharmacol Biochem Behav* 9: 91-98, 1978.

101. **Kreiskott H, Hofmann HP.** Stimulation of a specific drive (predatory behaviour) by *p*-chlorophenylalanine (pCPA) in the rat. *Pharmakopsychiatr Neuropsychopharmakol* 8: 136-140, 1975.

102. **Thurmond JB, Lasley SM, Conking AL, Brown JW.** Effects of dietary tyrosine, phenylalanine, and tryptophan on aggression in mice. *Pharmacol Biochem Behav* 6: 475-478, 1977.

103. **Jacobs BL, Mosko SS, Trulson ME.** The investigation of the role of serotonin in mammalian behavior, in *Neurobiology of Sleep and Memory*. Drucker-Colin RR, McGaugh JL. Eds., Academic Press, New York, 1977, 99.

104. **Miczek KA, Altmann JL, Appel JB, Boggan WO.** Parachlorophenylalanine, serotonin and killing behavior. *Pharmacol Biochem Behav* 3: 355-361, 1975.

105. **Paxinos G, Altrens DM.** 5,7-Dihydroxytryptamine lesions: effects on body weight, irritability, and muricide. *Aggressive Behav* 3: 107-118, 1977.

106. **Paxinos G, Burt J, Altrens DM, Jackson DM.** 5-Hydroxytryptamine depletion with para-chlorophenylalanine: effects on eating, drinking, irritability, muricide, and copulation. *Pharmacol Biochem Behav* 6: 439-447, 1977.

107. **Breese GR, Cooper BR, Grant LD, Smith RD.** Biochemical and behavioral alterations following 5,6-dihydroxytryptamine administration to brain. *Neuropharmacology* 13: 177-187, 1974.

108. **Dichiara G, Camba R, Spano PF.** Evidence for inhibition by brain serotonin of mouse killing behaviour in rats. *Nature* 233: 272-273, 1971.

109. **Waldbillig RJ.** The role of the dorsal and median raphe in the inhibition of muricide. *Brain Res* 160: 341-346, 1979.

110. **Jacobs BL, Asher R, Dement WC.** Electrophysiological and behavioral effects of electrical stimulation of the raphe nuclei in cats. *Physiol Behav* 11: 489-495, 1973.

111. **Vergnes M, Penot C, Kempf E, Mack G.** Lésion sélective des neurones sérotoninergiques du raphé par la 5,7-dihydroxytryptamine: effets sur le comportement d'agression interspécifique du rat. *Brain Res* 133: 167-171, 1977.

112. **Yamamoto T, Ueki S.** Characteristics in aggressive behavior induced by midbrain raphe lesions in rats. *Physiol Behav* 19: 105-110, 1977.

113. **Brutus M, Shaikh MB, Siegel H, Siegel A.** An analysis of the mechanisms underlying septal area control of hypothalamically-elicited aggression in the cat. *Brain Res* 310: 235-248, 1984.

114. **Latham EE, Thorne MB.** Septal damage and muricide: effects of strain and handling. *Physiol Behav* 12: 521-526, 1974.

115. **MacDonnell MFF, Stoddard-Apter S.** Effects of medial septal stimulation on hypothalamically-elicited intraspecific attack and associated hissing in cats. *Physiol Behav* 21: 679-683, 1978.

116. **Meiback RC, Siegel A.** Efferent connections of the septal area in the rat: an analysis utilizing retrograde and anterograde transport methods. *Brain Res* 119: 1-20, 1977.

117. **Stoddard-Apter SL, MacDonnell MF.** Septal and amygdalar efferents to the hypothalamus which facilitates hypothamically-elicited intraspecific aggression and associated hissing in the cat. An autoradiographic study. *Brain Res* 193: 19-32, 1980.

118. **Watson RE Jr, Edinger H, Siegel A.** An analysis of the mechanisms underlying hippocampal control of hypothalamically-elicited aggression in the cat. *Brain Res* 269: 327-345, 1983.

119. **Srebro B, Lorens SA.** Behavioral effects of selective midbrain raphe lesions in the rat. *Brain Res* 89: 303-325, 1975.

120. **Penot C, Vergnes M, Mack G, Kempf E.** Comportement d'agression interspécifique et réactivité chez le rat: étude comparative des effets de lésions électrolytiques du raphé et d'injections intraventriculaires de 5,7-DHT. *Biol Behav* 3: 71-85, 1978.

121. **Jones RSG, Broadbent J.** Further studies on the role of indoleamines in the responses of cortical neurones to stimulation of nucleus raphe medianus: effects of indoleamine precursor loading. *Neuropharmacology* 21: 1273-1277, 1982.

122. **Jones RSG.** Responses of cortical neurones to stimulation of the nucleus raphe medianus: a pharmacological analysis of the role of indoleamines. *Neuropharmacology* 21: 511-520, 1982.

123. **Sastry BSR, Phillis JW.** Inhibition of cerebral cortical neurones by a 5-hydroxytryptaminergic pathway from the median raphe nucleus. *Can J Physiol Pharmacol* 55: 737-743, 1977.

124. **Chronister RB, DeFrance JF.** Organization of projection neurons of the hippocampus. *Exp Neurol* 66: 509-523, 1979.

125. **Andersen P.** Organization of hippocampal neurons and their interconnections, in *The Hippocampus: A Comprehensive Treatise.* Isaacson RL, Pribram KH. Eds., Plenum Press, New York, 1975, 155.

126. **Sharma JN.** Microiontophoretic application of some mono-amines and their antagonists to cortical neurones of the rat. *Neuropharmacology* 16: 83-88, 1977.

127. **Blackshear MA, Steranka LR, Sanders-Bush E.** Multiple serotonin receptors: regional distribution and effect of raphe lesions. *Eur J Pharmacol* 76: 325-334, 1981.

128. **Peroutka SJ, Snyder SH.** Two distinct serotonin receptors: regional variations in receptor binding in mammalian brain. *Brain Res* 208: 339-344, 1981.

129. **Seeman P, Westman K, Coscina D, Warsh JJ.** Serotonin receptors in hippocampus and frontal cortex. *Eur J Pharmacol* 66: 179-184, 1980.

130. **Martin RF, Jordan LM, Willis WD.** Differential projections of cat medullary raphe neurons demonstrated by retrograde labelling following spinal cord lesions. *J Comp Neurol* 182: 77-88, 1978.

131. **Bobillier P, Seguin S, Petitjean F, Salvert D, Touret M, Jouve M.** The raphe nuclei of the cat brainstem: a topographical atlas of their efferent projections as revealed by autoradiography. *Brain Res* 113: 449-486, 1976.

132. **Brodal A, Walberg F, Taber E.** The raphe nuclei of the brainstem in the cat. III. Afferent connections. *J Comp Neurol* 14: 261-279, 1960.

133. **Taber-Pierce E, Foote WE, Hobson JA.** The efferent connection of the nucleus raphe dorsalis. *Brain Res* 107: 137-144, 1976.

134. **Consolo S, Ladinsky H, Forloni GL, Grombi P.** Modulation of the hippocampal-adrenoceptor population by lesion of the serotonergic raphe-hippocampal pathway in rats. *Life Sci* 30: 1113-1120, 1982.

135. **Roberts MHT, Straughan DW.** Excitation and depression of cortical neurones by 5-hydroxytryptamine. *J Physiol* 193: 269-294, 1976.

136. **Robinson SE.** Effect of specific serotonergic lesions on cholinergic neurons in the hippocampus, cortex and striatum. *Life Sci* 32: 345-353, 1982.

137. **Costa E, Panula P, Thompson HK, Cheney DL.** The trans-synaptic regulation of the septal-hippocampal cholinergic neurons. *Life Sci* 32: 165-179, 1982.

138. **Lamour Y, Rivot JP, Pointis D, Ory-Lavollee L.** Laminar distribution of serotonergic innervation in rat somato-sensory cortex, as determined by in vivo electrochemical detection. *Brain Res*, 259: 163-166, 1983.

139. **Hall RD, Lindholm EP.** Organization of motor and somato-sensory neocortex in the albino rat. *Brain Res* 66: 23-38, 1974.

140. **Kuhar MJ, Aghajanian GK, Roth RH.** Tryptophan hydroxylase activity and synaptosomal uptake of serotonin in discrete brain regions after midbrain raphe lesions: correlations with serotonin levels and histochemical fluorescence. *Brain Res* 44: 165-176, 1972.

141. **Lidov HG, Grzanna R, Molliver ME.** The serotonin innervation of the cerebral cortex in the rat. An immuno-histochemical analysis. *Neuroscience* 5: 207-227, 1980.

142. **Blaker WD, Cheney DL, Gandolfi O, Costa E.** Simultaneous modulation of hippocampal cholinergic activity and extinction by intraseptal muscimol. *J Pharmacol Exp Ther* 225: 361-365, 1983.

143. **Gray JA.** Effects of septal driving of the hippocampal theta rhythm on resistance to extinction. *Physiol Behav* 8: 481-490, 1972.

144. **Greene E, Stauff C.** Behavioral role of hippocampal connection. *Exp Neurol* 45: 141-160, 1974.

145. **Lewis PR, Shute CCD.** The cholinergic limbic system: projections to hippocampal formation, medial cortex, nuclei of the ascending cholinergic reticular system and the subfornical organ and supraoptic crest. *Brain* 90: 521-539, 1967.

146. **Bernardo LS, Prince DA.** Cholinergic pharmacology of mammalian hippocampal pyramidal cells. *Neuroscience* 7: 1703-1712, 1982.

147. **Bird SJ, Aghajanian GK.** The cholinergic pharmacology of hippocampal pyramidal cells: a microionto-phoretic study. *Neuropharmacology* 15: 273-282, 1976.

148. **Dodd J, Dingledine R, Kelly JS.** The excitatory action of acetylcholine on hippocampal neurones of the guinea pig and rat maintained in vitro. *Brain Res* 112: 413-419, 1976.

149. **Dutar P, Lamour Y, Jobert A.** Acetylcholine excites identified septo-hippocampal neurones in the rat. *Neurosci Lett* 43: 43-47, 1983.

150. **Krnjevic K, Ropert N.** Electrophysiological and pharmacologial characteristics of facilitation of hippocampal population spikes by stimulation of the medial septum. *Neuroscience* 7: 2165-2183, 1982.

151. **Lamour Y, Dutar P, Jobert A.** Excitatory effect of acetylcholine on different types of neurons in the first somatosensory neocortex of the rat: laminar distribution and pharmacological characteristics. *Neuroscience* 7: 1483-1494, 1982.

152. **Lynch G, Rose G, Gall C.** Anatomical and functional aspects of the septo-hippocampal projections,. in *Functions of the Septo-Hippocampal System*. Elsevier, Amsterdam, 1978, 5.

153. **Mesulam MM, Mufson EJ, Levey AI, Wainer BH.** Cholinergic innervation of cortex by the basal forebrain: cytochemistry and cortical connections of the septal area, diagonal band nuclei, nucleus basalis (substantia innominata) and hypothalamus in the rhesus monkey. *J Comp Neurol* 214: 170-197, 1983.

154. **Segal M.** Responses of septal nuclei neurons to micro-iontophoretically administered putative neurotransmitters. *Life Sci* 14: 1345-1351, 1974.

155. **Segal M.** Brain stem afferents to the rat medial septum. *J Physiol* 261: 617-631, 1976.

156. **Segal M, Weinstock M.** Differential effects of 5-hydroxytryptamine antagonists on behaviors resulting from activation of different pathways arising from the raphe nuclei. *Psychopharmacology* 79: 72-78, 1983.

157. **Pasquier DA, Kemper TL, Forbes WB, Morgane PJ.** Dorsal raphe, substantia nigra and locus coeruleus: inter-connections with each other and the neostriatum. *Brain Res Bull* 2: 323-329, 1977.

158. **Galindo-Mireles D, Meyer G, Castañeyra-Perdomo A, Ferres-Torres R.** Cortical projections of the nucleus centralis superior and the adjacent reticular tegmentum in the mouse. *Brain Res* 330: 343-348, 1985.

159. **Grossman SP.** An experimental "dissection" of the septal syndrome, in *Functions of the Septo-Hippocampal System*. Elliot K, Whelan J. Eds., Ciba Foundation Symposium, Elsevier/North-Holland, New York, 1978, 227.

160. **Mosko SS, Haubrich D, Jacobs BL.** Serotonergic afferents to the dorsal raphe nucleus: evidence from HRP and synaptosomal uptake studies. *Brain Res* 119: 269-290, 1977.

161. **Peroutka SJ, Snyder SH.** Multiple serotonin receptors: differential binding of [^{3}H]-5-hydroxytryptamine, [^{3}H] lysergic acid diethylamide and [^{3}H] spiroperidol. *Mol Pharmacol* 16: 687-690, 1979.

162. **Gallager DW, Pert A.** Afferents to brain stem nuclei (brain stem raphe, nucleus reticularis pontis caudalis and nucleus gigantocellularis) in the rat as demonstrated by microiontophoretically applied horseradish peroxidase. *Brain Res* 144: 257-275, 1978.

163. **Demontingy C, Aghajanian GK.** Preferential action of 5-methoxytryptamine and 5-methoxydimethyltryp-tamine on pre-synaptic serotonin receptors: a comparative iontophoretic study with LSD and serotonin. *Neuropharmacology* 16: 811-818, 1977.

164. **Rogawski MA, Aghajanian GK.** Serotonin autoreceptors on dorsal raphe neurons: structure-activity relationships of tryptamine analogs. *J Neurosci* 1: 1148-1154, 1981.

165. **Sakai K, Salvert D, Touret M, Jouvet M.** Afferent connections of the nucleus raphe dorsalis in the cat as visualized by the horseradish peroxidase technique. *Brain Res* 145: 1-25, 1977.

166. **Trulson ME, Preussler DW, Trulson VM.** Differential effects of hallucinogenic drugs on the activity of serotonin-containing neurons in the nucleus centralis superior and nucleus raphe pallidus in freely moving cats. *J Pharmacol Exp Ther* 228: 94-102, 1984.

167. **Hey MJ, Steinfels GF, Jacobs BL.** Medullary serotonergic neurons are insensitive to 5-MeODMT and LSD. *Eur J Pharmacol* 81: 667-680, 1982.

168. **Foldes A, Costa E.** Relationship of monoamine and locomotor activity in rats. *Biochem Pharmacol* 24: 1617-1625, 1975.

169. **Trulson ME, Heym J, Jacobs BL.** Dissociations between the effects of hallucinogenic drugs on behavior and raphe unit activity in freely moving cats. *Brain Res* 215: 275-293, 1981.

170. **Trulson ME, Jacobs BL.** Effects of 5-methoxy-*N,N*-dimethyltryptamine on behavior and raphe unit activity in freely-moving cats. *Eur J Pharmacol* 54: 43-50, 1979.

171. **Trulson ME, Jacobs BL.** Dissociations between the effects of LSD on behavior and raphe unit activity in freely moving cats. *Science* 205: 515-518, 1979.

172. **Chase TN, Murphy DL.** Serotonin and central nervous system function. *Annu Rev Pharmacol* 13: 181-197, 1973.

173. **McGinty DJ, Harper RM.** Dorsal raphe neurons: depression of firing during sleep in cats. *Brain Res* 101: 569-575, 1976.

174. **Mosko SS, Jacobs BL.** Midbrain raphe neurons: spontaneous activity and response to light. *Physiol Behav* 13: 589-593, 1974.

175. **Mosko SS, Jacobs BL.** Recording of dorsal raphe unit activity in vitro. *Neurosci Lett* 2: 195-200, 1976.

176. **Trulson ME, Jacobs BL.** Effects of LSD on behavior and raphe unit activity in freely-moving cats. *Fed Proc Fed Am Soc Exp Biol* 37: 346, 1978.

177. **Sheu Y-S, Nelson JP, Bloom FE.** Discharge patterns of cat raphe neurons during sleep and waking. *Brain Res* 73: 263-276, 1974.

178. **Steriade M, Hobson JA.** Neuronal activity during the sleep-waking cycle. *Prog Neurobiol* 6: 155-376, 1976.

179. **Trulson ME, Jacobs BL.** Raphe unit activity in freely moving cats: correlation with level of behavioral arousal. *Brain Res* 163: 135-150, 1979.

180. **Dyr W, Kostowski W, Zacharski B, Bidzinski A.** Differential clonidine effects on EEG following lesions of the dorsal and median raphe nuclei in rats. *Pharmacol Biochem Behav* 19: 177-185, 1983.

181. **Vandermaelen CP, Aghajanian GK.** Noradrenergic activation of serotonergic dorsal raphe neurons recorded in vitro. *Soc Neurosci Abstr* 8: 482, 1982.

182. **Anderson C, Pasquier D, Forbes W, Morgane P.** Locus coeruleus-to-dorsal raphe input examined by electrophysiological and morphological methods. *Brain Res Bull* 2: 209-221, 1977.

183. **Plaznik A, Danysz W, Kostowski W, Bidzinski A, Hauptmann M.** Interaction between noradrenergic and serotonergic brain systems as evidenced by behavioral and biochemical effects of microinjections of adrenergic agonists and antagonists into the median raphe nucleus. *Pharmacol Biochem Behav* 19: 27-32, 1983.

184. **Pujol J-F, Buguet A, Froment J-L, Jones B, Jouvet M.** The central metabolism of serotonin in the cat during insomnia: a neurophysiological and biochemical study after administration of *p*-chlorophenylalanine or destruction of the raphe system. *Brain Res* 29: 195-212, 1971.

185. **Ennis C.** Different adrenoceptors modulate the release of 5-hydroxytryptamine and noradrenaline in rat cortex. *Br J Pharmacol* 79: 279-283, 1983.

186. **Lechin F, van der Dijs B.** Slow wave sleep (SWS), REM Sleep (REMS) and depression. *Res Commun Psychol Psychiat Behav* 9: 227-262, 1984.

187. **Miliaressis E, Bouchard A, Jacobowitz DM.** Strong positive reward in median raphe: specific inhibition by parachlorophenylalanine. *Brain Res* 98: 194-201, 1975.

188. **Simon H, LeMoal M, Cardo B.** Mise en evidence du comportement d'autostimulation dans le noyau raphé median du rat. *C R Acad Sci* 277: 591-593, 1973.

189. **Costall B, Naylor RJ, Marsden CD, Pycock CJ.** Serotoninergic modulation of the dopamine response from the nucleus accumbens. *J Pharm Pharmacol* 28: 523-526, 1976.

190. **Lorens SA, Guldberg HC, Hole K, Kohler C, Srebro B.** Activity, avoidance learning and regional 5-hydroxytryptamine following intra-brain stem 5,7-dihydroxy-tryptamine and electrolytic midbrain raphe lesion in the rat. *Brain Res* 108: 97-113, 1976.

191. **Fibiger HC, Campbell BA.** The effect of parachloro-phenylalanine on spontaneous locomotor activity in the rat. *Neuropharmacology* 10: 25-32, 1971.

192. **Jacobs BL, Eubanks EE, Wise WD.** Effect of indolealkylamine manipulations on locomotor activity in rats. *Neuropharmacology* 13: 575-583, 1974.

193. **Heffner THG, Seiden S.** Possible involvement of serotonergic neurons in the reduction of locomotor hyperactivity caused by amphetamine in neonatal rats depleted of brain dopamine. *Brain Res* 244: 81-90, 1982.

194. **Green TK, Harvey JA.** Enhancement of amphetamine action after interruption of ascending serotonergic pathways *J Pharmacol Exp Ther* 190: 109-117, 1974.

195. **Mabry PD, Campbell BA.** Serotonergic inhibition of catecholamine-induced behavioral arousal. *Brain Res* 49: 381-391, 1973.

196. **Fuxe K, Hökfelt T, Agnati L, Johansson O, Ljungdahl A, Perez de La Mora M.** Regulation of the mesocortical dopamine neurons, in *Nonstriatal Dopaminergic Neurons* (Advances in Biochemical Psychopharmacology Series, Vol. 16). Costa E, Gessa GL. Eds., Raven Press, New York, 1977, 55.

197. **Thierry AM, Tassin JP, Blanc G, Glowinski J.** Selective activation of the mesocortical dopaminergic system by stress. *Nature* 263: 242-244, 1976.

198. **Bradley PB, Briggs I.** Further studies on the mode of action of psychomimetic drugs: antagonism of the excitatory actions of 5-hydroxytryptamine by methylated derivatives of tryptamine. *Br J Pharmacol* 50: 345-354, 1974.

199. **Kramarcy NR, Brown JW, Thurmond JB.** Effects of drug-induced changes in brain monoamines on aggression and motor behavior in mice. *Eur J Pharmacol* 99: 141-151, 1984.

200. **Thierry AM, Tassin JP, Blanc G, Glowinski J.** Topographic and pharmcological study of the mesocortical dopaminergic system, in *Brain Stimulation Reward*. Wauquier A, Rolls ET. Eds., Elsevier, New York, 1976, 290.

201. **Nai-Shin C.** Responses of midbrain raphe neurons to ethanol. *Brain Res* 311: 348-352, 1984.

202. **Ferron A, Thierry AM, Le Douarin C, Glowinski J.** Inhibitory influence of the mesocortical dopaminergic system on spontaneous activity or excitatory response induced from the thalamic mediodorsal nucleus in the rat medial prefrontal cortex. *Brain Res* 302: 257-265, 1984.

203. **Canedo A.** Subcortical influences upon prefrontal granular cortex. I. Patterns of focal field potentials evoked by stimulations of dorsomedial thalamus in conscious monkey. *Brain Res* 58: 401-414, 1973.

204. **Hwang EC, Van Woert MH.** Comparative effects of phenylethylamines on brain serotonergic mechanisms. *J Pharmacol Exp Ther* 213: 254-260, 1980.

205. **Korsgaard S, Gerlach J, Christensson E.** Behavioral aspects of serotonin-dopamine interaction in the monkey. *Eur J Pharmacol* 118: 245-252, 1985.

206. **French ED, Pilapil C, Quirion R.** Phencyclidine binding sites in the nucleus accumbens and phencyclidine induced hyperactivity are decreased following lesions of the mesolimbic dopamine system. *Eur J Pharmacol* 116: 1-9, 1985.

207. **Eisenstein NL, Lorio LC, Clody DE.** Role of serotonin in the blockade of muricidal behavior by tricyclic antidepressants. *Pharmacol Biochem Behav* 17: 847-849, 1982.

208. **Kostowski W, Valzelli L, Kozak W, Bernasconi S.** Activity of desipramine, fluoxetine and nomifensine on spontaneous and pCPA-induced muricidal aggression. *Pharmacol Res Commun* 16: 265-271, 1984.

209. **Kostowski W, Valzelli L, Kozak W.** Chlordiazepoxide antagonizes locus coeruleus-mediated suppression of muricidal aggression. *Eur J Pharmacol* 91: 329-336, 1983.

210. **Antelman S.** Stress and its timing: critical factors in determining the consequences of dopaminergic agents. *Pharmacol Biochem Behav* 17(Suppl. 1): 21-23, 1982.

211. **Antelman SM, Chiodo LA, DeGiovanni LA.** Antidepressant and dopamine autoreceptors: implications for both a novel means of treating depression and understanding bipolar illness, in *Typical and Atypical Antidepressants: Molecular Mechanisms*. Costa E, Racagni G. Eds., Raven Press, New York, 1982, 121.

212. **Van Kammen DP, Bunney WE Jr, Docherty JP, Jimerson DC, Post RM, Siris S, Ebert M, Gilin JC.** Amphetamine-induced catecholamine activation in schizophrenia and depression: behavioral and physiological effects, in *Nonstriatal Dopaminergic Neurons*. Costa E, Gessa GL. Eds., Raven Press, New York, 1977, 655.

213. **Gerner RH, Post RM, Bunney WE Jr.** A dopaminergic mechanism in mania. *Am J Psychiatry* 133: 1177-1179, 1976.

214. **Hollister LE.** Experiences with dopamine agonists in depression and schizophrenia, in *Apomorphine and Other Dopaminomimetics*, Vol. 2. Corsini GU, Gessa GL. Eds., Raven Press, New York, 1981, 57.

215. **Post RM, Gerner RH, Corman JS, Bunney WE Jr.** Effects of low doses of a dopamine-receptor stimulator in mania. *Lancet* 1: 203-204, 1976.

216. **Post RM, Cutler NR, Jimerson DC, Bunney WE Jr.** Dopamine agonists in affective illness. Implications for underlying receptor mechanisms, in *Apomorphine and Other Dopaminomimetics*, Vol. 2. Corsini GU, Gessa GL. Eds., Raven Press, New York, 1981, 77.

217. **Meltzer HY, Kolawoska T, Robertson A, Tricou BJ.** Effect of low-dose bromocriptine in treatment of psychosis: The dopamine autoreceptor-stimulation strategy. *Psychopharmacology* 81: 37-41, 1983.

218. **Colonna L, Peht M, Lepine JP.** Bromocriptine in affective disorders. *J Affect Disord* 1: 173-177, 1979.

219. **Dorr C, Sathananthan A.** Treatment of mania with bromocriptine. *Br Med J* 1: 1342-1343, 1976.

220. **Frye PE, Pariser SF, Kim MH, O'Shaughnessy RW.** Bromocriptine associated with symptom exacerbation during neuroleptic treatment of schizoaffective schizophrenia. *J Clin Psychiatry* 43: 252-253, 1982.

221. **Johnson JM.** Treated mania exacerbated by bromocriptine. *Am J Psychiatry* 138: 980-982, 1981.

222. **Smith AHW, Chambers C, Naylor GJ.** Bromocriptine in mania. A placebo-controlled double-blind trial. *Br Med J* 280: 86-90 1980.

223. **Trabucchi M, Andreoli VM, Frattola L, Spano PF.** Pre- and post-synaptic action of bromocriptine: its pharmacological effects in schizophrenia and neurological disease. *Adv Biochem Psychopharmacologia* 16: 661-665, 1977.

224. **Vlissides DN, Gill D, Castlelow J.** Bromocriptine-induced mania? *Br Med J* 1: 510-514, 1978.

225. **Thurmond JB, Kramarcy NR, Lasley SM, Brown JW.** Dietary amino acid precursors: effects on central monoamines, aggression and locomotor activity in the mouse. *Pharmacol Biochem Behav* 12: 525-531, 1980.

226. **Lechin F, van der Dijs B.** The effects of dopaminergic blocking agents on distal colon motility. *J Clin Pharmacol* 19: 617-625, 1979.

227. **Lechin F, Van Der Dijs B.** Intestinal pharmacomanometry and glucose tolerance: evidence for two antagonistic mechanisms in the human. *Biol Psychiatry* 16: 969-979 1981.

228. **Lechin F, van der Dijs.** *Clinical Pharmacology and Therapeutics.* Velazco M. Ed., Int Congr Ser No 604, Excerpta Medica, Amsterdam, 1982, 166.

229. **Lechin F, Gómez F, Acosta E, Arocha L, van der Dijs B.** Treatment of manic syndrome patients with dopaminergic antagonists. *Arch Venezolanos Farmacol Ter* 1: 150, 1982.

230. **Lechin F, van der Dijs B.** Antimanic effects of clonazepam. *Biol Psychiatry* 18: 1511, 1983.

231. **Pijnenburg AJJ, Van Rossum JM.** Stimulation of locomotor activity following injection of dopamine into the nucleus accumbens. *J Pharm Pharmacol* 25: 1003-1009, 1973.

232. **Pijnenburg AJJ, Woodruff GN, Van Rossum JM.** Ergometrine induced locomotor activity following intracerebral injection into the nucleus accumbens. *Brain Res* 59: 289-294, 1973.

233. **Stromberg U, Svensson TH.** L-Dopa induced effects on motor activity in mice after inhibition of dopamine-beta-hydroxylase. *Psychopharmacologia* 19: 53-58, 1971.

234. **Pijnenburg AJJ, Honig WMM, Van der Heyden JAM, Van Rossum JM.** Effects of chemical stimulation of the mesolimbic dopamine system upon locomotor activity. *Eur J Pharmacol* 35: 49-58, 1976.

235. **Costall B, Hui S-CG, Naylor RJ.** Hyperactivity induced by injection of dopamine into the accumbens nucleus: actions and interactions of neuroleptic, cholinomimetic and cholinolytic agents. *Neuropharmacology* 18: 661-665, 1979.

236. **Costall B, Naylor RJ.** A comparison of the abilities of typical neuroleptic agents and of thioridazine, clozapine, sulpiride and metoclopramide to antagonise the hyper-activity induced by dopamine applied intracerebrally to areas of the extrapyramidal and mesolimbic systems. *Eur J Pharmacol* 40: 9-19, 1976.

237. **Geyer MA, Puerto A, Menkes DB, Segal DS, Mandella AJ.** Behavioral studies following lesions of the mesolimbic and mesostriatal serotonergic pathways. *Brain Res* 106: 257-270, 1976.

238. **Miller FE, Heffner TG, Kotake C, Seiden L.** Magnitude and duration of hyperactivity following neonatal 6-hydroxy-dopamine is related to the extent of brain dopamine depletion. *Brain Res* 229: 123-132, 1981.

239. **Shaywitz RA, Klopper JH, Yager RD, Gordon JW.** Paradoxical response to amphetamine in developing rats treated with 6-hydroxydopamine. *Nature* 261: 153-155, 1976.

240. **Shaywitz RA, Klopper JH, Gordon JW.** Methylphenidate in 6-hydroxydopamine treated developing rats pups. *Child Neurol* 35: 463-469, 1978.

241. **Sorenson CA, Vayer JS, Goldberg CS.** Amphetamine reduction of motor activity in rats after neonatal administration of 6-hydroxydopamine. *Biol Psychiatry* 12: 133-137, 1977.

242. **Stoof JC, Dijkstra H, Hillegers JPM.** Changes in the behavioral responses to a novel environmental following lesioning of the central dopaminergic system in rat pups. *Psychopharmacology* 57: 163-166, 1978.

243. **Jones DL, Mogenson G, Wu M.** Injections of dopaminergic, cholinergic, serotonergic, and GABAergic drugs into the nucleus accumbens: effects of locomotor activity in the rat. *Neuropharmacology* 20: 29-36, 1981.

244. **Shaywitz RA, Yager RD, Klopper JH.** Selective brain dopamine depletion in developing rats: an experimental model of minimal brain dysfunction. *Science* 191: 305-308, 1976.

245. **Markowitsch HJ, Pritzel M.** Comparative analysis of prefrontal learning functions in rats, cats, and monkeys. *Psychol Bull* 84: 817-837, 1977.

246. **Carnoy P, Soubrie P, Puech AJ, Simon P.** Performance deficit induced by low doses of dopamine agonists in rats: toward a model for approaching the neurobiology of negative schizophrenic symptomatology? *Biol Psychiatry* 21: 11-22, 1986 .

247. **Mora F, Sweeney KF, Rolls ET, Sanguinetti AM.** Spontaneous firing rate of neurones in the prefrontal cortex of the rat: evidence for a dopaminergic inhibition. *Brain Res* 116: 516-522, 1976.

248. **Bevan P, Bradshaw CM, Pun RYK, Slater NT, Szabadi E.** Responses of single cortical neurones to noradrenaline and dopamine. *Neuropharmacology* 17: 611-617, 1978.

249. **Fuxe K, Hamberger B, Hökfelt T.** Distribution of noradrenaline nerve terminals in cortical areas of the rat. *Brain Res* 8: 125-131, 1968.

250. **Morrison JH, Grzanna R, Molliver ME, Coyles JT.** The distribution and orientation of noradrenergic fibers in neocortex of the rat: an immunofluorescence study. *J Comp Neurol* 181: 17-40, 1978.

251. **Rabey JM, Passeltiner P, Bystritsky A, Engel J, Goldstein M.** The regulation of striatal DOPA synthesis by A2-adrenoreceptors. *Brain Res* 230: 422-426, 1981.

252. **Shaywitz RA, Yager RD, Klopper JH.** Selective brain dopamine depletion in developing rats: an experimental model of minimal brain dysfunction. *Science* 191: 305-308, 1976.

253. **Thieme RE, Dijkstra H, Stoof JC.** An evaluation of the young dopamine-lesioned rat as an animal model for minimal brain dysfunction (MBD). *Psychopharmacology* 67: 165-169, 1980.

254. **Fink JS, Smith GP.** Mesolimbic and mesocortical dopaminergic neurons are necessary for normal locomotor and investigatory exploration in rats. *Neurosci Lett* 17: 61, 1980.

255. **Fink JS, Smith GP.** Mesolimbic-cortical dopamine terminal fields are necessary for normal locomotor and investigatory exploration in rats. *Brain Res* 199: 359-384, 1980.

256. **Fonseca JS, Gil MT, Figueira ML, Barata JG, Pego F, Pacheco MF.** How do normal subjects learn a simple adaptive task: how and why do paranoid schizophrenic patients fail? *Arch Psychiatr Nervenkr* 225: 31-43, 1978.

257. **Gaffori O, LeMoal M, Stinus L.** Locomotor hyperactivity and hypoexploration after lesion of the dopaminergic A-10 area in the ventral mesencephalic tegmentum (VMT) of rats. *Behav Brain Res* 1: 313-316, 1980.

258. **Hadfield MG.** Mesocortical vs. nigrostriatal dopamine up-take in isolated fighting mice. *J Neuropathol Exp Neurol* 40: 323-327, 1981.

259. **Levine MS, Hull CD, Villablanca JR, García-Rill E.** Effects of caudate nuclear or frontal cortical ablation in neonatal kittens or adults on the spontaneous firing of forebrain neurons. *Dev Brain Res* 4: 129-138, 1982.

260. **Levine MS, Hull CD, Buchwald NA, Villablanca JR.** Effects of caudate nuclei or frontal cortical ablations in kittens: motor activity and visual discrimination performance in neonatal and juvenile kittens. *Exp Neurol* 62: 555-569, 1978.

261. **Lidsky TI, Buchwald NA, Hull CD, Levine MS.** A neuro-physiological analysis of the development of cortico-caudate connections in the cat. *Exp Neurol* 50: 283-292, 1976.

262. **Morris R, Levine MS, Cherubini E, Buchwald NA, Hull CD.** Intracellular analysis of the development of responses of caudate neurons to stimulation of cortex, thalamus and substantia nigra in the kitten. *Brain Res* 173: 471-487, 1979.

263. **Villablanca JR, Olmstead CE, Levine MS, Marcus RJ.** Effects of caudate nuclei or frontal cortical ablations in kittens. I. Neurology and gross behavior. *Exp Neurol* 52: 389-420, 1976.

264. **Anden NE, Grabowska-Anden M, Wachtel H.** Effects of GABA and GABA-like drugs on the brain dopamine and on the motor activity of rats, in *GABA-Neurotransmitters*. Krogsgaard-Larsen P, Scheel-Kruger J, Kofod H. Eds., Academic Press, New York, 1979, 135.

265. **Anden NE, Stock G.** Inhibitory effect of gammahydroxy-butyric acid and gammaaminobutyric acid on the dopamine cells in the substantia nigra. *Naunyn-Schmiedeberg's Arch Pharmacol* 279: 89-92, 1973.

266. **Bartholini G, Keller H, Pieri L, Pletscher A.** The effect of diazepam on the turnover of cerebral dopamine, in *The Benzodiazepines*. Garattini S, Mussini E, Randall LO. Eds., Raven Press, New York, 1973, 235-240.

267. **Cheramy A, Nieoullon A, Glowinski J.** GABA-ergic processes involved in the control of dopamine release from nigro-striatal dopaminergic neurons in the cat. *Eur J Pharmacol* 48: 281-295, 1978.

268. **Corrodi H, Fuxe K, Lidbrink P, Olson L.** Minor tranquilizers, stress and central catecholamine neurons. *Brain Res* 29: 1-16, 1971.

269. **Fadda F, Argiolas A, Melis MR, Tissari AH, Ordi PL, Gessa GL.** Stress-induced increase in 3,4-dihydroxyphenyl-acetic acid (DOPAC) levels in the cerebral cortex and in n. accumbens: reversal by diazepam. *Life Sci* 23: 2219-2224, 1978.

270. **Reinhard JP Jr, Bannon MJ, Roth RH.** Acceleration by stress of dopamine synthesis and metabolism in prefrontal cortex: Antagonism by diazepam. *Naunyn-Schmiedeberg's Arch Pharmacol* 318: 374-380, 1982.

271. **Walters JR, Lakoski JM, Eng N, Waszczak BL.** Effect of muscimol, AOAA and Na valproate on the activity of dopamine neurons and dopamine synthesis, in *GABA-Neurotransmitters*. Krogsgaard-Larsen P, Scheel-Kruger J, Kofod H. Eds., Academic Press, New York, 1979, 118.

272. **Waszczak BL, Walters JR.** Effects of GABAergic drugs on single unit activity on A9 and A10 dopamine neurons. *Brain Res Bull* 5: 465-470, 1980.

273. **Costa E, Cheney DL.** Functional interactions of neuro-transmitter systems, in *Neuroactive Drugs in Endocrinology*. Muller EE. Ed., Elsevier/North-Holland, New York, 1980, 137.

274. **Waldmeier PC, Ortmann R, Bischoff S.** Modulation of dopaminergic transmission by alpha-noradrenergic agonists and antagonists: evidence for antidopaminergic properties of some alpha antagonists. *Experientia* 38: 1168-1176, 1982.

275. **Weinstock M, Zavadil AP III, Muth EA, Crowley WR, O'Donohue TL, Jacobowitz DM, Kopin IJ.** Evidence that noradrenaline modulates the increase in striatal dopamine metabolism induced by muscarine receptor stimulation. *Eur J Pharmacol* 68: 427-435, 1980.

276. **Fennessy MR, Lee JR.** The effect of benzodiazepines on brain amine of the mouse. *Arch Int Pharmacodyn Ther* 197: 37-44, 1971.

277. **Herman JP, Guilloneau D, Dantzer R, Scatton B, Semerdjian-Rouquier L, LeMoal M.** Differential effects of inescapable footshocks and of stimuli previously paired with inescapable footshocks on dopamine turnover in cortical and limbic areas of the rat. *Life Sci* 30: 2207, 1978.

278. **Unemoto H, Sasa M, Takaori S.** Inhibition from locus coeruleus of nucleus accumbens neurons activated by hippocampal stimulation. *Brain Res* 338: 376-379, 1985.

279. **Miller JD, Speciale SG, McMillen BA, German DC.** Naloxone antagonism of stress-induced augmentation of frontal cortex dopamine metabolism. *Eur J Pharmacol* 98: 437-439, 1984.

280. **McGeer PL, McGeer EG.** Chemistry of mood and emotion. *Rev Psychol* 31: 273-307, 1980.

281. **Breese GR, Cooper BR, Mueller RA.** Evidence for the involvement of 5-hydroxytryptamine in the actions of amphetamine. *Br J Pharmacol* 52: 307-314, 1974.

282. **Conrad LCA, Leonard CM, Pfaff DW.** Connections of the median and dorsal raphe nuclei in the rat: an autoradiographic and degeneration study. *J Comp Neurol* 156: 179-206, 1974.

283. **Dray A, Davies J, Oakley NR, Tongroach P, Vellucci S.** The dorsal and median raphe projections to the substantia nigra in the rat: electrophysiological biochemical and behavioral observations. *Brain Res* 151: 431-442, 1978.

284. **Jouvet M.** The role of monoamines and acetylcholine containing neurons in the regulation of the sleep-waking cycle. *Ergeb Physiol* 64: 166-307, 1972.

285. **Kostowski W.** Interactions between serotonergic and catecholaminergic systems in the brain. *Pol J Pharmacol Pharm Suppl* 27: 15-24, 1975.

286. **Kostowski W.** Noradrenergic interactions among central neurotransmitters, in *Neurotransmitters, Receptors and Drug Action.* Essman W. Ed., Spectrum, New York, 1980, 47.

287. **Sinha AK, Henricksen S, Dement WC, Barchas JD.** Cat brain amine content during sleep. *Am J Physiol* 224: 381-383, 1973.

288. **Hirschhorn ID, Hayes RL, Rosecrans JA.** Discriminative control of behavior by electrical stimulation of the dorsal raphe nucleus: generalization to lysergic acid diethylamide (LSD). *Brain Res* 86: 134-140, 1975.

289. **Jacobs B, Cohen A.** Differential behavioral effects of lesions of the median and dorsal raphe nuclei in rats: open field and pain elicited aggression. *J Comp Physiol Psychol* 46: 102-108, 1976.

290. **Raleigh MJ, Brammer GL, McGuire MT, Yuwiler A.** Dominant social status facilitates the behavioral effects of serotonergic agonists. *Brain Res* 348: 274-282, 1985.

291. **Kostowski W, Plaznik A, Pucilowski AO, Bidzinski A, Hauptmann M.** Lesion of serotonergic neurons antagonizes clonidine-induced suppression of avoidance behavior and locomotor activity in rats. *Psychopharmacologia* 73: 261-264, 1981.

292. **Kostowski W, Giacolono EW, Garattini S, Valzelli L.** Electrical stimulation of midbrain raphe: biochemical behavioral and bioelectric effects. *Eur J Pharmacol* 7: 170-178, 1969.

293. **Kovacevic R, Radulocvacki M.** Monoamine changes in the brain of cats during slow-wave sleep. *Science* 193: 1025-1027, 1976.

294. **Ogasahara S, Taguchi Y, Wada H.** Changes in serotonin in rat brain during slow-wave sleep and paradoxical sleep: application of the microwave fixation method to sleep research. *Brain Res* 189: 570-575, 1980.

295. **Krayniak PF, Meibach RC, Siegel A.** A projection from the entorhinal cortex to the nucleus accumbens in the rat. *Brain Res* 209: 427-431, 1981.

296. **Swanson LW.** The projections of the ventral tegmental area and adjacent regions. A combined fluorescent retrograde tracer and immunofluorescence study in the rat. *Brain Res Bull* 9: 321-353, 1982.

297. **Kalivas PW, Jennes L, Miller JS.** A catecholaminergic projection from the ventral tegmental area to the diagonal band of broca: modulation by neurotensin. *Brain Res* 326: 229-238, 1985.

298. **Kobayashi RM, Palkovits M, Jacobowitz DM, Kopin IJ.** Biochemical mapping of the noradrenergic projections from the locus coeruleus. *Neurology* 25: 223-233, 1975.

299. **Korf J, Aghajanian GK, Roth RH.** Increased turnover of norepinephrine in the rat cerebral cortex during stress: role of the locus coeruleus. *Neuropharmacology* 12: 933-938, 1973.

300. **Levitt P, Moore RY.** Origin and organization of brainstem catecholamine innervation in the rat. *J Comp Neurol* 186: 505-528, 1979.

301. **Lindvall O, Bjorklund A.** The organization of the ascending catecholamine neuron systems in the rat brain as revealed by the glyoxylic acid fluorescence method. *Acta Physiol Scand Suppl* 412: 1-48, 1974.

302. **McKellar S, Loewy, AD.** Efferent projections of the A1 catecholamine cell group in the rat: an autoradiographic study. *Brain Res* 241: 11-29, 1982.

303. **Kostowski W.** Two noradrenergic systems in the brain and their interactions with other monoaminergic neurons. *Pol J Pharmacol Pharm* 31: 425-436, 1979.

304. **Lindvall O, Björklund A.** Organization of catecholamine neurons in the rat central nervous system, in *Chemical Pathways in the Brain. Handbook of Psychopharmacology*, Vol. 9. Iversen L, Iversen S, Snyder SH. Eds., Plenum Press, New York, 1978, 139.

305. **Swanson LW, Cowan WM.** A note on the connections and development of the nucleus accumbens. *Brain Res* 92: 324-330, 1975.

306. **Anisman H, Ritch M, Sklar LS.** Noradrenergic and dopaminergic interactions in escape behavior: analysis of uncontrollable stress effects. *Psychopharmacology* 74: 263-268, 1981.

307. **Antelman SM, Black CA.** Dopamine-beta-hydroxylase inhibitors (DBHI) reverse the effects of neuroleptics under activating conditions: possible evidence for a norepinephrine (NE)-dopamine (DA) interaction. *Soc Neurosci Abstr* 1977; cited by **Anisman H, Ritch M, Sklar LS.** Noradrenergic and dopaminergic interactions in escape behavior: analysis of uncontrollable stress effects. *Psychopharmacology* 74: 263-268, 1981.

308. **O'Donohue TL, Crawley WR, Jacobowitz DM.** Biochemical mapping of the noradrenergic ventral bundle projection sites: evidence for a noradrenergic-dopaminergic interaction. *Brain Res* 172: 87-100, 1979.

309. **Fratta W, Biggio G, Gessa GL.** Homosexual mounting behavior induced in male rats and rabbits by a tryptophan-free diet. *Life Sci* 21: 379-384, 1977.

310. **Gessa GL, Tagliamonte A.** Role of brain serotonin and dopamine in male sexual behavior, in *Sexual Behavior: Pharmacology and Biochemistry*. Sandler M, Gessa GL. Eds., Raven Press, New York, 1975, 117.

311. **Day TA, Oliver JR, Nemadue MF, Davis B, Willoughby JO.** Stimulatory role for medial preoptic/anterior hypothalamic area neurones in growth hormone and prolactin secretion. A kainic acid study. *Brain Res* 238: 55-63, 1982.

312. **Kimura F, Kawakami M.** Reanalysis of the preoptic afferents and efferents involved in the surge of LH, FSH and prolactin release in the proestrous rat. *Neuroendocrinology* 27: 74-85, 1978.

313. **Willoughby JO, Terry LC, Brazeau P, Martin JB.** Pulsatile growth hormone, prolactin, and thyrotropin secretion in rats with hypothalamic deafferentation. *Brain Res* 127: 133-152, 1977.

314. **Brook NM, Cookson JB.** Bromocriptine-induced mania? *Br Med J* 1: 790-792, 1978.

315. **Brown GM , Friend WC, Chambers JW.** Neuropharmacology of hypothalamic pituitary regulation, in *Clinical Neuroendocrinology: A Pathological Approach*. Telis G, Labrie F, Martin JB, Naftolin F. Eds., Raven Press, New York, 1979, 47.

316. **Cooper BR, Black WC, Paolini RM.** Decreased septal-forebrain and lateral hypothalamic reward after alpha methyl-*p*-tyrosine. *Physiol Behav* 6: 425-429, 1968.

317. **Stein L.** Neurochemistry of reward and punishment: some implications for the ethiology of schizophrenia. *J Psychiatr Res* 8: 345-361, 1971.

318. **Rolls ET, Cooper SJ.** Activation of neurones in the prefrontal cortex by brain-stimulation reward in the rat. *Brain Res* 60: 351-368, 1973.

319. **Olds J.** Pleasure centers in the brain. *Sci Am* 195: 105-116, 1956.

320. **Olds J.** Commentary, in *Brain Stimulation and Motivation: Research and Commentary*. Valenstein E. Ed., Scott, Foresman, Glenview IL, 1973, 80.

321. **Olds J, Milner P.** Positive reinforcement produced by electrical stimulation of septal area and other regions of rat brain. *J Comp Physiol Psychol* 47: 419-427, 1954.

322. **Phillips AG, Fibiger HC.** The role of dopamine in maintaining intracranial self-stimulation in the ventral tegmentum, nucleus accumbens and medial prefrontal cortex. *Can J Psychol* 32: 58-66, 1978.

323. **Phillips AG, Mora F, Rolls ET.** Intracranial self-stimulation in orbitofrontal cortex and caudate nucleus of rhesus monkey: effects of apomorphine, pimozide, and spiroperidol. *Psychopharmacologia* 62: 79-82, 1979.

324. **Prado-Alcala RA, Wise RA.** Brain stimulation reward and dopamine terminal fields. I. Caudate putamen, nucleus accumbens and amygdala. *Brain Res* 297: 265-273, 1984.

325. **Robertson A, Lafarriere AD, Franklin KBJ.** Amphetamine and increases in current intensity modulate reward in the hypothalamus and substantia nigra but not in the prefrontal cortex. *Physiol Behav* 26: 809-813, 1981.

326. **Routtenberg A.** Self-stimulation pathways: origins and terminations — a three stage technique, in *Brain Stimulation Reward*. Wauquier A, Rolls ET. Eds., Elsevier, New York, 1976, 31.

327. **Routtenberg A, Sloan M.** Self-stimulation in the frontal cortex of Rattus norvegicus. *Behav Biol* 7: 567-572, 1972.

328. **Shizgal P, Bielajew C, Corbett D, Skelton R, Yeomans J.** Behavioral methods for inferring anatomical linkage between rewarding brain stimulation sites. *J Comp Physiol Psychol* 94: 227-237, 1980.

329. **Stein L.** Chemistry of reward and punishment, in *Psychopharmacology: A Review of Progress*. Efron DH. Ed., U.S. Government Printing Office, Washington, DC, 1968, 105.

330. **Wauquier A.** The pharmacology of catecholamine involvement in the neural mechanisms of reward. *Acta Neurobiol Exp* 40: 665-686, 1980.

331. **Wise RA.** Catecholamine theories of reward: a critical review. *Brain Res* 152: 215-247, 1978.

332. **Wise RA.** Action of drugs of abuse on brain reward system. *Pharmacol Biochem Behav* 13 (Suppl 1), 213-223, 1981.

333. **Yeomans JS.** The cells and axons mediating medial forebrain bundle reward, in *The Neural Basis of Feeding and Reward*. Hoebel BG, Novin D. Eds., Haer Institute, Brunswick, ME, 1982, 405.

334. **Zarevics P, Setler PE.** Simultaneous rate-independent and rate-dependent assesment of intracranial self-stimulation: evidence for the direct involvement of dopamine in brain reinforcement mechanisms. *Brain Res* 169: 499-512, 1979.

335. **Prado-Alcala R, Streather A, Wise RA.** Brain stimulation reward and dopamine terminal fields. II. Septal and cortical projections. *Brain Res* 301: 209-219, 1984.

336. **Collier TJ, Kurtzman S, Routtenberg A.** Intracranial self-stimulation derived from entorhinal cortex. *Brain Res* 137: 188-196, 1977.

337. **Corbett D, Wise RA.** Intracranial self-stimulation in relation to the ascending dopaminergic systems of the midbrain; a moveable electrode mapping study. *Brain Res* 185: 1-15, 1980.

338. **Fibiger HC.** Drugs and reinforcement mechanisms: a critical review of the catecholamine theory. *Annu Rev Pharmacol Toxicol* 18: 37-56, 1978.

339. **Fouriezos G, Wise RA.** Pimozide-induced extinction of intracranial self-stimulation: response patterns rule out motor or performance deficits. *Brain Res* 103: 377-380, 1976.

340. **Fouriezos G, Hansson P, Wise RA.** Neuroleptic-induced attenuation of brain stimulation reward in rats. *J Comp Physiol Psychol* 92: 661-667, 1978.

341. **Franklin KBJ.** Catecholamines and self-stimulation: reward and performance effects dissociated. *Pharmacol Biochem Behav* 9: 813-820, 1978.

342. **Franklin KB, McCoy SN.** Pimozide-induced extinction in rats: stimulus control of responding rules out motor deficits. *Pharmacol Biochem Behav* 11: 71-75, 1979.

343. **Gallistel CR, Boytim M, Gomita Y, Klebanoff L.** Does pimozide block the reinforcing effect of brain stimulation? *Pharmacol Biochem Behav* 17: 769-781, 1982.

344. **Gallistel CR, Shizgal P, Yeomans JS.** A portrait of the substrate for self-stimulation. *Psychol Rev* 88: 228-273, 1981.

345. **German DC, Bowden DM.** Catecholamine systems as the neural substrate for intracranial self-stimulation: a hypothesis. *Brain Res* 73: 381-419, 1974.

346. **Goodall EB, Carey RJ.** Effects of D- versus L-amphetamine, food deprivation, or current intensity on self-stimulation of the lateral hypothalamus, substantia nigra, and media frontal cortex of the rat. *J Comp Physiol* 89: 1029-1045, 1975.

347. **Nakajima S.** Serotonergic mediation of habenular self-stimulation in the rat. *Pharmacol Biochem Behav* 20: 859-862, 1984.

348. **Deakin JFW.** On the neurochemical basis of self-stimulation with midbrain raphe electrode placements. *Pharmacol Biochem Behav* 13: 525-530, 1980.

349. **Liebman JM.** Discriminating between reward and performance: a critical review of intracranial self-stimulation methodology. *Neurosci Behav Rev* 7: 45-72, 1983.

350. **Miliaressis E.** Serotonergic basis of reward in median raphe of the rat. *Pharmacol Biochem Behav* 7: 177-180, 1977.

351. **De Guchi T, Sinha AK, Barchas JD.** Biosynthesis of serotonin in raphe nuclei of rat brain: effect of *p*-chlorophenylalanine. *J Neurochem* 20: 1329-1336, 1973.

352. **Phillips AG, Carter DA, Fibiger HC.** Differential effects of para-chlorophenylalanine on self-stimulation in caudate-putamen and lateral hipothalamus. *Psychopharmacologia* 49: 23-27, 1976.

353. **Stark P, Fuller R.** Behavioral and biochemical effects of PCPA, 3-chlorotyrosine and 3-chlorotyramine: a proposed mechanism of inhibition of self-stimulation. *Neuropharmacology* 11: 261-272, 1972.

354. **Sutherland RJ.** The dorsal diencephalic conduction system: a review of the anatomy and functions of the habenular complex. *Neurosci Biobehav Rev* 6: 1-13, 1982.

355. **Sutherland RJ, Nakajima S.** Self-stimulation of the habenular complex in the rat. *J Comp Physiol Psychol* 95: 781-791, 1981.

356. **Van der Kooy D, Fibiger HC, Phillips AG.** Monoamine involvement in hippocampal self-stimulation. *Brain Res* 136: 119-130, 1977.

357. **Van der Kooy D, Fibiger HC, Phillips AG.** An analysis of dorsal and median raphe self-stimulation: effects of para-chlorophenylalanine. *Pharmacol Biochem Behav* 8: 441-445, 1978.

358. **White NM.** Strength-duration analysis of the organization of reinforcement pathways in the medial forebrain bundle of rats. *Brain Res* 110: 575-591, 1976.

359. **Oades RD.** Search strategies on a hole-board are impaired in rats with ventral tegmental damage: animal mode for tests of thought disorder. *Biol Psychiatry* 17: 243-258, 1982.

360. **Hökfelt T, Ljungdahl A, Fuxe K, Johansson O.** Dopamine nerve terminals in the rat limbic cortex: aspects of the dopamine hypothesis of schizophrenia. *Science* 184: 177-179, 1974.

361. **Takagi, H, Shiosaka S, Tohyama M, Senba E, Sakanaka H.** Ascending components of the medial forebrain bundle from the lower brainstem in the rat, with special reference to raphe and catecholamine cell groups. A study by the HRP method. *Brain Res* 193: 315-337, 1980.

362. **Speciale SG, Crowley WR, O'Donohue TL, Jacobowitz DM.** Forebrain catecholamine projections of the A5 cell group. *Brain Res* 154: 128-133, 1978.

363. **Geyer MA, Puerto A, Dawsey WJ, Knapp S, Bullard WP, Mandell AJ.** Histologic and enzymatic studies of the mesolimbic and mesostriatal serotonergic pathways. *Brain Res* 106: 241-256, 1976.

364. **Cedarbaum JM, Aghajanian GK.** Catecholamine receptors on locus coeruleus neurons: pharmacological characterization. *Eur J Pharmacol* 44: 375-385, 1977.

365. **Crawley JN, Roth RH, Maas JW.** Locus coeruleus stimulation increases noradrenergic metabolite levels in rat spinal cord. *Brain Res* 166: 180-184, 1979.

366. **Fuxe K, Hökfelt T, Goldstein M, Jonsson G, Lindbrink K, Ljungdahl A, Sachs CH.** Topography of central catecholamine pathways. *Symp Central Action of Drugs in the Regulation of Blood Pressure*. Royal Post Graduate Medical School, London, 1975.

367. **Silver MA, Soden W, Jacobowitz D, Bloom FE.** The functional organization of CNS noradrenergic neurons. *Anat Rev* 192: 684, 1979.

368. **Svensson TH, Thoren P.** Brain noradrenergic neurons in the locus coeruleus: inhibition by blood volume load through vagal afferents. *Brain Res* 172: 174-178, 1979.

369. **Swanson LW, Hartman BK.** The central adrenergic system. An immunofluorescent study of the location of cell bodies and their efferent connections in the rat utilizing dopamine-beta-hydroxylase as a marker. *J Comp Neurol* 163: 467-506, 1975.

370. **Takigawa M, Mogenson GJ.** A study of inputs to anti-dromically identified neurons of the locus coeruleus. *Brain Res* 135: 217-230, 1977.

371. **Kostowski W, Jerlicz M, Bidzinski A, Hauptmann M.** Evidence for existence of two opposite noradrenergic brain systems controlling behavior. *Psychopharmacology* 59: 311-312, 1978.

372. **Redmond DE Jr, Huang Y, Snyder DR, Maas JW.** Behavioral effects of stimulation of the nucleus locus coeruleus in the stump-tailed monkey *Macaca arctoides*. *Brain Res* 116: 502-512, 1976.

373. **Herve D, Blanc G, Glowinski J, Tassin JP.** Reduction of dopamine utilization in the prefrontal cortex but not in the nucleus accumbens after selective destruction of noradrenergic fibers innervating the ventral tegmental area in the rat. *Brain Res* 237: 510-516, 1982.

374. **Herve D, Simon H, Blanc G, LeMoal M, Glowinski J, Tassin JP.** Opposite changes in dopamine utilization in the nucleus accumbens and the frontal cortex after electrolytic lesion of the median raphe in the rat. *Brain Res* 216: 422-428, 1981.

375. **Lavielle S, Tassin JP, Thierry AM, Blanc G, Herve D, Barthelemy C, Glowinski J.** Blockade by benzodiazepines of the selective high increase in dopamine turnover induced by stress in mesocortical dopaminergic neurons of the rat. *Brain Res* 168: 585-594, 1978.

376. **Przuntek H, Guimaraes S, Philippu A.** Importance of adrenergic neurons of the brain for the rise of blood pressure evoked by hypothalamic stimulation. *Naunyn-Schmiedeberg's Arch Pharmacol* 271: 311-319, 1971.

377. **Przuntek H, Philippu A.** Reduced pressor responses to stimulation of the locus coeruleus after lesion of the posterior hypothalamus. *Naunyn-Schmiedeberg's Arch Exp Pathol Pharmakol* 276: 119-122, 1973.

378. **Andrade R, Aghajanian GK.** Single cell activity in the noradrenergic A-5 region: responses to drugs and peripheral manipulations of blood pressure. *Brain Res* 242: 125-135, 1982.

379. **Blessing WW, Reis DJ.** Inhibitory cardiovascular function of neurons in the caudal ventrolateral medulla of the rabbit: relationship to the area containing A1 noradrenergic cells. *Brain Res* 253: 161-171, 1982.

380. **Granata AR, Kumada M, Reis DJ.** Sympathoinhibition by A1-noradrenergic neurons is mediated by neurons in the C1 area of the rostral medulla. *J Auton Nerv Syst* 13: 387-395, 1985.

381. **Loewy DA, Mckellar S.** Serotonergic projections from the ventral medulla to the intermediolateral cell column in the rat. *Brain Res* 211: 146-152, 1981.

382. **Loewy DA, Mckellar S, Saper CB.** Direct projections from the A5 catecholamine cell group to the intermediolateral cell column. *Brain Res* 174: 309-314, 1979.

383. **Loewy DA, Neil IJ.** The role of descending monoaminergic systems in the central control of blood pressure. *Fed Proc Fed Am Soc Exp Biol* 40: 2778-2785, 1981.

384. **Tucker D, Saper C.** Specificity of spinal projections from hypothalamic and brainstem areas which innervate sympathetic preganglionic neurons. *Brain Res* 360: 159-164, 1985.

385. **Lechin F, Gómez F, van der Dijs B, Lechín E.** Distal colon motility in schizophrenic patients. *J Clin Pharmacol* 20: 459-465, 1980.

386. **Angrist BRJ.** Dopaminergic and non-dopaminergic elements in schizophrenia, in *Apomorphine and Other Dopaminomimetics,* Vol. 2. Corsini GV, Gessa GL. Eds., Raven Press, New York, 1981, 33.

387. **Bowers MB.** Central dopamine turnover in schizophrenic syndromes. *Arch Gen Psychiatry* 31: 50-54, 1974.

388. **Post RM, Fink E, Carpenter WT, Goodwin FK.** Cerebrospinal fluid amine metabolites in acute schizophrenia. *Arch Gen Psychiatry* 32: 1063-1069, 1975.

389. **Mackay AVP, Iversen LL, Rossor M, Spokes E, Bird E, Arregui A, Creese I, Snyder S.** Increased brain dopamine and dopamine receptors in schizophrenia. *Arch Gen Psychiatry* 39: 991-997, 1982.

390. **Lee T, Seeman P, Tourtelotte WW.** Binding of ^{3}H-neuroleptics and ^{3}H-apomorphine in schizophrenia brains. *Nature* 274: 897-900, 1978.

391. **Bagdy G, Perényi A, Frecska E, Révai K, Papp Z, Fekete MIK, Arató M.** Decrease in dopamine, its metabolites and noradrenaline in cerebrospinal fluid of schizophrenic patients after withdrawal of long-term neuroleptic treatment. *Psychopharmacology* 85: 62-64, 1985.

392. **Lee T, Seeman P.** Elevation of brain neuroleptics/dopamine receptors in schizophrenia. *Am J Psychiatry* 137: 191-197, 1980.

393. **Reisine TD, Rossor M, Spokes E.** Opiate and neuroleptic receptor alterations in human schizophrenic brain tissue, in *Receptors for Neurotransmitters and Peptide Hormones*. Pepeu G, Kuhar MJ, Enna SJ. Eds., Raven Press, New York, 1980, 443.

394. **Owen F, Crow TJ, Poulter M.** Increased dopamine-receptor sensitivity in schizophrenia. *Lancet* 2: 223-225, 1978.

395. **Bird ED, Crow TJ, Iversen LL.** Dopamine and homovanillic acid concentrations in the post-mortem brain in schizophrenia. *J Physiol* 293: 36-37, 1979.

396. **Cross AJ, Crow TJ, Longden A, Poulter M, Riley TJ.** Evidence for increased dopamine receptor sensitivity in post-mortem brains from patients with schizophrenia. *J Physiol* 28: 37-43, 1978.

397. **Crow TJ, Baker HF, Cross AJ, Joseph MH, Lofthouse R, Longden A, Owen F, Riley GJ, Glover V, Killpack WS.** Monoamine metabolism in chronic schizophrenia: postmortem neurochemical findings. *Br J Psychiatry* 134: 249-254, 1979.

398. **Farley IJ, Shannak KS, Hornykiewicz O.** Brain monoamine changes in chronic paranoid schizophrenia and their possible relation to increased dopamine receptor sensitivity, in *Receptors for Neurotransmitters and Peptide Hormones*. Pepeu G, Kuhar MJ, Enna SJ. Eds., Raven Press, New York, 1980, 427.

399. **Crow TJ.** Molecular pathology in schizophrenia: more than one disease process? *Br Med J* 280: 66-68, 1980.

400. **Iversen LL.** Biochemical and pharmacological studies: the dopamine hypothesis, in *Schizophrenia Towards a New Synthesis*. Wing J. Ed., Academic Press, New York, 1978, 89.

401. **Burt DR, Creese I, Snyder SH.** Antischizophrenic drugs: chronic treatment elevates dopamine receptor binding in brain. *Science* 196: 326-328, 1977.

402. **Mackay AVP, Bird ED, Spokes EG.** Dopamine receptors and schizophrenia: drug effect or illness? *Lancet* 2: 915-916, 1980.

403. **Seeman P.** Dopamine receptors in post-mortem schizophrenic brains. *Lancet* 2: 1130, 1981.

404. **Iversen LL, Mackay AVP.** Brain dopamine receptor densities in schizophrenics. *Lancet* 2: 149, 1981.

405. **Lechin F, van der Dijs B, Gómez F, Vall JM, Acosta E, Arocha L.** Pharmacomanometric studies of colonic motility as a guide to the chemotherapy of schizophrenia. *J Clin Pharmacol* 20: 664-171, 1980.

406. **Tamminga CA, De Fraites EG, Gotts MD, Chase TN.** Apomorphine and *N-n*-propylnorapomorphine in the treatment of schizophrenia, in *Apomorphine and Other Dopaminomimetics*, Vol. 2. Corsini GU, Gessa GL. Eds., Raven Press, New York, 1981, 49.

407. **Cutler NR, Jeste DV, Karoum F, Wyatt RJ.** Low-dose apomorphine reduces serum homovanillic acid concentrations in schizophrenic patients. *Life Sci* 30: 753-756, 1981.

408. **Anden NE, Strömbom U.** Adrenergic receptor blocking agents: effects on central noradrenaline and dopamine receptors and on motor activity. *Psychopharmacologia* 38: 91-103, 1974.

409. **Anden NE, Pauksens K, Svensson K.** Selective blockade of brain alpha2-autoreceptors by yohimbine: effects on motor activity and on turnover of noradrenaline and dopamine. *J Neural Transm* 55: 111-120, 1982.

410. **Donaldson IMcG, Dolphin A, Jenner P, Marsden CD, Pycock C.** The roles of noradrenaline and dopamine in contraversive circling behavior seen after unilateral electrolytic lesions of the locus coeruleus. *Eur J Pharmacol* 39: 179-191, 1976.

411. **Bacopoulos NC, Spokes EG, Bird ED, Roth RH.** Antipsychotic drug action in schizophrenic patients: effect on cortical dopamine metabolism after long term treatment. *Science* 205: 1405-1407, 1979.

412. **Bannon MJ, Reinhard JF Jr, Bunney EB, Roth RH.** Unique response to antipsychotic drugs is due to absence of terminal autoreceptors in mesocortical dopamine neurones. *Nature* 296: 444-446, 1982.

413. **Simon H, Scatton B, LeMoal M.** Dopaminergic A10 neurons are involved in cognitive functions. *Nature* 286: 150-151, 1980.

414. **Bunney BS, Aghajanian GK.** Mesolimbic and mesocortical dopaminergic systems: physiology and pharmacology, in *Psychopharmacology: A Generation of Progress*. Lipton MA, DiMascio A, Killam KF. Eds. Raven Press, New York, 1978, 221.

415. **Moore KE, Kelly PH.** Biochemical pharmacology of mesolimbic and mesocortical dopaminergic neurons, in *Psychopharmacology: A Generation of Progress*. Lipton MA, DiMascio A, Killam KF. Eds., Raven Press, New York, 1978, 221.

416. **Nicolaou NM, García-Munoz M, Arbuthnott G, Eccleston D.** Interactions between serotonergic and dopaminergic systems in rat brain demonstrated by small unilateral lesions of the raphe nuclei. *Eur J Pharmacol* 57: 295-305, 1979.

417. **Samanin R, Quattrone A, Consolo S, Ladinsky H, Algeri S.** Biochemical and pharmacological evidence of the interaction of serotonin with other aminergic systems in the brain, in *Interactions Between Putative Neurotransmitters*. Garattini S, Pujol JF, Samanin R. Eds., Raven Press, New York, 1978, 355.

418. **Phillipson OT.** Afferent projections to the ventral tegmental area of Tsay and interfascicular nucleus: a horseradish peroxidase study in the rat. *J Comp Neurol* 187: 117-144, 1979.

419. **Simon H, LeMoal M, Calas A.** Efferents and afferents of the ventral tegmental A10 region studies after local injection of (3H) leucine and horseradish peroxidase. *Brain Res* 178: 17-40, 1979.

420. **Wiklund L.** Studies on Anatomical, Functional, and Plastic Properties of Central Serotonergic Neurons. Doctoral dissertation, University of Lund, Sweden, 1980.

421. **Beart PM, McDonald D.** 5-Hydroxytryptamine and 5-hydroxy-tryptaminergic-dopaminergic interactions in the ventral tegmental area of rat brain. *J Pharm Pharmacol* 34: 591-593, 1982.

422. **Lyness WH, Moore KE.** Destruction of 5-hydroxy-tryptaminergic neurons and the dynamics of dopamine in nucleus accumbens septi and other forebrain regions of the rat. *Neuropharmacology* 20: 327-334, 1981.

423. **Herve D, Simon H, Blanc G, Lisoprawski A, LeMoal M, Glowinski J, Tassin JP.** Increased utilization of dopamine in the nucleus accumbens but not in the cerebral cortex after dorsal raphe lesion in the rat. *Neurosci Lett* 15: 127-134, 1979.

424. **Andrews DW, Patrick RL, Barchas JD.** The effects of 5-hydroxytryptophan and 5-hydroxytryptamine on dopamine synthesis and release in rat brain striatal synaptosomes. *J Neurochem* 30: 465-470, 1978.

425. **Cochran E, Robins E, Grote S.** Regional serotonin levels in brain: a comparison of depressive suicides and alcoholic suicides with controls. *Biol Psychiatry* 11: 283-294, 1976.

426. **Murphy DL, Campbell IC, Costa JL.** The brain serotonergic system in the affective disorders, *Prog Neuropsychopharmacol* 2: 1-31, 1978.

427. **Assaf SY, Miller JJ.** The role of a raphe serotonin system in the control of septal unit activity and hippocampal desynchronization. *Neuroscience* 3: 539-550, 1978.

428. **Lechin F, van der Dijs B.** Clonidine therapy for psychosis and tardive dyskinesia. *Am J Psychiatry* 138: 3, 1981.

429. **Lechin F, van der Dijs B.** Noradrenergic or dopaminergic activity in chronic schizophrenia? *Br J Psychiatry* 139: 472, 1981.

Chapter 5

CENTRAL NERVOUS SYSTEM CIRCUITRY INVOLVED IN BLOOD PRESSURE REGULATION

Fuad Lechin, Bertha van der Dijs, José Amat, Simón Villa, and Alex E. Lechin

TABLE OF CONTENTS

I. NORADRENERGIC SYSTEM

The pontine-medullary noradrenergic (NE) cell groups are the source of hypothalamic and spinal norepinephrine (NE). The action of this neurotransmitter at central level augments or decreases peripheral sympathetic activity which is dependent on: (1) NE released from sympathetic nerves and (2) norepinephrine (NE) + epinephrine (E) released from adrenal glands.[1-215]

A positive correlation exists between blood pressure and NE plasma levels. Furthermore, NE plasma levels during resting state and supine position are accepted as reflecting NE released from sympathetic terminals, only; whereas in standing and emotional states adrenal catecholamines are also stimulated.[131-223]

Sympathetic nerves are axons of postganglionic neurons. The NE postganglionic cells receive excitatory input from preganglionic sympathetic axons, located at central level, whose cell bodies (cholinergic in nature) are found in (1) intermediolateral spinal horn (IML), (2) lateral medullary formation, (3) the ventrolateral medullary nucleus known as nucleus reticularis lateralis (NRL), and (4) posterior hypothalamus.[148-215,220-324]

Adrenal catecholamines (NE, E, and dopamine, DA) are released by the adrenal glands. These glands receive two kinds of excitatory inputs, one cholinergic and the other β-adrenergic. Cholinergic input consists in plasma-circulating acetylcholine (ACh) as well as ACh released by sympathetic preganglionic axons whose cell bodies are located in the medulla oblongata and spinal cord. β-Adrenergic excitatory input consists in circulating catecholamines (NE and E) plus NE released by preganglionic sympathetic axons. Other excitatory input comes from circulating serotonin and histamine. Secretion of catecholamines by adrenal glands is inhibited by α-adrenergic agonists (NE preferentially). Thus, NE has a dual effect on adrenal secretion: both β-excitatory and α-inhibitory.[151-215,220-223,267-380]

Adrenal glands also receive excitatory serotonergic (5HT) and histaminergic inputs. With regard to this, it has been demonstrated that postganglionic sympathetic axons possess serotonin and histamine receptors.[198-215]

There is abundant evidence showing that central NE activity can stimulate or reduce peripheral sympathetic activity, depending on where in the central region NE is released or microinjected. In effect, NE reduces peripheral sympathetic activity and blood pressure (BP) when injected or released in anterior hypothalamus, IML spinal horn, nucleus tractus solitarius (NTS) and nucleus reticularis lateralis (NRL), lateral medullary formation, and C1 cell group. This cell consists in epinephrinergic neurons located rostrally to NE-A1 cell group. This latter NE nucleus sends inhibitory axons to E-C1 cell group. Norepinephrine released or microinjected in all these preganglionic sympathetic areas exerts peripheral sympathetic inhibition through stimulation of α_2- adrenoceptors located in the soma of these cholinergic neurons. On the other hand, α_1- and β-adrenergic stimulation of these preganglionic sympathetic areas induces peripheral sympathetic hyperactivity and BP increase. Moreover, electrical stimulation of all the above regions parallels NE stimulation. Finally, NE and electrical stimulation of discrete hypothalamic and medullary regions is able to increase or decrease adrenal catecholamines secretion.[14-134,138-140,145-150,155-223,232-284,288-300,303,306-324,334,335,340-345,350-356,362-366,373,376-403]

The anterior hypothalamus, posterior hypothalamus, IML spinal horn, NRL, lateral reticular formation, and C1 cell group are the projection areas of pontine-medullary NE cell groups. The fact that stimulation or inhibition of these NE and E neurons parallels α-adrenergic stimulation or blockade at their projection areas sheds light on the understanding of central NE circuitry involved in peripheral sympathetic activity. However, the overlapping in NE innervation and the interconnections between different NE nuclei prevent our complete comprehension of this matter. In all events, there exist enough well-documented findings to justify the postulation of some central NE circuit involvement in the regulation of BP and peripheral sympathetic activity.

A. Central Noradrenergic Sympathoexcitatory System (Figure 1)

A growing body of evidence suggests that central NE neurons are intimately involved in cardiovascular control. However, some suggest that NE input to central sympathetic neurons is excitatory, while others imply that the input is inhibitory.

The central acting antihypertensive drug clonidine has been postulated to produce its effect either by acting as an α_2-adrenergic autoreceptor agonist or by mimicking the effects of norepinephrine at postsynaptic α_2-receptor sites to decrease activity in central sympathetic pathways. This implies that central NE neurons either act directly to facilitate transmission in central sympathetic neurons or indirectly to antagonize an inhibitory input to sympathetic neurons.

Some data indicate that the specific α_1-antagonist prazosin acts centrally to reduce sympathetic nervous discharge (SND), when intraventricularly injected. Such observations suggest that NE neurons facilitate central sympathetic outflow and that clonidine acts by decreasing activity in noradrenergic pathways (disfacilitation). These findings are consistent with others showing that α_2-antagonists increase central NE turnover as well as the firing rate of some NE neurons. Further, a broad range of i.v. doses of piperoxane and rauwolscine (two α_2 blocking agents) increases sympathetic nervous discharge. These results are compatible with the hypothesis that at least some central NE neurons facilitate sympathetic nervous discharge.[3,6-28,33-41,46-50,53-60,65-77,82-107,112-116,121,122,127-130,148-199,206-228,232-237,244,251,254,257,260,262,265-271,275-284,288-297,303,306-311,316-324,328-335,342,343,354-356,363-366,373-380,400-420]

The pontine NE nucleus, locus coeruleus (LC), has a widely spread terminal distribution covering almost the entire central nervous system including the cerebral cortex, hippocampus, cerebellum, hypothalamus, medulla oblongata, pontis, and spinal cord. A number of physiological roles has been ascribed to the LC including maintenance of wakefulness and arousal and regulation or modulation of autonomic, e.g., cardiovascular functions. Electrical stimulation of LC elicits pressor response, tachycardia, and anxiety. An enhancement of NE release in all brain regions innervated by LC, as well as an increase of NE and its metabolites (MHPG and others) in the CSF, has also been registered after LC stimulation. The fact that CSF-NE and CSF-MHPG are closely correlated with plasma-NE and plasma-MHPG gives support to the presumption that LC activity is positively correlated with peripheral sympathetic activity. This hypothesis is reinforced by the finding that acute and chronic administration of drugs which reduce the firing of LC neurons induce a decrease of all the autonomic, behavioral, and biochemical changes provoked by LC stimulation.[248,385,391,418-420]

The pressor response induced by LC stimulation seems to be mediated through the posterior hypothalamus, since the same is abolished by destruction of this region. This response apparently results from stimulation of an ascending fiber system: LC-posterior hypothalamus. However, LC projects to the ventrolateral reticular formation and the dorsal medullar region levels where the NE-A1 and NE-A2 cell groups are respectively located. These NE cell groups, whose activity seems to be associated with sympathetic inhibition, have been postulated to possess α_2 inhibitory autoreceptors at the somatodendritic area; thus all NE input to these nuclei should induce reduction of NE-A1 and NE-A2 activity.[1-130,149-169,172-179,183-194,200-202,207,213,214,221,225,227,231,235,254,266-276,282-284,288,292,295,297,300,303,309,316,321-324,334,343,380,382]

Experiments demonstrate that LC axons (dorsal NE bundle = DNB) display opposite effects to NE-A1 (ventral NE bundle = VNB) and that an important NE fiber system arising from NE-A1 cell group reaches NE-A6 (LC). Other fiber systems reach LC arising from the dorsal vagal complex which is known to send inhibitory fiber systems to LC since stimulation of the vagus nerve produces inhibition of LC neurons. Similarly, it has been demonstrated that blood volume load clearly inhibits single units of the LC in rats, an effect which is mediated via vagal afferents. This blood volume-dependent effect lasted at least as long as recordings were maintained from LC neurons. Moreover, when the same amount of blood was withdrawn from the vein, LC neurons returned to their previous firing rates. Conversely, LC neurons are activated by blood

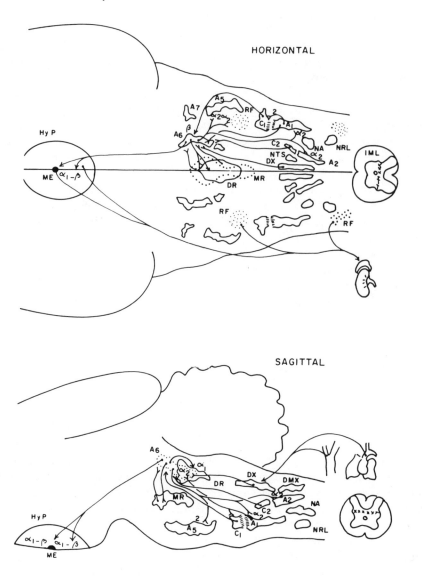

FIGURE 1. Sympathoexcitatory circuitry. The two main central sympathoexcitatory cell group are the NE neurons located in the dorsomedial part of the midbrain and pontis, e.g., locus coeruleus complex (LC) = A6 + A4 cell groups, and the adrenergic neurons located in the lateral ventromedullary C1 cell group rostral to NE-A1 cell group. NE-LC neurons exert sympathoexcitatory activity through direct monosynaptic as well as indirect polysynaptic mechanisms. Direct mechanisms consist in stimulation of sympathetic preganglionic neurons located in posterior hypothalamus inside the blood brain barrier (BBB), in median eminence (ME) outside the BBB, and in the reticular formation of the pontis (RF). The NE-LC axons stimulate these sympathetic preganglionic neurons through β- and α_1-postsynaptic adrenoceptors. Indirect mechanisms involve NE-LC axons which project to other NE cell groups (A1, A2, and A5). At these postsynaptic levels the release of norepinephrine by LC axons induces reduction of firing activity of these NE neurons. This inhibitory effect is mediated through α_2 adrenoceptors. NE neurons located in A1, A2 and A5 cell groups are able to display sympathoinhibitory activity; therefore bridling of these neurons results in sympathoexcitation. Another indirect mechanism is exerted by NE-LC neurons which stimulate 5HT DR neurons through α_1 adrenoceptors. In effect, stimulation of DR induces a brief increase in BP followed by longer lasting raised BP. The first increase is exerted through direct stimulation of sympathetic preganglionic neurons, whereas the second is triggered by plasma renin activity. E-C1 neurons display sympathoexcitatory activity through axons which bridle medullary as well as hypothalamic preganglionic parasympathetic neurons. E-C1 neurons are normally bridled by NE-A1 axons. Hence, inhibition of NE-A1 neurons occurring during NE-LC activity results in disinhibition of E-C1 neurons. HYP (hypothalamus), NTS (nucleus tractus solitarii), NA (nucleus ambiguus), NRL (nucleus reticularis lateralis), DX (sensory vagal neurons), DMX (dorsal motor vagal neurons), IML (intermediolateral column), ($\longrightarrow$) excitatory inputs, and ($\longrightarrow\!\!\!<$) inhibitory inputs.

loss. Since activation of LC in primates specifically seems to be associated with increased loss. apprehensiveness, restlessness, and anxiety reactions, the increased firing rate of LC units obtained in response to blood loss may provide part of the neurochemical basis for such reactions in man during hemorrhagic shock. Bilateral vagotomy readily reversed the load-induced inhibition of LC neurons to approximately their previous spontaneous firing rate. Moreover, after vagotomy LC neurons did not respond to all subsequent transfusions. Consequently, vagal afferents in all probability largely mediate the inhibition of LC neurons during increase of blood volume.[1-18,24-27,30-39,41-71,76-95,100-134,140,145,148-202,207-209,213-223,226,228-255,260-283,287-296,300-321,325,331,334,344,350,362-366,373,376-378,383-397,421-425]

Data obtained from hypertensive patients show that their elevated CSF-NE levels are reduced by clonidine, a drug which also sharply reduces their NE plasma levels. Elevated CSF and plasma levels suggest hypertensive patients have increased central NE activity. Since NE released at IML spinal horn, anterior hypothalamus, NRL, lateral reticular formation, C1 cell group, and other sympathetic preganglionic neurons exerts an inhibitory role, NE released in the CSF in hypertensive patients would be provided by NE axons other than those reaching these areas. The most probable candidates would be those NE axons innervating thelencephalic, mesencephalic, or diencephalic structures, i.e., LC axons.[2,3,6,9,15-21,27-30,35-46,51,52,62-79,83,84,89,94,100-105,109,112-130,136,141-143,149,150,155-165,180-188,192-194,198-202,207-214,233,239,254,264,266,272,273,279,307,310,323,324,345,380,400,403,420,426-431]

Summarizing, the pontine NE nucleus, LC, seems to be the most probable candidate for exerting central sympathoexcitatory influence among all the medullary-pontine NE nuclei. In effect, experiments show that spontaneously hypertensive rats have elevated CSF-NE levels and that clonidine not only reduces these levels but also BP. Moreover, electrical stimulation of LC provokes greater pressor response in normotensive than in hypertensive rats, a fact which might be related to greater basal activity of LC neurons in these hypertensive rats.

B. Central Noradrenergic-Sympathoinhibitory System (Figure 2)

It has been demonstrated that clonidine, an α_2-agonist, can reduce sympathetic nervous discharge in catecholamine-depleted animals. This suggests that endogenous NE is not required for the central cardiovascular action of clonidine and that this action is mediated by postsynaptic α_2-receptors. Together, these data suggest that NE input to central sympathetic neurons is inhibitory. The net effect of clonidine would result from direct inhibition of central sympathetic preganglionic neurons (postsynaptic effect) and from inhibition of sympathoexcitatory NE neurons (presynaptic effect). A third possibility would arise from the proven existence of polysynaptic pathways composed of two NE neurons and a third neuron (sympathetic preganglionic) of cholinergic nature. With respect to the third possibility, it should be remembered that NE-LC neurons send inhibitory axons to A1, A2, and A5-NE neurons which in turn send axons to sympathetic preganglionic neurons located in NRL, IML, posterior hypothalamus, lateral reticular formation, etc. Clonidine is able to inhibit all NE neurons as well as the final neuron: sympathetic preganglionic. Consequently, the effect of clonidine would depend on inhibition of that neuron in the chains of central sympathetic neurons: NE-A6 — NE-A1 — NE-A2 — NE-A6 — NE-A5; NE-A2 — NE-A5 — ACh-IML; NE-A1 — ACh-NRL; NE-A1 — ACh-RF; and NE-A1 — ACh-AH (IML: intermediolateral column, and NRL: nucleus reticularis lateralis = sympathetic preganglionic neurons; RF: reticular formation; and AH: anterior hypothalamus) which is most active at the moment of clonidine administration. This fact might explain the paradoxical pressor effect registered in some subjects after clonidine administration.

Although clonidine reduces sympathetic nervous discharge in catecholamine-depleted animals (postsynaptic effect), the dose required for sympathetic nervous discharge inhibition is at least three times greater than that in control animals. Therefore, only the presynaptic effect would be obtained in normal animals after the administration of low (therapeutic) doses of the drug.[1-130,148-215,221-226,235,248-260,265-324,331,334,335,342,343,354-356,362-366,373,376-406,409-422]

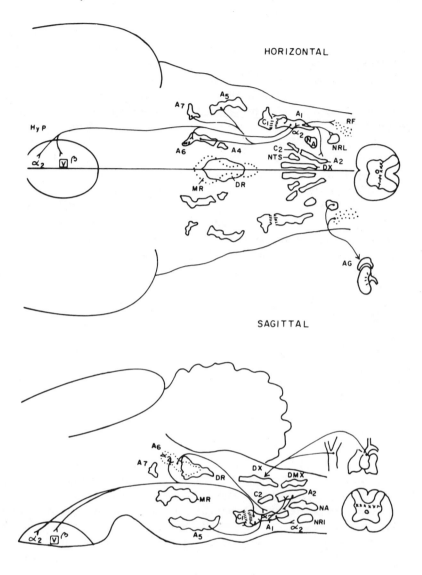

FIGURE 2.   Sympathoinhibitory circuitry. The main sympathoinhibitory system is centered in the NE lateral ventromedullary A1 cell group. These neurons send NE axons to NE-A6 cell group or LC, which is the main sympathoexcitatory nucleus. Norepinephrine released by NE-A1 axons at LC level reduces the NE-LC neuronal firing rate through α_2-adrenoceptors located in somato-dendritic area of the latter neurons. Other NE-A1 axons reach the dorsomedial medullary NE-A2 cell group which like LC, also possesses some sympathoexcitatory activity. Other indirect mechanisms are exerted through NE-A1 axons project to epinephrine (E) neurons in C1 cell group located just rostrally to A1 cell group. These E neurons display sympathoexcitatory activity through β-mediated stimulation of NE-LC neurons, as well as α_2-mediated inhibition of parasympathetic medullary preganglionic neurons. However, the main sympathoinhibitory activity exerted by NE-A1 cell group is mediated by its axons reaching sympathetic preganglionic neurons (vasopressure areas) in nucleus reticularis lateralis (NRL), medullary reticular formation (RF), and anterior hypothalamus. All these effects are mediated through α_2- adrenoceptors. Such sympathoinhibition by NE-A1 axons at medullary and hypothalamic areas is antagonized by serotonin (5HT) released at these levels by 5HT axons coming from 5HT neurons in the medulla oblongata, i.e., raphe magnus nucleus. In effect, although serotonin released in these areas is not able to provoke elevation of BP, it is able to annul the BP decrease induced by norepinephrine microinjected in the same areas. Another sympathoinhibitory effect displayed by NE-A1 neurons is exerted by axons reaching vasopressinergic neurons (magnocellular division) of the hypothalamic paraventricular nucleus; norepineph-rine released here by A1 axons reduces vasopressin plasma levels. This effect is mediated by β– adrenoceptors since it is blocked by β–adrenergic blocking agents. V (vasopressinergic neurons), NTS (nucleus tractus solitarii), NA (nucleus ambiguus), MR (median raphe), DR (dorsal raphe), DX (dorsal sensory vagal neurons), DMX (dorsal motor vagal neurons), AG (adrenal glands), HYP (hypothalamus), (——➤) excitatory inputs, and (——<) inhibitory inputs.

The existence of a central NE-sympathoinhibitory system is reinforced by findings showing that a reduction in NE content is reported in whole medulla oblongata during hypertension induced by DOCA salt. This NE reduction is not a consequence of elevated BP since it persists after BP returns to normal values following cervical cord transection. Similar findings registered in experimental renovascular hypertension strengthen the same hypothesis, namely, that lowered NE concentration in some central neurons is directly responsible for increased activity in peripheral sympathetic nerves. Compatible with this are the reduction in NE concentration registered in NRL, a pressor area receiving NE axons from A1 cell group and, further, the ability of α_2-agonists injected in NRL to provoke greatly lowered BP. Finally, taking into account the rise in BP and plasma NE in renovascular hypertension in rats, which is reversed by intracisternal 6-OHDA, it is clear that cells other than medullary NE neurons are responsible for the observed hypersympathetic activity. Thus, the assumption becomes logical that two NE systems coexist in these animals — hyperactive NE (LC) and hypoactive NE (medullary NE). Thus, peripheral sympathetic hyperactivity would result from the addition of two factors: (1) an enhancement of central NE sympathoexcitatory system and (2) a reduction of central NE sympathoinhibitory system.[3-130,148-223,226-275,280-355,362-366,373,376-378,401-422]

C. Anatomical, Physiological, and Pharmacological Basis Supporting the Existence of Two Central Noradrenergic Systems: Sympathoexcitatory and Sympathoinhibitory

The two medullary NE cell groups, A1 (ventrolateral) and A2 (dorsomedial) are associated with vasodepressor and vasopressor roles, respectively. Thus, this pair of NE nuclei may be postulated as two opposite poles in BP regulatory mechanisms.

The NE-A1 cell group sends axons to the medullary NRL, medullary reticular formation, medullary epinephrine cell group (C1), medullary NE-A2 cell group, anterior hypothalamus, hypothalamic paraventricular nucleus, and the pontine NE-A6 (LC) nucleus. Norepinephrine and other α_2-agonists such as clonidine, microinjected in these projection areas of NE-A1 axons, provoke strong reduction of BP and peripheral sympathetic activity. So, the NE-A1 cell group can be considered as a nucleus whose activity is positively correlated with vasodepression. In addition, NE-A1 axons provide the main NE innervation to magnocellular neurons of the hypothalamic paraventricular nucleus which are vasopressinergic; NE-A1 axons play a modulatory role, mainly inhibitory. In effect, destruction of NE-A1 cell group or of the VNB, which includes NE-A1 axons reaching paraventricular nucleus, induces an increase of plasma level vasopressin. β-Adrenergic receptors mediate the NE-A1 inhibition, since propranolol, a β-adrenergic blocking agent, interferes with it. In other words, NE-A1 activity can decrease vasopressinergic activity which is vasopressor in nature.[3-84,87-130,148,152-175,179-196,225,227,234,236,240-283,287-316,319-342,353-356,362-366,373-380,393,394,414]

The NE-A1 cell group receives NE axons from A5, A6, A7, and subcoeruleus cell groups. The activity of these pontine nuclei thus releases norepinephrine in NE-A1 nucleus and, because all NE cell bodies possess α_2-inhibitory autoreceptors, also lowers activity of NE-A1 neurons, which are sympathoinhibitory.

In addition, the NE-A1 cell group receives fibers from NE-A2 medullary cell group which is not NE in nature. Up to the present, no data have been published on the role displayed by this A2-A1 fiber system.

Some investigators suggest that NE-A1 axons reach sympathetic preganglionic neurons located in IML spinal horn; however, this postulation has been refuted by others. The catecholaminergic fibers which reach IML would appear to be epinephrinergic axons arising from medullary E-C1 cell group which is closely located to NE-A1 cell group.[3,6-41,46-90,94-137,146-215,220-255,259-292,296-327,334-338,343,353,363-366,373-382,414]

Located in the caudal part of NTS, the NE-A2 cell group sends axons to preganglionic vagal neurons in medulla oblongata (nuclei dorsal motor of vagus and ambiguus) and in lateral hypothalamus. Taking into account that NE released on parasympathetic preganglionic neurons exerts an inhibitory influence, NE-A2 activity would result in parasympathetic inhibition and vasopressor response. This vasopressor effect might also be mediated through inhibition of NE-A5 neurons which receive dopamine-β-hydroxylase (DBH)-positive axons (catecholaminergic) arising from NE-A2 nucleus. Since NE-A5 neurons reduce sympathetic activity through A5-IML spinal horn, the A2-A5 fiber system would display an indirect sympathoexcitatory activity.

We have mentioned that NE-A2 nucleus receives DBH-positive axons from NE-A1 nucleus. These NE axons along with another fiber system arising from NE-A5 nucleus constitute two important sources of NE input to NE-A2 nucleus. The finding that microinjection of NE in NE-A2 nucleus is able to induce reduction of BP and peripheral sympathetic activity is compatible with postulations that NE input to NE-A2 nucleus provokes vasodepression and that NE-A2 activity is positively correlated with vasopression, in this situation.

NE-A2 nucleus also sends axons to the parvocellular division of paraventricular nucleus in the hypothalamus, which is related to oxytocinergic neurons. These oxytocinergic neurons are also involved in BP and other autonomic function regulations.[1-255,259-293,296-303,306-310,314-394,401,407,408,414,422-429,432-435]

D. Possible Physiological Role of Pontine Noradrenergic Nuclei (A5 and A6)

The NE-A6 (LC) nucleus is located in the dorsoparamedial region of midbrain and pontine regions. It exerts a vasopressor effect mediated through projections to posterior hypothalamus. In turn, this effect seems to be mediated through postsynaptic α_1-adrenoceptors.

Locus coeruleus may also exert sympathoexcitatory effects through polysynaptic mechanisms. In effect, LC axons project to DR, amygdala, hippocampus, septum, and hypothalamic nuclei located outside the blood brain barrier (BBB), i.e., median eminence. All these central structures are involved in the regulation of autonomic functions.

Locus coeruleus axons also reach NE-A1 which exerts vasodepressor effects. Thus, NE released from LC terminals at NE-A1 nucleus would result in reduced NE-A1 activity and raised BP.

Despite postulations that LC sends axons to dorsal motor nucleus and nucleus ambiguus, both vagal motor nuclei, it has been amply demonstrated that LC does not send axons to motor nuclei, but only to sensory and association nuclei.[1-130,134,140-256,260,261,266-336,341-373,376-380,396-399,421-425,429,432-436]

Well-documented findings suggest that LC receives inhibitory fibers from sensory nucleus of the vagus, as well as excitatory input from NTS which in turn is part of the dorsal vagal complex. This excitatory fiber system seems to be epinephrinergic, arising from C2 cell group; the excitatory effect would be exerted through β-adrenergic receptors located in LC neurons, since propranolol interferes with excitation.[1-5,8-136,140-203,207-255,260,265-309,312-362,367-375,379-399,412-414,421-429]

According to the above findings, NE-A6 and NE-A1 cell groups would behave as antagonistic NE poles in such a way that hyperactivity of one results in hypoactivity of the other. In support of this, many experimental findings ratify the existence of a clearcut antagonism between DNB (collecting LC axons) and VNB (collecting non-LC axons).

The NE-A5 cell group is located ventrally and laterally to LC in the pontine region. The following structures are found to receive NE projections from A5 nucleus, based on unilateral or bilateral decreases of norepinephrine levels during A5 lesioning studies: caudate nucleus, piriform cortex, interstitiallis nucleus of stria terminalis, medial forebrain bundle, medial preoptica hypothalamic area, median eminence, etc.

Norepinephrine and other α_2- agonists such as clonidine induce BP reduction when directly microinjected into the anterior preoptic hypothalamic area which receives A1, A2, and A5 NE-axons but not A6 axons.[1-11,14-130,136,141-143,148-216,220-223,228,232-262,266-283,288-324,379-382,393,394,407,408,426-429,432,435]

The NE-A5 nucleus also sends axons to NE-A2 and NE-A1 nuclei as well as to dorsal motor nucleus of the vagus. In addition, NE-A5 axons represent the main NE innervation of 5HT raphe magnus nucleus and preganglionic sympathetic neurons located in IML spinal horn.[1-8,14-20,26-34,39,45-58,63-83,87,89,94-138,144-173,178-215,220-228,234,240-242,248,249,254,259,264,268-278,281,283,288-326,333-336,348,358,361,364,366,373-379,437-439]

Projections of the NE-A5 nucleus to both vasopressor and vasodepressor structures may explain contradictory findings which show that this NE cell group exerts sympathoexcitatory as well as sympathoinhibitory effects.

Stimulation of NE-A5 neurons activity produces vasodepressor effects following 30 mcl. microinjection in NE-A5 nucleus of L-glutamate, a strong firing-rate stimulant of all kinds of neurons which was found to contribute to a dose-related decrease in BP and heart rate. The bradycardic and depressor responses were markedly reduced in animals which had been sympathectomized with guanethidine. However, other experimental studies give contradictory results. NE-A5 neurons, like NE-LC, have been shown to receive inhibitory input from peripheral baroreceptors, while baroreceptor denervation increases turnover at IML spinal horn level, i.e., norepinephrine released from NE-A5 terminals. Such negative baroreceptor input to NE-A5 neurons combined with their massive projection to IML strongly suggests that these NE neurons normally contribute to the maintenance of BP. Moreover, demonstrations that clonidine inhibits NE-A5 neurons provide direct support for the hypothesis that these neurons are sympathoexcitatory. According to these paradoxical experimental data, it is possible to assume that NE-A5 axons are able to exert vasopressor or vasodepressor effects depending on the physiological circumstances . In other words, NE-A5 nucleus would act as a modulator (see Figure 3).[6-9,16-21,26-28,37-41,46-50,65-67,71-77,84,89,94,99-105,112-116,122,128,130,162,174,180-187,192,194,198,208,209,254,266-283,288-293,303,307,310,316,321-324,334,344,380,422]

Summarizing, NE released from A5 axons in anterior hypothalamus, IML spinal horn, 5HT raphe magnus nucleus, and NE-A2 neurons would result in vasodepression, whereas NE-A5 projections to NE-A1 nucleus and parasympathetic preganglionic neurons of the dorsal motor nucleus of the vagus and nucleus ambiguus would exert a vasopressor effect. This dual effect on BP regulation displayed by NE-A5 neurons is consistent with the identification of two types of NE-A5 axons, one showing high conduction velocity (about 2.5 m/s) = spinal axons and the other showing low conduction velocity (about 0.4 m/s) = anterior hypothalamus axons.

The above mechanisms of functioning suppose BP regulation to depend on the alternation of two opposite NE systems: one sympathoexcitatory (LC) and the other sympathoinhibitory (A1). During LC predominance (active waking and hyperarousal states) BP tends to rise because of NE release in posterior hypothalamus. Conversely, during A1 predominance (quiet waking and resting states) BP would tend to decrease because of NE release in anterior hypothalamus. This physiological alternation would be quantitatively exacerbated during pathological situations of stress, depression, and somatic disturbances. Other NE cell groups (A2 and A5) would act to modulate the activity of A6 and A1 cell groups by means of direct as well as indirect polysynaptic mechanisms. Finally, it is assumed that during LC activity there prevails a physiological deficiency of norepinephrine in anterior hypothalamus, whereas during A1 activity there would exist a physiological deficiency of NE in posterior hypothalamus.

II. SEROTONERGIC SYSTEM

An accumulation of recent evidence indicates that the central 5HT system participates in the regulation of BP. Anatomically, areas of the brain stem and spinal cord, namely the intermedio-lateral cell column (IML), NTS, and areas of the hypothalamus, are heavily innervated by 5HT-containing neurons. The turnover rate and/or content of 5HT in these areas is altered during experimental hypertension.[1-5,7,8,15-20,26-29,39,44-52,56,61-76,79-90,94,97,100,104-129,151-158,163,171,180-186,189,194-198,202,203,207-211,220-228,268-270,277-309,315-321,352-356,362,386,393-395]

Although pharmacological evidence suggests that central 5HT neurons also participate in

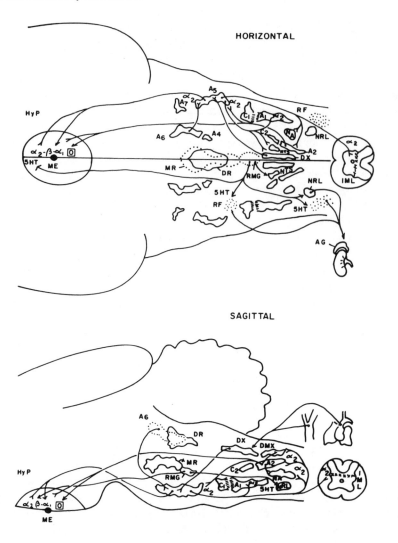

FIGURE 3. Sympathomodulatory circuitry. Ventrolateral (A5) and dorsomedial (A2) medullary NE cell groups are known to display both sympathoexcitatory and sympathoinhibitory activities. NE-A5 neurons are able to decrease sympathetic activity (BP) through projections to sympathetic preganglionic neurons located in anterior hypothalamus and intermediolateral spinal horn (IML). In effect, norepinephrine released at these levels inhibits sympathetic preganglionic neurons through α_2-adrenoceptors. 5HT released in these preganglionic sympathetic neurons annuls NE-induced inhibition. The medullary 5HT cell groups supply 5HT axons to hypothalamic and spinal preganglionic sympathetic areas. Conversely, NE-A5 axons may exert sympathoexcitatory influence through release of NE in median eminence (ME), a hypothalamic nucleus located outside the BBB (α_1- and β-adrenoceptors) as well as through release of NE on vagal motor neurons located in the dorsal motor nucleus vagii (DMX) and nucleus ambiguus (NA). These preganglionic parasympathetic neurons are inhibited by NE through α_2 adrenoceptors. In turn, NE-A5 cell group receives inhibitory input from sensory vagal neurons (DX) in such a way that stimulation of pressoreceptors located in the carotid sinus triggers a decrease of BP which is mediated through the DX — A5 pathway. A similar DX — A6 compensatory pathway has been found. Finally, NE-A5 axons project and inhibit sympathoinhibitory NE-A1 cell group. Activation of this pathway results in sympathoexcitation. NE-A2 cell group exerts sympathoexcitation, sending inhibitory NE axons to parasympathetic preganglionic neurons located in lateral hypothalamus, DMX, and NA (α_2 adrenoceptors). Other sympathoexcitatory effects are due to NE-A2 induced stimulation of oxitocinergic neurons located in parvocellular division of paraventricular hypothalamic nucleus. On the other hand, NE-A2 cell group displays sympathoinhibition through inhibitory NE axons projecting to NE-A5 cell group. Although the A2 cell group has been found to send axons to NE-A1, the neurotransmitters operating in this pathway are not known. An inhibitory pathway from A2 cell group to NE-A6 has been shown. However, this pathway is epinephrinergic in nature and probably arises from the closely related epinephrine C2 cell group. MR (median raphe), DR (dorsal raphe), RMG (raphe magnus), NTS (nucleus tractus solitarii), NRL (nucleus reticularis lateralis), RF (reticular formation), NA (nucleus ambiguus), AG (adrenal gland), Hyp (hypothalamus), ($\longrightarrow$) excitatory input, and ($\prec$) inhibitory input.

cardiovascular control, the nature of the interaction between 5HT pathways and central sympathetic networks is not well understood. For example, the 5HT precursor 5-hydroxytryptophan (5HTP) has been shown to produce a fall in BP in the conscious and anesthetized rat, cat and dog. Hypotension was accompanied by bradycardia and a decrease in splacnic sympathetic nerve activity.

The depressor response produced by i.v. or intracerebroventricular (i.c.v.) 5HTP was potentiated by monoamine oxydase inhibition and blunted by decarboxylase inhibition. In agreement with these observations, i.c.v. injection of 5HT decreases BP, heart rate, and sympathetic nervous discharge in cats and dogs. Further, several studies have shown that depletion of brain 5HT by the 5HT-synthesis inhibitor pCPA fails to alter BP. Collectively, these observations suggest that 5HT neurons normally inhibit transmission in central sympathetic pathways.[1-4,7,8,15-20,26-39,46-52,55,59-83,87,94,100-130,141-143,151-157,163,168-174,179-198,202-211,214,220-228,242,268-270, 277-325, 330, 334, 341-347, 352-354,357-362,367-372,382,386,395,422,425,429]

On the other hand, an equally impressive body of evidence suggests that 5HT facilitates central sympathetic nervous activity. First, i.c.v. injections of 5HT in the rat typically result in a pressure response. Similarly, microinjections in rats of 5HT into the anterior hypothalamus/ preoptic area or the NTS also increases BP (injection of NE in these areas provokes reduction of BP). The 5HT-induced pressor effects are enhanced by the 5HT-uptake inhibitor fluoxetine and blocked by the 5HT antagonist metergoline. Second, i.v. or i.c.v. administration of methysergide, a 5HT antagonist, decreases BP in the rat, cat, and dog. In the cat, the depressor effect produced by UML (a 5HT antagonist) is associated with lowered sympathetic nervous discharge. Third, pretreatment with the 5HT neurotoxins 5,6- or 5,7-dihydroxytryptamine (5,6-DHT or 5,7-DHT) typically decreases BP. Finally, electrical stimulation of 5HT nuclei (e.g., posterior portions of raphe pallidus and raphe obscurus, anterior portion of raphe magnus, and DR and MR) elicits pressor responses. All these data suggest that 5HT neurons facilitate activity in central sympathetic pathways.[1-8,15-20,26-39,45-56,61-89,94-130,141-143,150-157,163-174,179-203,207-214,220-228,253,266-270,277-307,312-318,323-325,330-334,341-362,367-372,379,380,386,395,422,437-440]

These extremely variable results undoubtedly reflect the complexity of 5HT neuronal pathways and their interaction with the central sympathetic system. In this regard, at least three types of 5HT receptors exist in the central nervous system (CNS). The first mediates an inhibition of neuronal activity in areas such as suprachiasmatic nucleus, ventrolateral geniculate and cortical and basolateral nuclei of the amygdala. A second type of postsynaptic 5HT receptor has been shown to facilitate excitatory input to motor nuclei. Finally, a third 5HT receptor mediates 5HT-induced inhibition of 5HT neurons in the DR (presynaptic receptor or autoreceptor).

Classical 5HT antagonists (e.g., UML, methysergide, cinaserin, cyproheptadine) fail to block either the presynaptic or postsynaptic inhibitory action of 5HT. In contrast, these antagonists act selectively on 5HT receptors to block the facilitating action of 5HT in the CNS.[180,268,270,274]

Recent studies reveal that 5HT agonists (lisuride and 5-methoxy-dimethyltryptamine) which act presynaptically to inhibit 5HT cell firing and 5HT antagonists which act postsynaptically to block the effect of synaptically released 5HT both mediate a central reduction of sympathetic nervous discharge. Therefore, it may be concluded that central 5HT neurons facilitate transmission in central sympathetic pathways.[180,182,186,291,292,307]

The sympathoexcitatory effect of the central 5HT system may be mediated through 5HT neurons projecting to IML, anterior hypothalamus, and NTS. 5HT injected in these areas induces a rise in BP. Anatomically, 5HT medullary neurons project to IML and NTS, whereas axons arising from DR, MR, and raphe magnus nuclei project to anterior hypothalamus.[186,227,276,279,282,309]

Taking into account the great deal of experimental evidence showing sympathoinhibitory as well as sympathoexcitatory roles for the central 5HT system, in order to explain these paradoxical findings, what is needed is not more experimental data but a logical integration of results into a coherent and operative mechanism. Using the accumulated information it may be possible to design a functioning hypothesis to reconcile apparently paradoxical results.

Microinjection of 5HT and NE in anterior hypothalamus has been shown to exert sympathoexcitatory and sympathoinhibitory influences, respectively. With respect to the source of 5HT released at this level, there are several anatomical possibilities: (1) 5HT-DR axons, (2) 5HT-MR axons, (3) 5HT-raphe magnus axons, and (4) 5HT-neurons located in hypothalamus.[116,292,328,329,352]

5HT-MR axons are involved in regulating hypothalamic hormonal secretion, rather than BP. In counterposition, 5HT-DR axons are certainly involved since electrical stimulation of DR nucleus raises BP, an increase which is blunted by the injection of 5HT antagonists in anterior hypothalamus. Although the vasopressor effect induced by DR stimulation is brief, the physiological role of this 5HT nucleus cannot be discarded because DR stimulation also provokes a late increase of plasma renin activity.[352,386]

What are the possibilities of a sudden increase of 5HT release from DR terminals occurring at anterior hypothalamic level? Although it is known that 5HT neurons are autoactive, they are capable of being stimulated and bridled by different physiological influences. For instance, NE released into DR nucleus exerts an excitatory influence on the firing rate of 5HT neurons, an influence which is mediated through α_1-adrenergic receptors. Taking into account that DR nucleus receives dense NE innervation arising from NE-LC neurons, it is logical to assume that enhancement of LC activity would be followed by hyperactivity of 5HT-DR neurons. In view of the fact that NE-LC displays strong sympathoexcitatory activity, the possible sympathoexcitatory role of 5HT-DR system fits well as part of a common central circuit addressed to this autonomic function. Such a vasopressor circuit would include NE-LC axons ending in posterior hypothalamus and median eminence. Furthermore, NE axons from non-LC nuclei (A1, A2, and A5) apparently interrupt their NE release in anterior hypothalamus during hyperactive NE-LC periods. This would be due to NE release by LC axons upon the somatodendritic areas of these non-LC NE nuclei. Thus, reduction of NE in anterior hypothalamus would be a reinforcing vasopressor factor.[8]

The finding that DR-induced BP increases are short lasting fits well with the fact that 5HT-DR sends heavy innervation to NE-LC nucleus, exerting a strong inhibitory influence. This effect is revealed through the decreased firing rate of NE neurons following 5HT microinjection at NE-LC. It follows that the 5HT-induced inhibitory influence on NE-LC activity would reduce NE release in posterior hypothalamus, median eminence, and NE-A1 nuclei, at which levels this neurotransmitter displays sympathoexcitatory activity.[1-8,15-20,26-35,39,45-56,61-83,87,89,94,100-130,141-143,151-157,163,168-174,179-195,202,203,208-210,214,220-228,236,268-270,277-334,341,344-354,357-362,367-372,386,395,422,437-440]

The sympathoexcitatory effect exerted by serotonin at anterior hypothalamus level has been postulated as indirect and phasic rather than direct and tonic. This hypothesis emerges from experiments showing that destruction of 5HT input at anterior hypothalamus (medial forebrain bundle) does not induce significant reduction of BP in rats. Therefore, it is logical to assume that the 5HT vasopressor mechanism is active mainly after vasodepression provoked by NE release in anterior hypothalamus. 5HT is known to act as a bridle and modulator of NE transmission at different peripheral and central levels. For example, (1) NE and 5HT antagonize postsynaptically at hippocampal pyramidal cells, the anterior hypothalamic areas, and IML spinal horn; (2) α_2-presynaptic receptors located on both NE and 5HT terminals regulate release of the two neurotransmitters on many central levels, thus supporting other observations that these terminals are frequently intermingled in structures receiving mixed innervation; and (3) NE and 5HT cell bodies interchange axons in such a way that NE neurons receive 5HT axons, and 5HT neurons receive NE axons. NE stimulates 5HT-DR neurons through α_1-receptors located on 5HT cell bodies, while inhibiting 5HT-MR neurons, also through α_1-receptors. However, all 5HT input reaching NE neurons has an inhibitory influence. Such 5HT-induced reductions in the NE-LC sympathoexcitatory system would explain experimental results associating the 5HT system with sympathoinhibition.[1-8,15-20,26-39,45-56,61-83,87,89,94,100-130,141-143,151-157,163,168-174,179-198,202,203,208-210,214,220-234,240,242,247,250-255,261,268-270,277-309,312-325,330-334,341-362,367-372,386,395,422,425]

The association of 5HT system with sympathoinhibitory activity emerges from several animal and human studies. For example, it has been shown that administration of L-tryptophane exerts a vasodepressor effect in hypertensive but not normotensive rats, suggesting that 5HT would bridle a hyperactive, but not a normoactive, NE-LC sympathoexcitatory system.[198,283,317,318,334]

Accordingly, the administration of 5HT agonists would induce vasodepressor effect in those animals or humans showing central sympathoexcitatory predominance, whereas it would induce vasopressor effect when administered during predominance of the central sympathoinhibitory system.

5HT agonists are able to inhibit the firing rate not only of NE neurons but also of 5HT neurons located in DR nucleus. In effect, only these 5HT neurons are provided with 5HT-inhibitory autoreceptors in their cell bodies. On the contrary, 5HT agonists microinjected in other 5HT cell groups provoke firing rate increase of these neurons. These data, along with anatomical findings showing the existence of a 5HT-MR → 5HT-DR fiber system, are coherent with studies showing that L-tryptophane or 5HTP are able to reduce hypertensive syndrome. In these circumstances the increase of central 5HT, secondary to administration of 5HT precursors, would act through inhibition of both NE-LC and 5HT-DR nuclei.[8,73,74,122,151,185,186,225,274,436-444]

Such findings on the complex mechanisms underlying central NE and 5HT functioning should provide clues to the apparently paradoxical relation between the central 5HT system and BP regulation. For instance, it has been shown that injection of 5HT into the lateral or the fourth cerebral ventricle of rats produces a pressor response sometimes followed by a prolonged depressor effect. On the other hand, administration of 5HT into the cisterna magna evoked a slow, progressive, and long-lasting depressor effect with or without an initial pressor effect. These demonstrations that 5HT injected centrally produces an increase as well as decrease in arterial BP would depend on varying anatomical, physiological, and pharmacological experimental conditions.

III. DOPAMINERGIC SYSTEM

During the past few years evidence has grown supporting the role of dopamine (DA) as a neurotransmitter substance in the CNS.

It has been suggested recently that the central DA system is involved in central regulation of BP. Supporting this are pharmacological experiments in which direct injection of DA into the i.c.v. elicited changes in systemic BP. However, experimental studies give some conflicting results. The i.c.v. injection of DA has been shown to decrease BP in anesthetized cats and rats, whereas in unanesthetized cats and dogs a pressure increase was observed.[4,5,9,11,19-28,37,40-50,54-61,65-76,82,85-106,112-116,121,128,149,156-162,165,166,170,173,174,177,179,182,184,187,189-191,194,328,329,393,394]

The mechanisms whereby centrally applied DA produces a cardiovascular response are similarly controversial. Apart from acting on DA receptors, DA may stimulate α- and β–receptors, directly or after being converted to norepinephrine. However, inhibition of DBH was not observed to affect the decrease of BP caused by centrally applied DA, suggesting a direct effect of DA on the CNS. Furthermore, in anesthetized and unanesthetized cats and dogs, the central cardiovascular effects of DA are interfered by i.c.v. administration of dopaminergic blocking agents (DBA) such as haloperidol, pimozide, etc., but not by i.c.v. administration of phentolamine, an α-adrenoceptor antagonist. On the other hand, i.c.v. administration of metoclopramide, a DBA, increased BP in conscious rats. Pretreatment with hexamethonium almost completely abolished the pressor effect of metoclopramide acting at a site in the CNS.[24,59,216,260,261,331,393]

In order to explain the role of DA system in regulating BP we may invoke direct and indirect effects of DA. There is ample proof that DA axons arising in ventral tegmental area (VTA), where DA cell group A10 is located, end in NE-LC and NE-A2 nuclei, both of which are strongly involved in BP regulation. In turn, NE-LC axons reach VTA and exert inhibitory influence on DA activity.[3,6-8,11,16,18,26-29,39-46,57-67,71,76-84,88-93,99-116,122-130,156,161-184,188-206,211-2334,340,366,373,380,407,433,439]

Other findings demonstrate that NE axons in VNB, which collect A1, A2, and A5 axons, exert inhibitory influence on nigrostriatal DA system.[432]

NE-DA systems interact at the level of NE and DA terminals. In effect, inhibitory DA receptors are known to exist at NE terminals, just as inhibitory NE receptors exist at DA terminals. Finally, NE has been observed to act on postsynaptic DA receptors and conversely, DA may act on postsynaptic NE receptors.[86]

In view of the above NE-DA interactions and other experimentally demonstrated levels of DA-5HT interaction, the DA system would necessarily be involved in all functions depending on NE and 5HT activities.[3,4,7,8,15-20,26-39,46-55,61,65-71,74-83,87-105,109-123,126-130,141-143,151-157,163,168-174,179,183-195,198,202-214,220-223,228,268-270,277-283,287-292,298-305,312-321,334,345,353,358,362]

IV. CHOLINERGIC SYSTEM

The various cardiovascular responses obtained after central administration of acetylcholine (ACh) may arise from species differences in the animals studied and in the route of administration of ACh. In cats, i.c.v. administration of ACh or superfusion of ACh in cardiovascular loci of the posterior hypothalamus elicits pressor effects accompanied by a tachycardia; superfusion of ACh in lateral medullary reticular pressor areas evokes biphasic cardiovascular response. In dogs, ACh administered into the lateral cerebral ventricle causes an increase of BP and heart rate; yet in rats ACh injected by the same route evokes a rise in BP accompanied by slight and variable changes in heart rate. Carbachol administered by i.c.v. produces a pressor response in dogs, but in cats, whether i.c.v. injected or applied to ventral surface of the brainstem, it induces a fall in BP, providing further evidence that vascular responses to central injection of cholinergic drugs vary according to species. A good example of this is the reversal of pressor response to i.c.v. injection of ACh in rats after central application of physostigmine; the resulting depressor response, with or without a secondary rise of BP, is accompanied by bradycardia.[1-3,8-34,39,46-52,55,56,61-71,75-83,87-93,99-129,134-145,151-166,171,176,179-189,192-223,227,228,241,246,267-288,295,297-322,328-334,339-353,357,378,383,386,390,396,412-414,421-429,433,434]

Other studies show that cardiovascular responses to ACh can be initiated in a number of cardiovascular loci situated at various levels of the rat brain. Apparently, the cardiovascular loci of all these brain areas possess muscarinic cholinoceptors mediating pressor responses, as well as those mediating depressor responses, but not in equal proportion. It seems that muscarinic cholinoceptors mediating pressor effects are more sensitive to ACh and functionally are more competent than those mediating depressor effects, the latter being markedly greater in number.[11,24,25,51,91,157,189,190,264,298,300,330,333,339,358]

ACh-sympathetic preganglionic neurons exist in the IML spinal horn, lateral reticular pressor medullary area, the nucleus reticularis lateralis (NRL), and posterior hypothalamus. In addition, ACh-parasympathetic preganglionic neurons exist in the dorsal motor nucleus of vagus nerve, nucleus ambiguus, and medullary medial reticular formation, as well.[1-21,26-53,56,61-84,82-95,113,212,252,321,325,331,334,362,383,385,390-392,396,397,412,413,421]

All these preganglionic neurons send axons to postganglionic neurons of both divisions of the autonomic nervous system. In general, it is accepted that ACh and 5HT input to preganglionic neurons is excitatory while NE input is inhibitory, e.g., the well-known existence of an ACh excitatory input from the anterior to posterior hypothalamus.

The ACh neurons located in lateral reticular formation of the medullary and spinal regions send excitatory axons to adrenal glands, which then secrete epinephrine, norepinephrine, and dopamine.[3,7,8,12-18,22,26-35,41,46-50,55,65,67,70,71,76-89,98-105,111-123,127-140,144-150,159-175,179-208,212,2329,334,352,380]

ACh input has likewise been demonstrated to NE and 5HT neurons of the pontine and medullary regions.[1-3,7,8,15-20,26-29,39,46,51,52,61-83,87-94,99-129,136,141-143,15,1-159,163,171,174,179-2334,357-429,433]

Sensory neurons of vagus nerve, also cholinergic in nature, likewise receive excitatory peripheral input, probably glutamatergic in nature.[264,358,361,364,396]

All the above findings give an idea of the complex role played by the central ACh pathways in BP and autonomic regulation.

REFERENCES

1. **Adair JR, Hamilton BL, Scappaticci KA, Kelke CJ, Gillis RA.** Cardiovascular responses to electrical stimulation of the medullary raphe area of the cat. *Brain Res* 128: 141-145, 1977.
2. **Ader JP, Sebens JB, Korf J.** Central levels of noradrenaline, 3-methoxy-4-hydroxyphenylethyleneglycol and cyclic AMP in the rat after activation of locus coeruleus neurons: influence of single and repeated neuroleptic treatment. *Psychopharmacology* 70: 239-245, 1980.
3. **Aghajanian GK, Cedarbaum JM, Wang RY.** Evidence of norepinephrine mediated collateral inhibition of locus coeruleus neurons. *Brain Res* 136: 570-577, 1977.
4. **Anden NE, Dahlstrom A, Fuxe K, Larsson K.** Mapping out of catecholamine and 5-hydroxytryptamine neurons innervating the telencephalon and diencephalon. *Life Sci* 4: 1275-1279, 1965.
5. **Anden NE, Dahlstrom A, Fuxe K, Olson L, Ungerstedt U.** Ascending noradrenaline neurons from the pons and medulla oblongata. *Experientia* 22: 44-45, 1966.
6. **Andrade R, Aghajanian GK.** Single cell activity in the noradrenergic A-5 region: responses to drugs and peripheral manipulations of blood pressure. *Brain Res* 242: 125-135, 1982.
7. **Aston-Jones G, Segal M, Bloom FE.** Brain aminergic axons exhibit marked variability in conduction velocity. *Brain Res* 195: 215-222, 1980.
8. **Baraban JM, Aghajanian GK.** Suppression of firing activity of 5-HT neurons in the dorsal raphe by alpha-adrenoceptor antagonists. *Neuropharmacology* 19: 355-363, 1980.
9. **Blessing WW, Chalmers JP, Howe PRC.** Distribution of catecholamine-containing cell bodies in the rabbit central nervous system. *J Comp Neurol* 179: 407-424, 1978.
10. **Blessing WW, Costa M, Furness JB, West MJ, Chalmers JP.** Projection from A1 neurons towards the nucleus solitarius in rabbit. *Cell Tissue Res* 220: 27-40, 1981.
11. **Blessing WW, Reis DJ.** Inhibitory cardiovascular function of neurons in the caudal ventrolateral medulla of the rabbit: relationship to the area containing A1 noradrenergic cells. *Brain Res* 253: 161-171, 1982.
12. **Bloch R, Feldman J, Bousquet P, Schwartz J.** Relationship between the ventromedullary clonidine-sensitive area and the posterior hypothalamus. *Eur J Pharmacol* 45: 55-59, 1977.
13. **Bousquet P, Feldman J, Bloch R, Schwartz J.** The nucleus reticularis lateralis: A region highly sensitive to clonidine. *Eur J Pharmacol* 69: 389-392, 1981.
14. **Bunag RD, Eferakeya AE.** Immediate hypotensive after-effects of posterior hypothalamic lesions in awake rats with spontaneous, renal or DOCA hypertension. *Cardiovasc Res* 10: 663-670, 1976.
15. **Carey HM, Dacey RG, Jane JA, Winn HR, Ayers CR, Tyson GW.** Production of sustained hypertension by lesion in the nucleus tractus solitarii of the American foxhound. *Hypertension* 1: 246-264, 1979.
16. **Cedarbaum JM, Aghajanian GK.** Catecholamine receptors on locus coeruleus neurons: pharmacological characterization. *Eur J Pharmacol* 44: 375-385, 1977.
17. **Cedarbaum JM, Aghajanian GK.** Activation of locus coeruleus neurons by peripheral stimuli: modulation of collateral inhibitory mechanisms. *Life Sci* 23: 1383-1392, 1978.
18. **Chung JM, Chung K, Wurster RD.** Sympathetic preganglionic neurons of the cat spinal cord: horseradish peroxidase study. *Brain Res* 91: 126-131, 1975.
19. **Coote JH, Fleetwood-Walker SM, Martin IL.** The origin of the catecholamine innervation of the sympathetic lateral column. *J Physiol* 295: 57-58P, 1979.
20. **Coote JH, MacLeod VH.** The influence of bulbospinal monoaminergic pathways on sympathetic nerve activity. *J Physiol* 241: 453-475, 1974.
21. **Coote JH, MacLeod VH.** The effect of intraspinal micro-injections of 6-hydroxydopamine on the inhibitory influence exerted on spinal sympathetic activity by the baroreceptors. *Pfluegers Arch Gesamte Physiol Menschen Tiere* 371: 271-277, 1977.

22. **Cubeddu LX, Hoffman IS, Davila J, Barbella, YR, Ordaz P.** Clonidine reduces elevated cerebrospinal fluid catecholamine levels in patients with essential hypertension. *Life Sci* 35: 1365-1371, 1984.

23. **Dahlstrom A, Fuxe K.** Evidence for the existence of monoamine containing neurons in the central nervous system. I. Demonstration of monoamines in the cell bodies of brain stem neurons. *Acta Physiol Scand Suppl* 232: 1, 1964.

24. **Dampney RAL.** Brain stem mechanisms in the control of arterial pressure. *Clin Exp Hyperten* 3: 379-391, 1981.

25. **Dampney RAL, Moon EA.** Role of ventrolateral medulla in vasomotor response to cerebral ischemia. *Am J Physiol* 239: H349-358, 1980.

26. **Day TA, Blessing W, Willoughby JO.** Noradrenergic and dopaminergic projections to the medial preoptic area of the rat. A combined horseradish peroxidase catecholamine fluorescence study. *Brain Res* 193: 543-548, 1980.

27. **Degroat WC, Ryall RW.** An excitatory action of 5-hydroxytryptamine on sympathetic preganglionic neurones. *Exp Brain Res* 3: 299-303, 1967.

28. **Dejong W.** Noradrenaline: central inhibitory control of blood pressure and heart rate. *Eur J Pharmacol* 29: 179-186, 1974.

29. **Dejong W, Nijkamp FP, Bohus B.** Role of noradrenaline and serotonin in the central control of blood pressure in normotensive and spontaneously hypertensive rats. *Arch Int Pharmacodyn* 213: 272-284, 1975.

30. **Dejong W, Palkovits M.** Hypertension after localized transection of brainstem fibers. *Life Sci* 18: 61-64, 1976.

31. **Dejong W, Zandberg P, Bohus B.** Central inhibitory noradrenergic cardiovascular control. *Prog Brain Res* 42: 285-298, 1975.

32. **Dequattro V, Eide I, Myers MR, Eide K, Kolloch R, Whigham H.** Enhanced hypothalamic noradrenaline biosynthesis in Goldblatt I renovascular hypertension. *Clin Sci Mol Med* 55: 109-115, 1978.

33. **Dietl H, Sinha JN, Philippu A.** Presynaptic regulation of the release of catecholamine in the cat hypothalamus. *Brain Res* 208: 213-218, 1981.

34. **Doba N, Reis DJ.** Acute fulminating neurogenic hypertension produced by brainstem lesions in the rat. *Circ Res* 32: 584-593, 1973.

35. **Eide I, Kolloch R, Dequattro V, Miano L, Dugger R, Van Der Muelen J.** Raised cerebrospinal fluid norepinephrine in some patients with primary hypertension. *Hypertension* 1: 255-260, 1979.

36. **Eide I, Myers MR, Dequattro V, Kolloch R, Eide K, Whigham M.** Increased hypothalamic noradrenergic activity in one-kidney, one clip renovascular hypertensive rats. *J Cardiovasc Pharmacol* 2: 833-839, 1980.

37. **Fleetwood-Walker SM.** Catecholamine Systems Descending from the Lower Brainstem: Their Contribution to the Innervation of the Sympathetic Lateral Column. Ph.D. thesis, Birmingham University, Birmingham, England, 1979.

38. **Fleetwood-Walker SM, Coote JH.** The contribution of brain stem catecholamine cell groups to the innervation of the sympathetic lateral cell column. *Brain Res* 205: 141-155, 1981.

39. **Folkow BUG, Hallback MIL.** Physiopathology of spontaneous hypertension in rats, in *Hypertension*. Genest J, Koiw E, Kuehel O. Eds., McGraw-Hill, New York, 1977, 507.

40. **Fuxe K.** Evidence for existence of monoamine containing neurons in the central nervous system. IV. Distribution of monoamine nerve terminals in the central nervous system. *Acta Physiol Scand Suppl* 247: 36, 1965.

41. **Fuxe K, Ganten D, Jonsson G, Agnati LF, Andersson K, Hökfelt T, Bolme P, Goldstein M, Hallman H, Unger T, Rascher W.** Catecholamine turnover changes in hypothalamus and dorsal midline area of the central medulla oblongata of spontaneously hyper-tensive rats. *Neurosci Lett* 15: 283-288, 1979.

42. **Fuxe K, Hökfelt T, Goldstein M, Jonsson G, Lindbrink K, Ljungdahl A, Sachs CH.** Topography of central catecholamine pathways. Symp Central Action of Drugs in the Regulation of Blood Pressure. Royal Post Graduate Medical School, London, 1975.

43. **Gagnon DJ, Melville KI.** Centrally mediated cardiovascular response to isoprenaline. *Int J Neuropharmacol* 6: 245-251, 1967.

44. **Gordon EK, Perlow M, Oliver J, Ebert M, Kopin IJ.** Origins of catecholamine metabolites in monkey cerebrospinal fluid. *J Neurochem* 25: 347-349, 1975.

45. **Gunn CG, Sevelius G, Puiggari J.** Vagal cardiomotor mechanisms in the hindbrain of the dog and cat. *Am J Physiol* 214: 258-262, 1968.

46. **Guyenet PG, Cabot JB.** Inhibition of sympathetic preganglionic neurons by catecholamines and clonidine: mediation by an alpha-adrenergic receptor. *J Neurosci* 1: 908-917, 1981.

47. **Hancock MB, Fougerousse CL.** Spinal projections from the nucleus locus coeruleus and nucleus subcoeruleus in the cat and monkey as demonstrated by the retrograde transport of horseradish peroxidase. *Brain Res Bull* 1: 229-234, 1976.

48. **Henry JL, Calaresu FR.** Excitatory and inhibitory inputs from medullary nuclei projecting to spinal cardioacceleratory neurons in the cat. *Exp Brain Res* 20: 485-504, 1974.

49. **Hilton SM, Spyer KM.** The hypothalamic depressor area and the baroreceptor reflex. *J Physiol* 200: 107P, 1969.

50. **Hilton SM, Spyer KM.** Participation of the anterior hypothalamus in the baroreceptor reflex. *J Physiol* 218: 271-277, 1971.

51. **Hoffman WE, Phillips MI.** A pressor response to intraventricular injections of carbachol. *Brain Res* 105: 157-162, 1976.

52. **Hoffman WE, Schmid PG, Phillips MI.** Central cholinergic and noradrenergic stimulation in spontaneously hypertensive rats. *J Pharmacol Exp Ther* 206: 644-651, 1978.

53. **Johnson AK, Buggy J, Fink GD, Brody MJ.** Prevention of renal hypertension and of the central pressor effect of angiotensin by ventromedial hypothalamic ablation. *Brain Res* 205: 255-260, 1981.

54. **Jonsson G, Fuxe K, Hökfelt T.** On the catecholamine innervation of the hypothalamus, with special reference to the median eminence. *Brain Res* 40: 271-278, 1972.

55. **Jordan D, Spyer KM.** Studies on the termination of sinus nerve afferents. *Pfluegers Arch* 369: 65-73, 1977.

56. **Julius S, Esler MD.** *The Nervous System in Arterial Hypertension*. Charles C Thomas, Springfield, IL, 1976, 3.

57. **Kawamura H, Gunn CG, Frohlich ED.** Modified cardiovascular responses by nuclei tractus solitarius and locus coeruleus in spontaneously hypertensive rat (SHR). *Circulation* 54, II: 143, 1976.

58. **Kawamura H, Gunn CG, Frohlich ED.** Cardiovascular alteration by nucleus locus coeruleus in spontaneously hypertensive rat. *Brain Res* 140: 137-147, 1978.

59. **Kobayashi RM, Palkovits M, Jacobowitz DM, Kopin IJ.** Biochemical mapping of the noradrenergic projections from the locus coeruleus. *Neurology* 25: 223-233, 1975.

60. **Kobayashi RM, Palkovits M, Kopin IJ, Jacobowitz DM.** Biochemical mapping of noradrenergic nerves arising from the locus coeruleus. *Brain Res* 77: 269-276, 1974.

61. **Konig JFR, Klippel RA.** *The Rat Brain: A Stereotaxic Atlas*. Williams & Wilkins, Baltimore, 1963.

62. **Korf J, Aghajanian GK, Roth RH.** Stimulation and destruction of the locus coeruleus: opposite effects on 3-methoxy-4-hydroxyphenylglycol sulfate levels in the rat cerebral cortex. *Eur J Pharmacol* 21: 305-310, 1973.

63. **Korf J, Roth RH, Aghajanian GK.** Alterations in turnover and endogenous levels of norepinephrine in cerebral cortex following electrical stimulation and acute axotomy of cerebral noradrenergic pathways. *Eur J Pharmacol* 23: 276-282, 1973.

64. **Krstic MK, Djurkovic D.** Analysis of cardiovascular responses to central administration of 5-hydroxytryptamine in rats. *Neuropharmacology* 19: 455-463, 1980.

65. **Laubie M, Schmitt H.** Sites of action of clonidine: centrally mediated increase in vagal tone, centrally mediated hypotensive and sympatho-inhibitory effects, in *Hypertension and Brain Mechanisms*. (Progress in Brain Research Series, Vol. 47), DeJong W, Provoost AP, Shapiro AP. Eds., 1977. 337.

66. **Levitt P, Moore RY.** Noradrenaline neurons innervation of the neocortex of the rat. *Brain Res* 139: 219-231, 1978.

67. **Levitt P, Moore RY.** Origin and organization of brainstem catecholamine innervation in the rat. *J Comp Neurol* 186: 505-528, 1979.

68. **Lewander T, Joh TH, Reis DJ.** Prolonged activation of tyrosine hydroxylase in noradrenergic neurons of rat brain by cholinergic stimulation. *Nature* 258: 440-441, 1975.

69. **Lindvall O, Bjorklund A.** The organization of the ascending catecholamine neuron systems in the rat brain as revealed by the glyoxylic acid fluorescence method. *Acta Physiol Scand Suppl* 412: 1-48, 1974.

70. **Lipski J, Przybylski J, Solnicka E.** Reduced hypotensive effect of clonidine after lesions of area of nucleus tractus solitarii in rats. *Eur J Pharmacol* 38: 19-29, 1976.

71. **Loewy AD, Gregorie EM, McKellar S, Baker RP.** Electrophysiological evidence that the A5 catecholamine cell group is a vasomotor center. *Brain Res* 178: 196-200, 1979.

72. **Loewy AD, McKellar S, Saper CB.** Direct projections from the A5 catecholamine cell group to the intermediolateral cell column. *Brain Res* 174: 309-314, 1979.

73. **Maruyama S.** Inhibition by topically applied clonidine and guanfacine on the pressor response to stimulation of the locus coeruleus in cats. *Jpn J Pharmacol* 31: 586-589, 1981.

74. **McKellar S, Loewy AD.** Spinal projections of norepinephrine-containing neurons in the rat. *Neurosci Abstr* 5: 344, 1979.

75. **Mizuno N, Nakamura Y.** Direct hypothalamic projections to the locus coeruleus. *Brain Res* 19: 160-162, 1970.

76. **Moore RY, Bloom FE.** The central catecholamine neuron systems: anatomy and physiology of the norepinephrine and epinephrine system. *Annu Rev Neurosci* 2: 113-168, 1976.

77. **Morris MJ, Woodcock EA.** Central alpha-adrenoceptors and blood pressure regulation in the rat. *Clin Exp Pharmacol Physiol* 9: 303-307, 1982.

78. **Mullen PE, Lightman S, Linsel C, McKeon P, Sever PS, Todd K.** Rhythms of plasma noradrenaline in man. *Psychoneuroendocrinology* 6: 213-222, 1981.

79. **Nakamura K, Nakamura K.** Role of brainstem and spinal noradrenergic and adrenergic neurons in the development and maintenance of hypertension in spontaneously hypertensive rats. *Naunyn-Schmiedeberg's Exp Pathol Pharmakol* 305: 127-133, 1978.

80. **Nathan MA, Reis DJ.** Chronic labile hypertension produced by lesions of the nucleus tractus solitarii in the cat. *Circ Res* 40: 72-81, 1977.

81. **Neumayr RJ, Hare BD, Franz DN.** Evidence for bulbospinal control of sympathetic preganglionic neurons by monoaminergic pathways. *Life Sci* 14: 793-806, 1974.

82. **Nijkamp FP, Dejong W.** Methylnoradrenaline induced hypotension and bradycardia after administration into the area of the nucleus tractus solitarii. *Eur J Pharmacol* 32: 361-370, 1975.

83. **Ogawa M, Fujita Y, Niwa M, Takami N, Ozaki M.** Role on blood pressure regulation of noradrenergic neurons originating from the locus coeruleus in the Wistar-Kyoto rat. *Jpn Heart J* 18: 586-587, 1977.

84. **Palkovits M.** Catecholamines in the hypothalamus: an anatomical review. *Neuroendocrinology* 33: 123-128, 1981.

85. **Palkovits M.** Distribution of neuroactive substances in the dorsal vagal complex of the medulla oblongata (Critique). *Neurochem Int* 7: 213-219, 1985.

86. **Palkovits M, Brownstein M, Saavedra JM, Axelrod J.** Norepinephrine and dopamine content of hypothalamic nuclei of the rat. *Brain Res* 77: 137-141, 1974.

87. **Palkovits M, Zaborszky L.** Neuroanatomy of central cardiovascular control. Nucleus tractus solitarii: afferent and efferent neuronal connections in relation to the baroreceptor reflex arc. *Prog Brain Res* 47: 9-34, 1978.

88. **Palkovits M, Zaborszky L.** Neuronal connections of the hypothalamus, in *Handbook of the Hypothalamus.* Vol. 1, Morgane PJ, Panksepp J. Eds., Marcel Dekker, New York, 1979, 379.

89. **Palkovits M, Zaborszky L, Feminger A, Mezey E, Fekete MIK, Herman JP, Kanicska B, Szabo D.** Noradrenergic innervation of the rat hypothalamus: experimental biochemical and electron microscopic studies. *Brain Res* 191: 161-171, 1980.

90. **Paxinos G, Watson C.** *The Rat Brain.* Academic Press, New York, 1982.

91. **Philippu A, Dietl H, Sinha JN.** In vivo release of endogenous catecholamines in the hypothalamus. *Naunyn-Schmiedeberg's Arch Pharmacol* 308: 137-142, 1979.

92. **Philippu A, Kittel E.** Presence of beta-adrenoceptors in the hypothalamus: their importance for the pressor response to hypothalamic stimulation. *Naunyn-Schmiedeberg's Arch Pharmacol* 297: 219-225, 1977.

93. **Philippu A, Rosenberg W, Przuntek H.** Effects of adrenergic drugs on pressor responses to hypothalamic stimulation. *Naunyn-Schmiedeberg's Arch Pharmacol* 278: 373-386, 1973.

94. **Poitras D, Parent A.** Atlas of the distribution of monoamine-containing nerve cell bodies in the brainstem of the rat. *J Comp Neurol* 179: 699-718, 1978.

95. **Przuntek H, Philippu A.** Reduced pressor responses to stimulation of the locus coeruleus after lesion of the posterior hypothalamus. *Naunyn-Schmiedeberg's Arch Exp Pathol Pharmakol* 276: 119-122, 1973.

96. **Ross CA, Reis DJ.** Effect of lessions of locus coeruleus on regional distribution of dopamine-beta-hydroxylase activity in rat brain. *Brain Res* 73: 161-166, 1974.

97. **Ross CA, Ruggiero DA, Joh TH, Park DH, Reis DJ.** Adrenaline synthesizing neurons in the rostral ventrolateral medulla: a possible role in tonic vasomotor control. *Brain Res* 273: 356-361, 1983.

98. **Sakumoto T, Tohyama M, Satoh K, Kimoto Y, Kinugasa T, Tanizawa O, Kurachi K, Shimizu N.** Afferent fiber connections from lower brain stem to hypothalamus studied by the horseradish peroxidase method with special reference to noradrenaline innervation. *Exp Brain Res* 31: 81-94, 1978.

99. **Saper CB, Loewy AD, Swanson LW, Cowan WM.** Direct hypothalamo-autonomic connections. *Brain Res* 117: 305-312, 1978.

100. **Satoh K, Tohyama M, Yamamoto K, Sakumoto T, Shimizu N.** Noradrenaline innervation of the spinal cord studied by the horseradish peroxidase method combined with monoamine oxidase staining. *Exp Brain Res* 30: 175-186, 1977.

101. **Schmitt H, Laubie M.** Destruction of the nucleus tractus solitarii in dogs: acute effects on blood pressure and haemodynamics chronic effects on blood pressure. Importance of the nucleus for the effects of drugs, in *Nervous System and Hypertension.* Meyer P, Schmitt H. Eds., Wiley-Flammarion, New York, 1979, 173.

102. **Scriabine A, Clineschmidt BV, Sweet CS.** Central noradrenergic control of blood pressure. *Annu Rev Pharmacol* 16: 113-123, 1978.

103. **Sharma JN, Sandrew BB, Wang SC.** CNS site of clonidine induced hypotension: a microiontophoretic study of bulbar cardiovascular neurons. *Brain Res* 151: 127-132, 1978.

104. **Silver MA, Jacobowitz D, Crowley W, O'Donohue T.** Retrograde transport of dopamine-beta hydroxylase antibody (ADBH) by CNS noradrenergic neurons: hypothalamic noradrenergic innervations. *Anat Rec* 190: 541, 1978.

105. **Silver MA, Soden W, Jacobowitz D, Bloom FE.** The functional organization of CNS noradrenergic neurons. *Anat Rec* 192: 684, 1979.

106. **Sinha JN, Dhawan KN, Chandra O, Gupta GP.** Role of acetylcholine in central vasomotor regulation. *Can J Physiol Pharmacol* 45: 503-507, 1967.

107. **Sinha JN, Dietl H, Philippu A.** Effect of a fall of blood pressure on the release of catecholamines in the hypothalamus. *Life Sci* 26: 1751-1760, 1980.

108. **Sinha JN, Tangri KK, Bhargava KP, Schmitt H.** Central sites of sympatho-inhibitory effects of clonidine and L-dopa, in *Recent Advances in Hypertension*, Vol. 1. Millez P, Safar M. Eds., Boehringer-Ingelheim, Reims, 1975, 97.

109. **Sladek CD, Knigge KM.** Cholinergic stimulation of vasopressin release from the rat hypothalamic-neurohypophyseal system in organ culture. *Endocrinology* 101: 411-420, 1977.

110. **Smits JF, Struyker-Boudier HA.** Intrahypothalamic serotonin and cardiovascular control in rats. *Brain Res* 111: 422-427, 1976.

111. **Snyder DW, Nathan MA, Reis DJ.** Chronic lability of arterial pressure produced by selective destruction of catecholaminergic innervation of the nucleus tractus solitarii in the rat. *Circ Res* 43: 662-671, 1978.

112. **Speciale SG, Crowley WR, O'Donohue TL, Jacobowitz DM.** Forebrain catecholamine projections of the A5 cell group. *Brain Res* 154: 128-133, 1978.

113. **Svensson TH, Bunney BS, Aghajanian GK.** Inhibition of both noradrenergic and serotonergic neurons in brain by the alpha-adrenergic agonist clonidine. *Brain Res* 92: 291-306, 1975.

114. **Svensson TH, Thoren P.** Brain noradrenergic neurons in the locus coeruleus: inhibition by blood volume load through vagal afferents. *Brain Res* 172: 174-178, 1979.

115. **Swanson LW, Hartman BK.** The central adrenergic system. An immunofluorescent study of the location of cell bodies and their efferent connections in the rat utilizing dopamine-beta-hydroxylase as a marker. *J Comp Neurol* 163: 467-506, 1975.

116. **Takagi H, Shiosaka S, Tohyama M, Senba E, Sakanaka H.** Ascending components of the medial forebrain bundle from the lower brainstem in the rat, with special reference to raphe and catecholamine cell groups. A study by the HRP method. *Brain Res* 193: 315-337, 1980.

117. **Takigawa M, Mogenson GJ.** A study of inputs to anti-dromically identified neurons of the locus coeruleus. *Brain Res* 135: 217-230, 1977.

118. **Talman WT, Perrone MH, Reis DJ.** Acute hypertension after the local injection of kainic acid into the nucleus tractus solitarii of rats. *Circ Res* 48: 292-298, 1981.

119. **Talman WT, Snyder D, Reis DJ.** Chronic lability of arterial pressure produced by destruction of A2 catecholaminergic neurons in rat brainstem. *Circ Res* 46: 842-853, 1980.

120. **Turton MD, Deagan T.** Circadian variations of plasma catecholamine, cortisol and immunoreactive insulin concentration in supine subjects. *Clin Chem Acta* 55: 389-397, 1974.

121. **Ungerstedt U.** Stereotaxic mapping of the monoamine pathways in the rat brain. *Acta Physiol Scand Suppl* 367: 1-11, 1971.

122. **Van Ameringen M-R, De Champlain J, Imbeault S.** Participation of central noradrenergic neurons in experimental hypertension. *Can J Physiol Pharmacol* 55: 1246-1251, 1977.

123. **Vlachakis ND, Lampano C, Alexander N, Maronde RF.** Catecholamines and their major metabolites in plasma and cerebrospinal fluid of man. *Brain Res* 229: 67-74, 1981.

124. **Ward DG, Gunn CG.** Locus coeruleus complex: elicitation of a pressor response and a brain stem region necessary for its occurrence. *Brain Res* 107: 401-406, 1976.

125. **Ward DG, Leftcourt AM, Gunn CG.** Responses of neurons in the locus coeruleus to hemodynamic changes. *Fed Proc Fed Am Soc Exp Biol* 37: 743, 1978.

126. **Wing LMH, Chalmers JP.** Participation of central serotonergic neurons in the control of circulation of the unanesthetized rabbits. *Circ Res* 35: 504-513, 1974.

127. **Winternitz SR, Katholi RE, Oparil S.** Decrease in hypothalamic norepinephrine content following renal denervation in the one-kidney, on clip goldblatt hypertensive rat. *Hypertension* 4: 369-373, 1982.

128. **Wolf WA, Kuhn DM, Lovenberg W.** Blood pressure responses to local application of serotonergic agents in the nucleus tractus solitarii. *Eur J Pharmacol* 69: 291-299, 1981.

129. **Wolf WA, Kuhn DM, Lovenberg W.** Pressor effects of dorsal raphe stimulation and intrahypothalamic application of serotonin in the spontaneously hypertensive rat. *Brain Res* 208: 192-197, 1981.

130. **Yukimaru T, Fuxe K, Ganten D, Andersson K, Harfstrand A, Unger T, Agnati LF.** Acute sino-aortic denervation in rats produces a selective increase of adrenaline turnover in the dorsal midline area of the caudal medulla oblongata and a reduction of adrenaline levels in the anterior and posterior hypothalamus. *Eur J Pharmacol* 69: 361-365, 1981.

131. **Augustine SJ, Buckley JP, Tachikawa S, Lokhandwala MF.** Involvement of central noradrenergic mechanisms in the rebound hypertension following clonidine withdrawal. *J Cardiovasc Pharmacol* 4:449-455, 1982.

132. **Cleroux J, Peronnet F, Cousineau D, De Champlain J.** Plasma catecholamines and local modifications of sympathetic nervous activity. *J Auton Nerv System* 11: 323-327, 1984.

133. **Dominiak P, Kees F, Grobecker H.** Sympathoadrenal dysfunction in rats with chronic neurogenic hypertension. *Eur J Pharmacol* 107: 263-266, 1985.

134. **Gavras H, Bain GT, Bland L, Vlahakos D, Gavras I.** Hypertensive response to saline micro-injection in the area of the nucleus tractus solitarii of the rat. *Brain Res* 343: 113-119, 1985.

135. **Goldstein DS, McCarty R, Polinsky RJ, Kopin IJ.** Relationship between plasma norepinephrine and sympathetic neural activity. *Hypertension* 5: 552-559, 1983.

136. **Lake CR, Ziegler MG, Kopin IJ.** Use of plasma norepinephrine for evaluation of sympathetic neuronal function in man. *Life Sci* 18: 1315-1326, 1976.

137. **Martin PR, Ebert MH, Gordon EK, Weingartner H, Kopin IJ.** Catecholamine metabolism during clonidine withdrawal. *Psychopharmacology* 84: 58-63, 1984.

138. **Matsui H.** Adrenal medullary secretory response to stimulation at the medulla oblongata in the cat. *Neuroendocrinology* 29: 385-390, 1979.

139. **Matsui H.** Adrenal medullary secretory response to pontine stimulation in the rat. *Neuroendocrinology* 33: 84-87, 1981.

140. **Reiner PB.** Clonidine inhibits central noradrenergic neurons in unanesthetized cats. *Eur J Pharmacol* 115: 249-257, 1985.

141. **Reis DJ, Weinbren M, Covelli A.** A circadian rhythm of norepinephrine regionally in cat brain. Its relationship to environmental lighting and to regional diurnal variations in brain serotonin. *J Pharmacol Exp Ther* 164: 135-145, 1968.

142. **Reis DJ, Wurtman RJ.** Diurnal changes in brain noradrenaline. *Life Sci* 7: 91-98, 1968.

143. **Renton GH, Weis-Malherbe H.** Adrenaline and noradrenaline in human plasma during sleep. *J Physiol* 131: 170-175, 1965.

144. **Robinson RL, Culberson JL, Carmichael SW.** Influence of hypothalamic stimulation on the secretion of adrenal medullary catecholamines. *J Auton Nerv Syst* 8: 89-96, 1983.

145. **Sourkes TL.** Neurotransmitters and central regulation of adrenal functions. *Biol Psychiatry* 20: 182-191, 1985.

146. **Svensson TH, Strombom U.** Discontinuation of chronic clonidine treatment: evidence for facilitated brain noradrenergic neurotransmission. *Naunyn-Schmiedeberg's Arch Pharmacol* 299: 83-87, 1977.

147. **Tang SW, Helmeste DM, Stancer HC.** The effect of clonidine withdrawal on total 3-methoxy-4-hydroxy-phenylglycol in rat brain. *Psychopharmacology* 61: 11-12, 1979.

148. **Gauthier P, Reader TA.** Adrenomedullary secretory response to midbrain stimulation in rat: effects of depletion of brain catecholamines or serotonin. *Can J Physiol Pharmacol* 6: 1464-1474, 1982.

149. **Yamaguchi I, Kopin IJ.** Plasma catecholamine and blood pressure responses to sympathetic stimulation in pithed rats. *Am J Physiol* 237: H305-H310, 1979.

150. **Young JG, Cohen DJ, Hattox SE, Kavanagh ME, Anderson GM, Shaywitz BA, Maas JW.** Plasma free MHPG and neuroendocrine responses to challenge doses of clonidine in Tourette's syndrome: preliminary report. *Life Sci* 29: 1467-1475, 1981.

151. **Azmitia EC, Segal M.** An autoradiographic analysis of the differential ascending projections of the dorsal and median raphe nuclei in the rat. *Comp Neurol* 179: 641-668, 1978.

152. **Beaudet A, Descarries L.** Radiographic characterization of a serotonin-accumulating nerve cell group in adult rat hypothalamus. *Brain Res* 160: 231-243, 1979.

153. **Bobillier P, Petitjean F, Salvert D, Leger M, Seguin S.** Differential projections of the nucleus raphe dorsalis and nucleus raphe centralis as revealed by autoradiography. *Brain Res* 85: 205-210, 1975.

154. **Bobillier P, Seguin S, Petitjean F, Salvert D, Touret M, Jouvet M.** The raphe nuclei of the cat brain stem: topographical atlas of their efferent projection as revealed by autoradiography. *Brain Res* 113: 449-486, 1976.

155. **Brezenoff HE, Rusin J.** Brain acethylcholine mediates by hypertensive response to physostigmine in the rat. *Eur J Pharmacol* 29: 262-266, 1974.

156. **Brownstein MJ, Palkovits M, Tappaz M, Saavedra JM, Kizer JS.** Effect of surgical isolation of the hypothalamus on its neurotransmitter content. *Brain Res* 117: 287-295, 1976.

157. **Buccafusco JJ, Brezenoff HE.** The hypertensive response to injection of physostigmine into the hypothalamus of the unanesthetized rat. *Clin Exp Hypertension* 1: 219-227, 1978.

158. **Campese VM, Myers MR, Dequattro V.** Neurogenic factors in low renin essential hypertension. *Am J Med* 60: 83-91, 1980.

159. **Cedarbaum JM, Aghajanian GK.** Noradrenergic neurons of the locus coeruleus: inhibition by epinephrine and activation by the alpha-antagonist piperoxane. *Brain Res* 112: 413-419, 1976.

160. **Cedarbaum JM, Aghajanian GK.** Afferent projections to the rat locus coeruleus as determined by a retrograde tracing technique. *J Comp Neurol* 178: 1-5, 1978.

161. **Crawley JN, Hattox SE, Maas JW, Roth RH.** 3-Methoxy-4-hydroxy-phenethyleneglycol increase in plasma after stimulation of the nucleus locus coeruleus. *Brain Res* 131: 380-384, 1978.

162. **Crawley JN, Maas JW, Roth RH.** Biochemical evidence for simultaneous activation of multiple locus coeruleus efferents. *Life Sci* 26: 1373-1378, 1980.

163. **Crofton JT, Share L, Shade RE, Allen C, Tarnowski D.** Vasopressin in the rat with spontaneous hypertension. *Am J Physiol* 235: H361-H366, 1978.

164. **Dechamplain J, Farley L, Cousineau D, Van Ameringen MR.** Circulating catecholamine levels in human and experimental hypertension. *Circ Res* 38: 109-114, 1976.

165. **Elsworth JD, Redmond DE Jr, Roth RH.** Plasma and cerebrospinal fluid 3-methoxy-4-hydroxyphenylethylene glycol (MHPG) as indices of brain norepinephrine metabolism in primates. *Brain Res* 235: 115-124, 1982.

166. **Elsworth JD, Roth RH, Stogin JM, Leahy DJ, Moore MR, Redmond DE.** Peripheral correlates of central noradrenergic activity. *Neurosci Abstr* 6: 140, 1980.

167. **Esler M, Jackman G, Bobik A, Leonard P, Keleher D, Skews H, Jennings G, Korner P.** Norepinephrine kinetics in essential hypertension. Defective neuronal uptake of norepinephrine in some patients. *Hypertension* 3: 149-156, 1981.

168. **Faiers AA, Calaresu FR, Mogenson GJ.** Factors affecting cardiovascular responses to stimulation of hypothalamus in the rat. *Exp Neurol* 51:188-206, 1976.

169. **Folkow BUG, Von Euler US.** Selective activation of noradrenaline and adrenaline producing cells in the cat's adrenal gland by hypothalamic stimulation. *Circ Res* 2: 191-195, 1954.

170. **Francke PF, Culberson JL, Carmichael SW, Robinson RL.** Bilateral secretory responses of the adrenal medulla during stimulation of hypothalamic or mesencephalic sites. *J Neurosci Res* 8: 1-6, 1982.

171. **Fuller RW, Snoddy HD.** Effect of serotonin-releasing drugs on serum corticosterone concentration in rats. *Neuroendocrinology* 31: 96-100, 1980.

172. **Gauthier P.** Pressor responses and adrenomedullary catecholamine release during brain stimulation in the rat. *Can J Physiol Pharmacol* 59:485-492, 1981.

173. **Gauthier P, Reis DJ, Nathan MA.** Arterial hypertension elicited either by lesions or by electrical stimulations of the rostral hypothalamus in the rat. *Brain Res* 211: 91-105, 1981.

174. **Gilbey MP, Coote JH, Fleetwood-Walker S, Peterson DF.** The influence of the paraventriculo-spinal pathway, and oxytocin and vasopressin on sympathetic preganglionic neurones. *Brain Res* 251: 283-290, 1982.

175. **Grimm M, Weidmann P, Keusch G, Meier A, Gluck Z.** Norepinephrine clearance and pressor effect in normal and hypertensive man. *Klin Wochenschr* 58: 1175-1181, 1980.

176. **Haskins JT, Moyer JA, Muth EA, Sigg EB.** DMI, WY-45,030, WY-45,881, and ciramadol inhibit locus coeruleus neuronal activity. *Eur J Pharmacol* 115: 139-146, 1985.

177. **Hjemdahl P, Sjoquist B, Daleskog M.** A comparison of noradrenaline MHPG and VMA in plasma as indicators of sympathetic nerve activity in man. *Acta Physiol Scand* 115: 507-509, 1982.

178. **Kopin IJ, Blombery P, Ebert MH, Gordon EK, Jimerson DC, Markey SP, Polinsky RJ.** Disposition and metabolism of MHPG-CD3 in humans: plasma MHPG as the principal pathway of norepinephrine metabolism and as an important determinant of CSF levels of MHPG, in *Frontiers in Biochemical and Pharmacological Research in Depression*. Sjoqvist F, Usdin E. Eds., Raven Press, New York, 1985.

179. **Korf J, Aghajanian GK, Roth RH.** Increased turnover of norepinephrine in the rat cerebral cortex during stress: role of the locus coeruleus. *Neuropharmacology* 12: 933-938, 1973.

180. **Lin MT, Tsay BL, Fan YC.** Effects of 5-hydroxytryptamine, fluoxetine and chlorimipramine on reflex bradycardia in rats. *J Pharm Pharmacol* 32: 493-496, 1980.

181. **Loewy AD, McKellar S.** The neuroanatomical basis of central cardiovascular control. *Proc Fed Am Soc Exp Biol* 39: 2495-2503, 1980.

182. **Loewy AD, McKellar S.** Serotonergic projections from the ventral medulla to the intermediolateral cell column in the rat. *Brain Res* 211: 146-152, 1981.

183. **Loewy AD, Neil IJ.** The role of descending mono-aminergic systems in the central control of blood pressure. *Fed Proc Fed Am Soc Exp Biol* 40: 2778-2785, 1981.

184. **Maas JW, Hattox SE, Greene NM, Landis DH.** 3-Methoxy-4-hydroxy-phenethyleneglycol production by human brain in vivo. *Science* 205: 1025-1027, 1979.

185. **Martin GF, Humbertson AO, Laxson C, Panneton M.** Evidence for direct bulbospinal projections to lamine IX, X and the intermediolateral cell column. Studies using axonal transport technique in the North American opossum. *Brain Res* 170: 165-171, 1979.

186. **McCall RB, Humphrey SJ.** Central serotoninergic neurons facilitate sympathetic nervous discharge (SND). *Soc Neurosci Abstr* 7: 365.121.8, 1981.

187. **Nilaver G, Zimmerman EA, Wilkins J, Michaels J, Hoffman D, Silverman AJ.** Magnocellular hypothalamic projections to the lower brain stem and spinal cord of the rat. Immunohistochemical evidence for predominance of the oxytocin-neurophysin system compared to the vasopressin-neurophysin system. *Neuroendocrinology* 30: 150-158, 1980.

188. **Pedersen EB, Christensen NJ.** Catecholamines in plasma and urine in patients with essential hypertension determined by double-isotope derivative techniques. *Acta Med Scand* 198: 373-377, 1975.

189. **Philippu A, Demmler R, Rosenberg G.** Effects of centrally applied drugs on pressor responses to hypothalamic stimulation. *Naunyn-Schmiedeberg's Arch Pharmacol* 282: 389-400, 1974.

190. **Philippu A, Dietl H, Strohl U, Truc VT.** Adrenoceptors of the hypothalamus: their importance for the regulation of the arterial blood pressure, in *Catecholamines: Basic and Clinical Frontiers*. Usdin E, Kopin IJ, Bashes J. Eds., Pergamon Press, Elmsford, NY, 1979, 1428.

191. **Philippu A, Heyd G, Burger A.** Release of noradrenaline from the hypothalamus in vivo. *Eur J Pharmacol* 9: 52-58, 1970.

192. **Przuntek H, Guimaraes S, Philippu A.** Importance of adrenergic neurons of the brain for the rise of blood pressure evoked by hypothalamic stimulation. *Naunyn-Schmiedeberg's Arch Pharmacol* 271: 311-319, 1971.

193. **Sauerbier I, Von Mayersbach H.** Circadian variation of catecholamine in human blood. *Horm Metab Res* 9: 529-530, 1977.

194. **Svensson TH, Elan M, Yao T, Thoren P.** Parallel regulation of brain norepinephrine (NE) neurons and peripheral, splanchnic NE nerves by chemoreceptors, baroreceptors and blood volume receptors. *Neurosci Abstr* 6: 234, 1980.

195. **Van De Kaar LD, Wilkinson CW, Shrobik Y, Brownfield MS, Ganong WF.** Evidence that serotonergic neurons in the dorsal raphe nucleus exert a stimulatory effect on the secretion of renin but not of corticosterone. *Brain Res* 235: 233-243, 1982.

196. **Wallin BG, Sundlof G, Eriksson BM, Dominiak P, Grobecker H, Lindblad E.** Plasma noradrenaline correlates to sympathetic muscle nerve activity in normotensive man. *Acta Physiol Scand* 111: 69-73, 1981.

197. **Yamane Y, Nakai M, Yamamoto J, Umeda Y, Ogino K.** Release of vasopressin by electrical stimulation of the intermediate portion of the nucleus of the tractus solitarius in rats with cervical spinal cordotomy and vagotomy. *Brain Res* 324: 358-360, 1984.

198. **Coote JH, MacLeod VH, Martin IL.** Bulbospinal tryptaminergic neurons: a search for the role of bulbospinal tryptaminergic neurons in the control of sympathetic activity. *Pfluegers Arch Eur J Physiol* 377: 109-116, 1978.

199. **Culberson JL, Robinson RL, Carmichael SW, Francke PF.** CNS control of secretion by the adrenal medulla in the cat. *Anat Rec* 187: 559-560, 1977.

200. **Dequattro V, Campese V, Miura Y, Meijer D.** Increased plasma catecholamines in high renin hypertension. *Am J Cardiol* 38: 801-804, 1976.

201. **Esler M, Zweifler A, Randall O, et al.** Suppression of sympathetic nervous function in low-renin essential hypertension. *Lancet* 2: 115-118, 1976.

202. **Feniuk W, Hare J, Humphrey PPA.** An analysis of the mechanism of 5-hydroxytryptamine-induced vasopressor responses in ganglion-blocked anesthetized dogs. *J Pharm Pharmacol* 33: 155-160, 1981.

203. **Geyer MA, Puerto A, Dawsey WJ, Knapp S, Bullard WP, Mandell AJ.** Histologic and enzymatic studies of the mesolimbic and mesostriatal serotonergic pathways. *Brain Res* 106: 241-256, 1976.

204. **Goldstein DS.** Plasma norepinephrine during stress in essential hypertension. *Hypertension* 3: 551-556, 1981.

205. **Goldstein DS.** Plasma norepinephrine in essential hypertension: a study of the studies. *Hypertension* 3: 48-52, 1981.

206. **Goldstein DS, Horwitz D, Keiser HR, Polinsky RJ, Kopin IJ.** Plasma L-(3H) norepinephrine, D-(14C) norepinephrine, and d-l-(3H) isoproterenol kinetics in essential hypertension. *J Clin Invest* 72: 1748-1758, 1983.

207. **Gurtu S, Pant KK, Sinha JN, Bhargava KP.** An investigation into the mechanism of cardiovascular responses elicited by electrical stimulation of locus coeruleus and subcoeruleus in the cat. *Brain Res* 301: 59-64, 1984.

208. **Howe PRC, Stead BH, Chalmers JP.** Central serotonin nerves in spontaneously hypertensive and DOCA-salt hypertensive rats, in Proc 5th Int Symp SHR and Related Studies; cited by **Howe PRC et al.** *Clin Exp Pharmacol Physiol* 9: 335-339, 1982.

209. **Howe PRC, Stead BH, Lovenberg W, Chalmers JP.** Effects of central serotonin nerve lesions on blood pressure in normotensive and hypertensive rats. *Clin Exp Pharmacol Physiol* 9: 335-339, 1982.

210. **Jacobs BL, Wise WD, Taylor KM.** Differential behavioral and neurochemical effects following lesions of the dorsal or median raphe nuclei in rats. *Brain Res* 79: 353-361, 1974.

211. **Kent DL, Sladek JR Jr.** Histochemical, pharmacological and microspectrofluorometric analysis of new sites of serotonin localization in the rat hypothalamus. *J Comp Neurol* 180: 221-236, 1978.

212. **Kopin IJ, Goldstein DS, Feuerstein GZ.** The sympathetic nervous system and hypertension, in *Frontiers in Hypertension Research*. Laragh JJ, Buhler FR, Seldin DW. Eds., Springer-Verlag, New York, 1981, 283.

213. **Louis WJ, Doyle AE, Anavekar S.** Plasma norepinephrine levels in essential hypertension. *N Engl J Med* 288: 599-601, 1973.

214. **Martinez AA, Lokhandwala MF.** Evidence for a presynaptic inhibitory action of 5-hydroxytryptamine on sympathetic neurotransmission to the myocardium. *Eur J Pharmacol* 53: 303-311, 1980.

215. **Robinson RL, Culberson JL, Carmichael SW, Francke PF.** Selective CNS control of epinephrine and norepinephrine release by the adrenal medulla. *Fed Proc Fed Am Soc Exp Biol* 36: 381, 1977.

216. **Beckstead RM, Morse JR, Norgren R.** The nucleus of the solitary tract in the monkey: projections of the thalamus and brainstem nuclei. *J Comp Neurol* 190: 259-282, 1980.

217. **Lechin F, Van Der Dijs B, Jakubowicz D, Camero RE, Lechin S, Villa S, Reinfeld B, Lechin ME.** Role of stress in the exacerbation of chronic illness: effects of clonidine administration on blood pressure and plasma norepinephrine, cortisol, growth hormone and prolactin concentrations. *Psychoneuroendocrinology* 12: 117-129, 1987.

218. **Lechin F, Van Der Dijs B, Jakubowicz D, Camero RE, Villa S, Arocha L, Lechin AE.** Effects of clonidine on blood pressure, noradrenaline, cortisol, growth hormone, and prolactin plasma levels in high and low intestinal tone depressed patients. *Neuroendocrinology* 41: 156-162, 1985.

219. **Lechin F, Van Der Dijs B, Jakubowicz D, Camero RE, Villa S, Lechin E, Gomez F.** Effects of clonidine on blood pressure, noradrenaline, cortisol, growth hormone, and prolactin plasma levels in high and low intestinal tone subjects. *Neuroendocrinology* 40: 253-261, 1985.

220. **Antonaccio MJ, Robson RD.** Cardiovascular effects of 5-hydroxytryptophan in anesthetized dogs. *J Pharm Pharmacol* 25: 495-497, 1973.

221. **Crawley JN, Maas JW, Roth RH.** Role of the nucleus locus coeruleus in sympathetic and central noradrenergic activation as reflected by changes in norepinephrine metabolite 3-methoxy-4-hydroxy-pheneth-ylene-glycol (MHPG) in rats, 4th Int. Catecholamine Symposium, Pacific Grove, CA, 1978.

222. **Sofroniew MV.** Projections from vasopressin, oxytocin and neurophysin neurons to neural targets in the rat and human. *Histochem Cytochem* 28: 475-478, 1980.

223. **Sved AF, Van Itallie CM, Fernstrom JD.** Studies on the antihypertensive action of L-tryptophan. *J Pharmacol Exp Ther* 221: 329-333, 1982.

224. **Aghajanian GK.** Regulation of central noradrenergic cell firing: role of alpha-2 adrenoceptors and opiate receptors, in *Chemical Neurotransmission, 75 Years*. Stjarne L, Hedqvist P, Lagererantz H, Wennmaln A. Eds., Academic Press, London, 1981, 273.

225. **Aghajanian GK, Wang RY.** Physiology and pharmacology of central serotonergic neurons, in *Psychopharmacology: A Generation of Progress*. Lipton MA, DiMascio A, Killam KF. Eds., Raven Press, New York, 1978, 171.

226. **Amendt K, Czachursky K, Sellar H.** Bulbospinal projections to the intermediolateral cell column: a neuroanatomical study. *J Auton Nerv Syst* 1: 103-117, 1979.

227. **Antonaccio MJ, Kelly E, Halley J.** Centrally mediated hypotension and bradycardia by methysergide in anesthetized dogs. *Eur J Pharmacol* 33:107-117, 1975.

228. **Antonaccio MJ, Robson RD.** Centrally mediated cardio-vascular effects of 5-hydroxytryptophan in MAO-inhibited dogs: Modification by autonomic antagonists. *Arch Int Pharmacodyn Ther* 213: 200-210, 1975.

229. **Banerji TK, Quay WB.** Twenty-four hour rhythm in plasma dopamine-beta-hydroxylase activity: evidence of age and strain differences and an adrenomedullary contribution. *Chronobiologia Suppl* 1: 6, 1975.

230. **Baum T, Shropshire AT.** Susceptibility of spontaneous sympathetic outflow and sympathetic reflexes to depression by clonidine. *Eur J Pharmacol* 44: 121-129, 1977.

231. **Calza L, Giardino L, Grimaldi R, Rigoli M, Steinbusch HWM, Tiengo M.** Presence of 5-HT-positive neurons in the medial nuclei of the solitari tract. *Brain Res* 347: 135-139, 1985.

232. **Chalmers J.** Brain amines and models of experimental hypertension. *Circ Res* 36: 469-480, 1975.

233. **Chase TN, Gordon EK, Ng LK.** Norepinephrine metabolism in the central nervous system of man: studies using 3-methoxy-4-hydroxy-phenylethyleneglycol levels in cerebrospinal fluid. *J Neurochem* 21: 581-587, 1973.

234. **Ciriello J, Caverson M, Ancalresu FR.** Lateral hypothalamic and peripheral cardiovascular afferent inputs to ventrolateral medullary neurons. *Brain Res* 347: 173-176, 1985.

235. **Daiguji M, Mikuni M, Okada F, Yamashita I.** The diurnal variations of dopamine-beta-hydroxylase activity in the hypothalamus and locus coeruleus of the rat. *Brain Res* 155: 409-412, 1978.

236. **Dampney RAL, Goodchild AK, Tan E.** Vasopressor neurons in the rostral ventrolateral medulla of the rabbit. *J Auton Nerv Syst* 14: 239-254, 1985.

237. **Dechamplain J, Van Ameringen MR.** Role of sympathetic fibres and of adrenal medulla in the maintenance of cardiovascular homeostasis in normotensive and hypertensive rats, in *Frontiers in Catecholamine Research*. Usdin E, Snyder S. Eds., Pergamon Press, Oxford, 1973, 859.

238. **Dembrowsky K, Czachurski J, Amendt K, Seller H.** Tonic descending inhibition of the spinal-sympathetic reflex from the lower brainstem. *J Auton Nerv Syst* 2: 157-182, 1980.

239. **Diraddo J, Kellog C.** In vivo rates of tyrosine hydroxylation in regions of rat brain at four times during the lightdark cycle. *Naunyn-Schmiedeberg's Arch Exp Pathol Pharmakol* 286: 389-394, 1975.

240. **Edery H, Berman HA.** Yohimbine antagonism of the vasodepression elicited by organophosphates applied on ventral medulla oblongata. *J Auton Nerv Syst* 14: 229-238, 1985.

241. **Egan TM, North RA.** Acetylcholine acts on M2-muscarinic receptors to excite rat locus coeruleus neurones. *Br J Pharmacol* 85: 733-735, 1985.

242. **Elam M, Svensson TH, Thoren P.** Differentiated cardiovascular afferent regulation of locus coeruleus neurons and sympathetic nerves. *Brain Res* 358: 77-84, 1985.

243. **Glazer EJ, Ross LL.** Localization of noradrenergic terminals in sympathetic nuclei of the rat: demonstration by immunocytochemical localization of dopamine-beta-hydroxylase. *Brain Res* 185: 39-49, 1980.

244. **Granata AR, Kumada M, Reis DJ.** Sympathoinhibition by A1-noradrenergic neurons is mediated by neurons in the C1 area of the rostral medulla. *J Auton Nerv Syst* 13: 387-395, 1985.

245. **Guyenet PG, Stornette RL.** Inhibition of sympathetic preganglionic discharges by epinephrine and alpha-epinephrine. *Brain Res* 235: 271-283, 1982.

246. **Helke CJ, Muth EA, Jacobowitz DM.** Changes in central cholinergic neurons in the spontaneously hypertensive rats. *Brain Res* 188:425-436, 1980.

247. **Howe PRC.** Blood pressure control by neurotransmitters in the medulla oblongata and spinal cord. *J Auton Nerv Syst* 12: 95-115, 1985.

248. **Illert M, Gabriel M.** Descending pathways in the cervical cord of cats affecting blood pressure and sympathetic activities. *Arch Gen Physiol* 335: 109-124, 1982.

249. **Kobinger W, Pichler L.** Centrally induced reduction in sympathetic tone — a postsynaptic alpha-adrenoceptors-stimulating action of imidazolines. *Eur J Pharmacol* 40: 311-320, 1976.

250. **Leslie RA.** Neuroactive substance in the dorsal vagal complex of the medulla oblongata: nucleus of the tractus solitarius, area postrema, and dorsal motor nucleus of the vagus. *Neurochem Int* 7: 191-211, 1985.

251. **Lipski J, McAllen RM, Spyer KM.** The sinus nerve and baroreceptor input to the medulla of the cat. *J Physiol* 251: 61-78, 1975.

252. **Manschardt T, Wurtman JR.** Daily rhythm in noradrenaline content of the rat hypothalamus. *Nature* 217: 574-575, 1968.

253. **McCall RB, Humphrey SJ.** Evidence of GABA mediation of sympathetic inhibition evoked from midline medullary depressor sites. *Brain Res* 339: 356-360, 1985.

254. **Neil JJ, Loewy AD.** Decrease in blood pressure in response to L-glutamate microinjection into the A5 catecholamine cell group. *Brain Res* 241: 271-278, 1982.

255. **Palkovits M, Saavedra JM, Brownstein M.** Serotonin content of the rat brain stem nuclei. *Brain Res* 80: 237-243, 1974.

256. **Perlow M, Ebert MH, Gordon EK, Ziegler MG, Lake CR, Chase TN.** The circadian variation of catecholamine metabolism in the subhuman primate. *Brain Res* 139: 101-113, 1978.

257. **Petty MA, Reid JL.** Changes in noradrenaline concentration in brain stem and hypothalamic nuclei during the development of renovascular hypertension. *Brain Res* 136: 376-380, 1977.

258. **Reis DJ, Ross RA.** Dynamic changes in brain dopamine-beta-hydroxylase activity during anterograde and retrograde reactions to injury of central noradrenergic axons. *Brain Res* 57: 307-326, 1973.

259. **Ross CA, Armstrong DM, Ruggiero DA, Pickel VM, Joh TH, Reis DJ.** Adrenaline neurons in the rostral ventro-lateral medulla innervate thoracic spinal cord: a combined immunocytochemical and retrograde transport demonstration. *Neurosci Lett* 25: 257-262, 1981.

260. **Shagerberg G, Bjorklund A, Lindvall O, Schmidt RH.** Origin and termination of the diencephalo-spinal dopamine system in the rat. *Brain Res Bull* 9: 237-244, 1982.

261. **Tucker D, Saper C.** Specificity of spinal projections from hypothalamic and brainstem areas which innervate sympathetic preganglionic neurons. *Brain Res* 360: 159-164, 1985.

262. **Versteeg DHG, Palkovits M, Van Der Gugten J, Wijen HLSM, Smeets GWM, DeJong W.** Catecholamine content of individual brain regions of spontaneously hypertensive rats. *Brain Res* 112: 429-434, 1976.

263. **Vlahakos D, Gavras I, Gavras H.** Alpha-adrenoceptors agonists applied in the area of the nucleus tractus solitarii in the rat: effect of anesthesia on cardiovascular responses. *Brain Res* 347: 372-375, 1985.

264. **Willette RN, Barcas PP, Krieger AJ, Sapru HN.** Vasopressor and depressor areas in the rat medulla: identification by microinjection of L-glutamate. *Neuropharmacology* 22: 1071-1079, 1983.

265. **Willete RN, Punnen S, Krieger AJ, Sapru HN.** Interdependence of rostral and caudal ventrolateral medullary areas in the control of blood pressure. *Brain Res* 321: 169-174, 1984.

266. **Zandberg P, Palkovits M, DeJong W.** Effect of various lesions in the nucleus tractus solitarii of the rat on blood pressure, heart rate and cardiovascular reflex responses. *Clin Exp Hypertension* 1: 355-361, 1978.

267. **Bousquet P, Feldman J, Velly J, Bloch R.** Role of the ventral surface of the brain stem in the hypotensive action of clonidine. *Eur J Pharmacol* 34: 151-157, 1975.

268. **Bowker RM, Westlund KN, Coulter JD.** Origins of serotonergic projections to the spinal cord in the rat: an immunocytochemical-retrograde transport study. *Brain Res* 226: 187-199, 1981.

269. **Browning RA, Bundman MC, Smith ML, Myers JH.** Effects of *p*-chlorophenylalanine (PCPA) and 5,7-dihydroxy-tryptamine (5,7-DHT) on blood pressure in normo-tensive and spontaneously hypertensive (SH) rats. *Fed Proc Fed Am Soc Exp Biol* 36: 1042, 1977.

270. **Cartens E, Klumpp D, Randic M, Simmermann M.** Effect of iontophoretically applied 5-hydroxytryptamine on the excitability of single primary afferent C- and A-fibres in the cat spinal cord. *Brain Res* 220: 151-158, 1981.

271. **Chan SHH, Koo A.** The participation of medullary reticular formation in clonidine-induced hypotension in rats. *Neuropharmacology* 17: 367-373, 1978.

272. **Chemerinski E, Ramirez AJ, Enero MA.** Sinoaortic denervation induced changes in central serotonergic neurons. *Eur J Pharmacol* 64: 195-202, 1980.

273. **Crawley JN, Roth RH, Maas JW.** Locus coeruleus stimulation increases noradrenergic metabolite levels in rat spinal cord. *Brain Res* 166: 180-184, 1979.

274. **Demontigny C, Aghajanian GK.** Preferential action of 5-methoxytryptamine and 5-methoxydimethyltryptamine on presynaptic serotonin receptors: a comparative iontophoretic study with LSD and serotonin. *Neuropharmacology* 16: 811-818, 1977.

275. **Feldberg W.** The ventral surface of the brain stem: a scarcely explored region of pharmacological sensitivity. *Neuroscience* 1: 427-433, 1976.

276. **Felpel LP, Huffman RD.** Supersensitivity to norepinephrine and serotonin in the intermediolateral cell column following chronic spinal transection. *Soc Neurosci Abstr* 8: 988, 1982.

277. **Finch L.** The cardiovascular effects of intraventricular 5,6-dihydroxytryptamine in conscious hypertensive rats. *Clin Exp Pharmacol* 2: 503-508, 1975.

278. **Florez J, Armijo JA.** Effect of central inhibition of the L-amino acid decarboxylase on the hypotensive action of 5HT precursor in cats. *Eur J Pharmacol* 26: 108-110, 1974.

279. **Franz DN, Madsen PW, Peterson RG, Sangdee C.** Functional roles of monoaminergic pathways to sympathetic preganglionic neurons. *Clin Exp Hypertension* 4: 543-562, 1982.

280. **Fuller RW, Holland DR, Yen TT, Bemis KG, Stamm NB.** Antihypertensive effects of fluoxetine and L-5-hydroxy-tryptophan in rats. *Life Sci* 25: 1237-1242, 1979.

281. **Fuller RW, Yen TT, Stamm NB.** Lowering of blood pressure by direct and indirect-acting serotonin agonists in spontaneously hypertensive rats. *Clin Exp Hypertension* 3: 497-508, 1981.

282. **Giarcovich-Martinez S, Fernandez M, Chemerinski E, Enero MA.** Central serotonergic activity after neurogenic hypertension. *Eur J Pharmacol* 86: 337-345, 1983.

283. **Gilbey MP, Coote JH, MacLeod VH, Peterson DF.** Inhibition of sympathetic activity by stimulating in the raphe nuclei and the role of 5-hydroxytryptamine in this effect. *Brain Res* 226: 131-142, 1982.

284. **Gurtu S, Sharma DK, Sinha JN, Bhargava KP.** Evidence of the involvement of alpha-adrenoceptors in the nucleus ambiguus in baroreflex mediated bradicardia. *Naunyn-Schmiedeberg's Arch Pharmacol* 323: 199-204, 1983.

285. **Gurtu S, Sinha JN, Bhargava KP.** Receptors in the medullary cardioinhibitory loci. I. Nucleus tractus solitarius: catecholaminergic modulation of baroreflex induced bradycardia. *Ind J Pharmacol* 14: 37-45, 1982.

286. **Heinricher MM, Rosenfeld JP.** Microinjection of morphine into the nucleus reticularis paragiganto-cellularis of the rat suppresses spontaneous activity in nucleus raphe magnus neurons. *Brain Res* 272: 382-386, 1983.

287. **Henning M, Rubenson A.** Effects of 5-hydroxy-tryptophan on arterial blood pressure, body temperature and tissue monoamines in the rat. *Acta Pharmacol Toxicol* 29: 145-154, 1971.

288. **Kadzielawa K.** Inhibition of the spinal sympathetic preganglionic neurons by alpha-methylnorepinephrine. *Pharmacologist* 20: 189-228, 1978.

289. **Kadzielawa K.** Inhibition of the sympathetic preganglionic neurons by catecholamines. *Soc Neurosci Abstr* 4: 274.857, 1978.

290. **Kadzielawa K.** Alpha-methylnorepinephrine inhibition of spinal sympathetic preganglionic neurons (SPGN) mediated by catecholamine receptors of the alpha type. *Pharmacologist* 22: 162-229, 1980.

291. **Kadzielawa K.** Antagonism of 5-hydroxytryptamine (5HT) excitatory effects on sympathetic preganglionic neurons (SPGN). *Soc Neurosci Abstr* 6: 606.208.1, 1980.

292. **Kadzielawa K.** Antagonism of the excitatory effects of 5-hydroxy-tryptamine on sympathetic preganglionic neurones and neurones activated by visceral afferents. *Neuropharmacology* 22: 19-27, 1983.

293. **Kadzielawa K.** Inhibition of the activity of sympathetic preganglionic neurones and interneurones activated by visceral afferents by alpha-methylnoradrenaline and endogenous catecholamines. *Neuropharmacology* 22: 3-17, 1983.

294. **Kahn N, Mills E.** Centrally evoked sympathetic discharge: a functional study of medullary vasomotor areas. *J Physiol* 191: 339-352, 1967.

295. **Kawamura H, Gunn CG, Frohlich ED.** Altered cardiovascular modulation by locus coeruleus in spontaneously hypertensive rat. *Circulation* 52, II: 122, 1975.

296. **Klemfuss H, Seiden LS.** Water deprivation increases anterior hypothalamic norepinephrine metabolism in the rat. *Brain Res* 341: 222-227, 1985.

297. **Koss M, Bernthal PJ, Chandler MJ.** Use of a sympathetic-cholinergic system in the analysis of sympatho-inhibitory produced by clonidine and some congeneric derivatives of clonidine. *Eur J Pharmacol* 87: 301-308, 1983.

298. **Krstic MK.** Cardiovascular response to intracerebro-ventricular administration of acetylcholine in rats treated with physostigmine. *Neuropharmacology* 17: 1003-1008, 1978.

299. **Krstic MK, Djurkovic D.** Hypertension mediated by the activation of the rat brain 5-hydroxytryptamine receptor sites. *Experientia* 32: 1187-1188, 1976.

300. **Krstic MK, Djurkovic D.** Cardiovascular response to intracerebro-ventricular administration of acetylcholine in rats. *Neuropharmacology* 17: 341-347, 1978.

301. **Kuhn DM, Wolf WA, Lovenberg W.** Pressor effects of electrical stimulation of the dorsal and median raphe nuclei in anesthetized rats. *J Pharmacol Exp Ther* 214: 403-409, 1980.

302. **Kuhn DM, Wolf WA, Lovenberg W.** Review of the role of the central serotonergic neuronal system in blood pressure regulation. *Hypertension* 213: 243-255, 1980.

303. **Laguzzi R, Talman WT, Reis DJ.** Serotonergic mechanisms in the nucleus tractus solitarius may regulate blood pressure and behavior in the rat. *Clin Sci* 63: 323s-326s, 1982.

304. **Lambert GA, Friedman E, Buchweitz E, Gershon S.** Involvement of 5-hydroxytryptamine in the central control of respiration, blood pressure and heart rate in the anesthetized rat. *Neuropharmacology* 17: 807-813, 1978.

305. **Lambert G, Friedman E, Gershon S.** Centrally-mediated cardiovascular response to 5-HT. *Life Sci* 17: 915-920, 1975.

306. **Langer SZ.** Presynaptic regulation of catecholamine release. *Biochem Pharmacol* 23: 1793-1800, 1974.

307. **Loewy AD.** Raphe pallidus and raphe obscurus projections to the intermediolateral cell column in the rat. *Brain Res* 222: 129-133, 1981.

308. **Mathias CJ, Reid JL, Wing LMH, Frankel HL, Christenson NJ.** Antihypertensive effects of clonidine in tetraplegic subjects devoid of central sympathetic control. *Clin Sci* 5: 325s-428s, 1979.

309. **McCall RB.** Serotonergic excitation of sympathetic preganglionic neurons: a microiontophoretic study. *Brain Res* 289: 121-127, 1983.

310. **McCall RB, Schuette MR, Humphrey SJ, Lahti RA, Barsuhn C.** Evidence for a central sympatho-excitatory action of alpha$_2$-adrenergic agonists. *J Pharmacol Exp Ther* 224: 501- 507, 1983.

311. **Nathan MA.** Pathways in medulla oblongata of monkeys mediating splanchnic nerve activity. Electrophysiological and anatomical evidence. *Brain Res* 45: 115-126, 1972.

312. **Nolan PL.** The effects of serotonin precursors on the pressor response to intravenous clonidine in conscious rats. *Clin Exp Pharmacol Physiol* 4: 579-583, 1979.

313. **Ogawa M.** Interaction between noradrenergic and serotonergic mechanisms on the central regulation of blood pressure in the rat. *Jpn Circ J* 42: 581-597, 1978.

314. **Pant KK, Gurtu S, Sharma DK, Sinha JN, Bhargava KP.** Cardiovascular effects of microinjection of morphine into nucleus locus coeruleus in the cat. *Jpn J Pharmacol* 33: 253-256, 1983.

315. **Smits JF, Van Essen H, Struyker-Boudier HAJ.** Serotonin-mediated cardiovascular responses to electrical stimulation of the raphe nuclei in the rat. *Life Sci* 23: 173-178, 1978.

316. **Steinbusch HWM.** Distribution of serotonin immuno-reactivity in the central nervous system of the rat-cell bodies and terminals. *Neuroscience* 6: 557-618, 1981.

317. **Sved AF, Fernstrom JD.** Tryptophan administration lowers blood pressure in spontaneously hypertensive rats. *Fed Proc Fed Am Soc Exp Biol* 39: 608, 1980.

318. **Sved AF, Fernstrom JD, Wurtman RJ.** Tyrosine administration reduces blood pressure and enhances brain norepinephrine release in spontaneously hypertensive rats. *Proc Natl Acad Sci USA* 76: 3511-3514, 1979.

319. **Ward DG, Baertschi AJ, Gann DS.** Activation of solitary nucleus neurons from the locus coeruleus and vicinity. *Neurosci Abstr* 1: 658, 1975.

320. **Ward DG, Gunn CG.** Locus coeruleus complex: differential modulation of depressor mechanisms. *Brain Res* 107: 407-411, 1976.

321. **Wikberg JES.** The pharmacological classification of adrenergic alpha$_1$- and alpha$_2$-receptors and their mechanisms of action. *Acta Physiol Scand Suppl* 468: 1-99, 1979.

322. **Wolf DL, Mohrland JS.** Lateral reticular formation as a site for morphine- and clonidine-induced hypotension. *Eur J Pharmacol* 98: 93-98, 1984.

323. **Zandberg P, DeJong W.** Alpha-methylnoradrenaline-induced hypotension in the nucleus tractus solitarii of the rat: a localization study. *Neuropharmacology* 16: 219-225, 1977.

324. **Zandberg P, DeJong W, DeWied D.** Effect of catecholamine receptor stimulating agents on blood pressure after local application in the nucleus tractus solitarii of the medulla oblongata. *Eur J Pharmacol* 55: 43-55, 1979.

325. **Aghajanian GK, Vandermaelen CP.** Alpha-adrenoceptor mediated hyperpolarization of locus coeruleus neurons: intracellular studies in vivo. *Science* 215: 1394-1396, 1982.

326. **Anden NE, Golembiowska-Nikitin K, Thormstrom U.** Selective stimulation of dopamine and noradrenaline autoreceptors by B-HT 920 and B-HT 933, respectively. *Naunyn-Schmiedeberg's Arch Pharmacol* 321: 100-104, 1982.

327. **Anden NE, Nilsson H, Ros E, Thormstrom U.** Effect of B-HT 920 and B-HT 933 on dopamine and noradrenaline autoreceptors in the rat brain. *Acta Pharmacol Toxicol* 52: 51-56, 1983.

328. **Benarroch EE, Balda MS, Finkielman S, Nahmod VE.** Neurogenic hypertension after depletion of norepinephrine in anterior hypothalamus induced by 6-hydroxy-dopamine administration into the ventral pons: role of serotonin. *Neuropharmacology* 22: 29-34, 1983.

329. **Benarroch EE, Pirola CJ, Alvarez AL, Nahmod VE.** Serotonergic and noradrenergic mechanisms involved in the cardiovascular effects of angiotensin II injected into the anterior hypothalamic preoptic regions of rats. *Neuropharmacology* 20: 9-13, 1981.

330. **Bhargava KP.** Role of cholinergic and tryptaminergic mechanisms in cardiovascular control, in *Proc 6th Int Cong Pharmacology*. Tuomisto J, Paasonen, MK. Eds., Forssa, Finland, 1975, 69.

331. **Blessing WW, Sved AF, Reis DJ.** Destruction of noradrenergic neurons in the rabbit brainstem elevates plasma vasopressin, causing hypertension. *Science* 217: 661-663, 1982.

332. **Brezenoff HE.** Cardiovascular response to intrahypothalamic injection of carbachol and certain cholinesterase inhibitors. *Neuropharmacology* 11: 637-644, 1972.

333. **Brezenoff HE, Giuliano R.** Cardiovascular control by cholinergic mechanisms in the central nervous system. *Annu Rev Pharmacol Toxicol* 22: 341-350, 1982.
334. **Cabot JB, Wild JM, Cohen DH.** Raphe inhibition of sympathetic preganglionic neurons. *Science* 203: 184-186, 1979.
335. **Calaresu FR, Ciriello J.** Projection to the hypothalamus from buffer nerves and nucleus tractus solitarius in the cat. *Am J Physiol* 239: R 126-129, 1980.
336. **Cavero I, Lefevre-Borg F, Gomeni R.** Blood pressure lowering effects of N,N-di-n-propyl-dopamine in rats: evidence for stimulation of peripheral dopamine receptors leading to inhibition of sympathetic vascular tone. *J Pharmacol Exp Ther* 218: 515-524, 1981.
337. **Cavero I, Lefevre-Borg F, Gomeni R.** Heart rate lowering effects of N,N-di-n-propyl-dopamine in rats: evidence for stimulation of central dopamine receptors leading to inhibition of sympathetic tone and enhancement of parasympathetic outflow. *J Pharmacol Exp Ther* 219: 510-519, 1981.
338. **Clapham JC, Hamilton TC.** Presynaptic dopamine receptors mediate the inhibitory action of the dopamine agonists on stimulation-evoked pressor responses in the rat. *J Auton Pharmacol* 3: 181-188, 1982.
339. **Criscione L, Reis DJ, Talman WT.** Cholinergic mechanisms in the nucleus tractus solitarii and cardiovascular regulation in the rat. *Eur J Pharmacol* 88: 47-55, 1983.
340. **Day MD, Roach AG.** Central alpha and beta-adrenoceptors modifying arterial blood pressure and heart rate in conscious cats. *Br J Pharmacol* 51: 325-333, 1974.
341. **Echizen H, Freed CR.** Altered serotonin and norepinephrine metabolism in rat dorsal raphe nucleus after drug-induced hypertension. *Life Sci* 34: 1581-1589, 1984.
342. **Folkow BUG, Rubinstein EH.** Cardiovascular effects of acute and chronic stimulations of hypothalamic defense area in rats. *Acta Physiol Scand* 68: 48-57, 1966.
343. **Gurtu S, Sinha JN, Bhargava KP.** Involvement of alpha-adrenoceptors of the nucleus tractus solitarius in baroreflex mediated bradycardia. *Naunyn-Schmiedeberg's Arch Pharmacol* 321: 38-43, 1982.
344. **Guyenet PG.** Baroreceptor-mediated inhibition of A5 noradrenergic neurons. *Brain Res* 303: 31-40, 1984.
345. **Hiller JG, Martin PR, Redfern PH.** A possible interaction between the 24 hour rhythms in catecholamine and 5-hydroxytryptamine concentration in the rat brain. *J Pharm Pharmacol* 27, (Suppl.): 400, 1975.
346. **Hwa JY, Chan SHH.** Suppression of bradycardia induced by gigantocellular reticular nucleus by clonidine and morphine in the cat. *Neurosci Abstr* 6: 755, 1980.
347. **Kalia M, Mesulam MM.** Brainstem projections of sensory and motor components of the vagus complex in the cat. I. Cervical vagus and nodose ganglion. *J Comp Neurol* 193: 435-465, 1980.
348. **Kubo T, Misu Y.** Changes in arterial blood pressure after microinjections of nicotine into the dorsal area of the medulla oblongata of the rat. *Neuropharmacology* 20: 521-530, 1981.
349. **Kubo T, Misu Y.** Pharmacological characterization of the alpha-adrenoceptors responsible for a decrease in blood pressure in the nucleus tractus solitarius of the rat. *Naunyn-Schmiedeberg's Arch Pharmacol* 317: 120-125, 1981.
350. **Lumb BM, Wolstencroft JH.** Electrophysiological studies of a rostral projection from the nucleus raphe magnus to the hypothalamus in the rat and cat. *Brain Res* 327: 336-339, 1985.
351. **McAllen RM, Spyer KM.** The location of cardiac vagal preganglionic motoneurones in the medulla of cat. *J Physiol* 258: 187-192, 1976.
352. **McCall RB.** Evidence for a serotonergically mediated sympathoexcitatory response to stimulation of medullary raphe nuclei. *Brain Res* 311: 131-139, 1984.
353. **Moore RY.** The anatomy of central serotonin neuron systems in the rat brain, in *Serotonin Neurotransmission and Behavior*. Jacobs BL, Galperin, A. Eds., MIT Press, Cambridge, 1981, 35.
354. **Moore RY, Halavis AE, Jones BE.** Serotonin neurons of the midbrain raphe: ascending projections. *J Comp Neurol* 180: 471-488, 1978.
355. **Morris MJ, Devynck MA, Woodcock EA, Johnston CI, Meyer P.** Specific changes in hypothalamic alpha-adrenoceptors in young spontaneously hypertensive rats. *Hypertension* 3: 516-520, 1981.
356. **Nicholson G, Greely G, Humm J, Youngblood W, Kiser JS.** Lack of effect of noradrenergic denervation of the hypothalamus and medial preoptic area on the feed back regulation of gonadotropin secretion and the estrous cycle of the rat. *Endocrinology* 103: 556-566, 1978.
357. **Ono TH, Nishino H, Sasaka K, Muramoto K, Yano I, Simpson A.** Paraventricular connections to spinal cord and pituitary. *Neurosci Lett* 10: 141-146, 1978.
358. **Philippu A.** Review: involvement of cholinergic systems of the brain in the central regulation of cardiovascular functions. *J Auton Pharmacol* 1: 321-332, 1981.
359. **Renaud LP, Day TA.** Excitation of supraoptic putative vasopressin neurons following electrical stimulation of the A1 catecholamine cell group region of the rat medulla. *Soc Neurosci Abstr* 8: 422, 1982.
360. **Rogers RC, Nelson DO.** Neurons of the vagal division of the solitary nucleus activated by the paraventricular nucleus of the hypothalamus. *J Auton Nerv Syst* 10: 193-197, 1984.
361. **Ross CA, Ruggiero DA, Ries DJ.** Afferent projections to cardiovascular portions of the nucleus of the tractus solitarius in the rat. *Brain Res* 223: 402-410, 1981.

362. **Saavedra JM, Palkovits M, Brownstein MJ, Axelrod J.** Serotonin distribution in the nuclei of the rat hypothalamus and preoptic region. *Brain Res* 77: 157-165, 1974.

363. **Sawchenko PE, Swanson LW.** Central noradrenergic pathways for the integration of hypothalamic neuroendocrine and autonomic responses. *Science* 214: 685-687, 1981.

364. **Sawchenko PE, Swanson LW.** Anatomic relationships between vagal preganglionic neurons and aminergic and peptidergic neural systems in the brainstem of the rat. *Soc Neurosci Abstr* 8: 427, 1982.

365. **Sawchenko PE, Swanson LW.** Immunohistochemical identification of paraventricular hypothalamic neurons that project to the medulla or to the spinal cord in the rat. *J Comp Neurol* 205: 260-272, 1982.

366. **Sawchenko PE, Swanson LW.** The organization of noradrenergic pathways from the brainstem to the paraventricular and supraoptic nuclei in the rat. *Brain Res Rev* 4: 275-325, 1982.

367. **Sharma DK.** Further Analysis of the Central Receptors Involved in the Baroreceptor Reflex. M.D. thesis, University of Lucknow, India, 1982.

368. **Sharma DK, Gurtu S, Sinha JN, Bhargava KP.** Receptors in the medullary cardioinhibitory loci. II. Nucleus ambiguus: changes in heart rate and blood pressure following microinjection of adrenergic and cholinergic agents. *Ind J Pharmacol* 13: 38-51, 1982.

369. **Sinha JN, Gurtu S, Bhargava KP.** Effects of microinjection of alpha-adrenoceptors agonists and antagonists into medullary cardioinhibitory loci, Proc 8th Int Conf IUPHAR, Tokyo, Abstr. 374, 1981.

370. **Sinha JN, Gurtu S, Bhargava KP.** Characterization of the receptors of the nucleus tractus solitarius (NTS) involved in regulation of heart rate. *Neurosci Abstr* (Suppl.) 7: S195, 1982.

371. **Sinha JN, Gurtu S, Sharma DK, Bhargava KP.** An investigation of the regulation of heart rate. *Naunyn-Schmiedeberg's Arch Pharmacol* (Suppl.) 319: R47, 1982.

372. **Sinha JN, Sharma DK, Gurtu S, Pant KK, Bhargava KP.** Nucleus locus coeruleus: evidence for alpha$_1$-adrenoceptor mediated hypotension in the cat. *Naunyn-Schmiedeberg's Arch Pharmacol* 326: 193-197, 1984.

373. **Sofroniew MV, Schrell U.** Evidence for a direct projection from oxytocin and vasopressin neurons in the hypothalamic paraventricular nucleus to the medulla oblongata. *Neurosci Lett* 22: 211-217, 1981.

374. **Starke K.** Regulation of noradrenaline release by presynaptic receptor systems. *Rev Physiol Biochem Pharmacol* 77: 1-124, 1977.

375. **Starke K.** Presynaptic receptors. *Annu Rev Pharmacol Toxicol* 21: 7-30, 1981.

376. **Swanson LW, Kuypers HGJM.** The paraventricular nucleus of the hypothalamus: cytoarchitectonic subdivisions and the organization of projections to the pituitary, dorsal vagal complex and spinal cord as demonstrated by retrograde fluorescence double labeling methods. *J Comp Neurol* 194: 555-570, 1980.

377. **Swanson LW, Sawchenko PE.** Hypothalamic integration: organization of paraventricular and supra-optic nuclei. *Annu Rev Neurosci* 6: 269-324, 1983.

378. **Swanson LW, Sawchenko PE, Berod A, Hartman BK, Helle KB, Van Orden DE.** An immunohistochemical study of the organization of catecholaminergic cells and terminals fields in the paraventricular and supraoptic nuclei of the hypothalamus. *J Comp Neurol* 196: 271-285, 1981.

379. **Wilffert B, Smit G, DeJong A, Thoolen MJMC, Timmermans PBMWM, Van Zwieten PA.** Inhibitory dopamine receptors on sympathetic neurons innervating the cardiovascular system of the pithed rat. Characterization and role in relation to presynaptic alpha-2-adrenoceptors. *Naunyn-Schmiedeberg's Arch Pharmacol* 326: 91-98, 1984.

380. **Zandberg P, DeJong W.** Localization of catecholaminergic receptor sites in the nucleus tractus solitarii involved in the regulation of arterial blood pressure, in *Hypertension and Brain Mechanisms*. DeJong W, Provoost AP, Shapiro AP. Eds., Elsevier, Amsterdam, 1977, 117.

381. **Cabot JB, Edwards E, Bogan N, Schechter N.** Alpha-2-adrenergic receptors in avian spinal cord: increases in apparent density associated with the sympathetic preganglionic cell column. *J Auton Nerv Syst* 11: 77-89, 1984.

382. **Chu NS, Bloom FE.** The catecholamine containing neurons in the cat dorsolateral pontine tegmentum: distribution of the cell bodies and some axonal projections. *Brain Res* 66: 1-21, 1974.

383. **Contreras RJ, Gomez MM, Norgren R.** Central origins of cranial nerve parasympathetic neurons in the rat. *J Comp Neurol* 169: 373-394, 1980.

384. **Doxey JC, Everitt J.** Inhibitory effects of clonidine on responses to sympathetic nerve stimulation in the pithed rat. *Br J Pharmacol* 61: 559-566, 1977.

385. **Gebber GL, Taylor DG, Weaver LC.** Electrophysiological studies on organization of central vasopressor pathways. *Am J Physiol* 224: 470-481, 1973.

386. **Howe PRC, Kuhn DM, Minson JB, Stead BH, Chalmers JP.** Evidence for a bulbospinal serotonergic pressor pathway in the rat brain. *Brain Res* 270: 29-36, 1983.

387. **Kalia M, Mesulam MM.** Brainstem projections of sensory and motor components of the vagus complex in the cat. II. Laryngeal, tracheobronchial, pulmonary, cardiac, and gastrointestinal branches. *J Comp Neurol* 193: 467-508, 1980.

388. **Kalia M, Sullivan JM.** Brainstem projections of sensory and motor components of the vagus nerve in the rat. *J Comp Neurol* 211: 248-264, 1982.

389. **Kobinger W.** Central alpha-adrenergic systems as targets for hypotensive drugs. *Rev Physiol Biochem Pharmacol* 81: 39-100, 1978.

390. **Loewy AD, Burton H.** Nuclei of the solitary tract: efferent projections to the lower brain stem and spinal cord of the cat. *J Comp Neurol* 181: 421-450, 1978.

391. **Miura M, Reis DJ.** The paramedian reticular nucleus: a site of inhibitory interaction between projections from fastigial nucleus and carotid sinus nerve acting on blood pressure. *J Physiol* 216: 441-460, 1971.

392. **Miura M, Reis DJ.** The role of the solitary and paramedian reticular nuclei in mediating cardiovascular reflex responses from carotid baro- and chemoreceptors. *J Physiol* 223: 525-548, 1972.

393. **O'Donohue TL, Crawley WR, Jacobowitz DM.** Biochemical mapping of the noradrenergic ventral bundle projection sites: evidence for a noradrenergic-dopaminergic interaction. *Brain Res* 172: 87-100, 1979.

394. **Olson L, Fuxe K.** Further mapping out of central noradrenaline neuron systems: projection of the sub-coeruleus area. *Brain Res* 43: 289-295, 1972.

395. **Rochette L, Bralet J.** Effect of norepinephrine receptor stimulating agent "clonidine" on the turnover of 5-hydroxytryptamine in some areas of the rat brain. *J Neural Transm* 37: 259, 1975.

396. **Sawchenko PE.** Central connections of the sensory and motor nuclei of the vagus nerve. *J Auton Nerv Syst* 9: 13-26, 1983.

397. **Snyder DW, Gebber GL.** Relationships between medullary depressor region and central vasopressor pathways. *Am J Physiol* 225: 1129-1137, 1973.

398. **Takahashi H, Bunag RD.** Augmentation of centrally induced alpha-adrenergic vasodepression in spontaneously hypertensive rats. *Hypertension* 2: 198-202, 1980.

399. **Takahashi H, Takeda K, Yoneda S, Inoue A, Yoshimura A, Nakagawa M, Ijichi H.** Dysfunction of supramedullary alpha-adrenergic mechanisms following sino-aortic denervation in Kyoto Wistar rats. *Life Sci* 32: 1539-1545, 1983.

400. **Idowu OA, Zar MA.** Inhibitory effect of clonidine on a peripheral adrenergic synapse. *Br J Pharmacol* 58: 278P, 1976.

401. **Schmitt H.** Influence of adrenergic and cholinergic mechanisms on the central cardiovascular structures and their interactions, in *Drugs and Central Synaptic Transmission*. MacMillan, London, 1976, 63.

402. **Starke K, Borowski E, Endo T.** Preferential blockade of presynaptic alpha-adrenoceptors by yohimbine. *Eur J Pharmacol* 34: 384-388, 1975.

403. **Starke K, Montel H, Endo T.** Relative potencies of sympathomimetic drugs on pre and postsynaptic adrenoceptors. *Naunyn-Schmiedeberg's Arch Pharmacol* 287, Suppl. 5, 1975.

404. **Anden NE, Strombon U.** Stimulation of central adrenergic alpha-receptors by L-dopa, alpha-methyl-dopa and clonidine, in *Central Action of Drugs in Blood Pressure Regulation*. Davies DS, Reid JL. Eds., Pitman Medical, Tunbridge Wells, Kent, England, 1975, 225.

405. **Anderson C, Stone TW.** On the mechanism of action of clonidine. Effects on single central neurons. *Br J Pharmacol* 51: 359-365, 1974.

406. **Bogaievsky D, Bogaievsky Y, Tsoucaris-Kupfer D, Schmitt H.** Blockade of the central hypotensive effect of clonidine by alpha-adrenoceptor antagonists in rats, rabbits and dogs. *Clin Exp Pharmacol Physiol* 1: 527-534, 1974.

407. **DeJong A, Van Den Berg G, Qian JQ, Wilffert B, Thoole MJMC, Timmermans PBMWM, Van Zwieten PA.** Inhibitory effect of alpha-1 adrenoceptor stimulation on cardiac sympathetic neurotransmission in Pithed normotensive rats. *J Pharmacol Exp Ther* 236: 500-504, 1986.

408. **Elliot JM, Stead BH, West MJ, Chalmers J.** Cardiovascular effects of intracisternal 6-hydroxydopamine and of subsequent lesions of the ventrolateral medulla coinciding with the A1 group of noradrenaline cells in the rabbit. *J Auton Nerv Syst* 12: 117-130, 1985.

409. **Kobinger W.** Central cardiovascular actions of clonidine, in *Central Action of Drugs in Blood Pressure Regulation*. Davies DS, Reid IL. Eds., Pitman Medical, Tunbridge Wells, Kent, England, 1975, 181.

410. **Kobinger W, Pichler L.** Localization in the CNS of adrenoceptors which facilitate a cardioinhibitory reflex. *Naunyn Schmiedeberg's Arch Pharmacol* 286: 371-380, 1975.

411. **Kobinger W, Pichler L.** The central modulatory effect of clonidine on the cardio depressor reflex after suppression of synthesis and storage of noradrenaline. *Eur J Pharmacol* 30: 56-64, 1975.

412. **Kobinger W, Walland A.** Involvement of adrenergic receptors in central vagus activity. *Eur J Pharmacol* 16: 120-128, 1971.

413. **Kobinger W, Walland A.** Facilitation of vagal reflex bradycardia by an action of clonidine on central alpha-receptors. *Eur J Pharmacol* 19: 210-222, 1972.

414. **McKellar S, Loewy AD.** Efferent projections of the A1 catecholamine cell group in the rat: an autoradiographic study. *Brain Res* 241: 11-29, 1982.

415. **Srimal RC, Gulati K, Dhawan BN.** On the mechanism of central hypotensive action of clonidine. *Can J Physiol Pharmacol* 55: 1007-1014, 1977.

416. **Struyker-Boudier H, Smeets G, Brouwer G, Van Rossum J.** Central and peripheric alpha-adrenergic activity of imidazoline derivatives. *Life Sci* 15: 887-895, 1974.

417. **Struyker-Boudier HAK, Van Rossum JM.** Clonidine-induced cardiovascular effects after stereotaxic application in the hypothalamus of rats. *J Pharm Pharmacol* 24: 410-418, 1972.

418. **Enero MA, Langer SZ, Rothlin RP, Stefano FJE.** Role of the alpha-adrenoceptor in regulating noradrenaline overflow by nerve stimulation. *Br J Pharmacol* 44: 672-679, 1972.

419. **Koss MC.** Studies on the site of action of clonidine utilizing a sympathetic-cholinergic system. *Eur J Pharmacol* 37: 381-384, 1976.

420. **Starke K, Altman KP.** Inhibition of adrenergic neurotransmission by clonidine: an action on prejunctional alpha-receptors *Neuropharmacology* 12: 339-341, 1973.

421. **Bhargava KP, Jain IP, Saxena AK, Sinha JN, Tangri KK.** Central adrenoceptors and cholinoceptors in cardiovascular control. *Br J Pharmacol* 74: 842P, 1978.

422. **Byrum CF, Stornetta RL, Guyenet PG.** Electro-physiological properties of spinally projecting A5 not adrenergic neurons. *Brain Res* 303: 15-29, 1984.

423. **Eriksson E, Eden S, Modigh K.** Up- and down-regulation of central postsynaptic alpha$_2$-receptors reflected in the growth hormone response to clonidine in reserpine pre-treated rats. *Psychopharmacology* 77: 327-331, 1982.

424. **Hamilton TC, Hunt AAE, Poyser RH.** Involvement of central alpha$_2$-adrenoceptors in the mediation of clonidine-induced hypotension in the cat. *J Pharmacol* 32: 788-789, 1980.

425. **Moore SD, Guyenet PG.** Effect of blood pressure on A2 noradrenergic neurons. *Brain Res* 338: 169-172, 1985.

426. **Jimerson DC, Gordon EK, Post RM, Goodwin FK.** Central noradrenergic function in man: vanyll-mandelic acid in CSF. *Brain Res* 99: 434-439, 1975.

427. **Jouvet M.** The role of monoamines and acetylcholine containing neurons in the regulation of sleep-waking cicle. *Ergeb Physiol* 64: 168-342, 1972.

428. **Lake CR, Ziegler MG, Kopin IJ.** Human plasma norepinephrine. I. Variations in normal subjects. *Neurosci Abstr* 1: 413, 1975.

429. **Morgan WW, McFadin LS, Harvey CY.** A daily rhythm in norepinephrine content in regions of the hamster brain. *Comp Gen Pharmacol* 1: 47-52, 1973.

430. **Vogt M.** Metabolites of cerebral transmitters entering the cerebrospinal fluid: their value as indicators of brain function, in *Fluid Environment of the Brain*. Cserr HF, Fenstermacher JD, Vencl V. Eds., Academic Press, New York, 1975, 225.

431. **Ziegler MG, Lake CR, Foppen FH, Shoulson I, Kopin I.** Norepinephrine in cerebrospinal fluid. *Brain Res* 108: 436-440, 1976.

432. **Kobayashi RM, Palkovits M, Kizer JS, Jacobowitz DM, Kopin IJ.** Selective alterations of catecholamines and tyrosine hydroxylase activity in the hypothalamus following acute and chronic stress, in *Catecholamines and Stress*. Usdin E, Kvetnansky R, Kopin IJ. Eds., Pergamon Press, Oxford, 1976, 29.

433. **Norgren R.** Projection from the nucleus of the solitary tract in the rat. *Neuroscience* 3: 207-218, 1978.

434. **Ricardo JA, Koh ET.** Anatomical evidence of direct projections from the nucleus of the solitary tract to the hypothalamus, amygdala, and other forebrain structures in the rat. *Brain Res* 153: 1-26, 1978.

435. **Weiner RY, Shryne JE, Gorski RA, Sawyer CH.** Changes in the catecholamine content of the rat hypothalamus following deafferentation. *Endocrinology* 90: 867-877, 1971.

436. **Bobillier P, Seguin S, Degueurce A, Lewis BD, Pujol JF.** The efferent connection of the nucleus raphe centralis in the rat as revealed by autoradiography. *Brain Res* 166: 1-8, 1979.

437. **Mosko SS, Jacobs BL.** Electrophysiological evidence against negative neuronal feedback from the forebrain controlling midbrain raphe unit activity. *Brain Res* 119: 291-303, 1977.

438. **Sakai K, Salvert D, Touret M, Jouvet M.** Afferent connections of the nucleus raphe dorsalis in the cat as visualized by the horseradish peroxidase technique. *Brain Res* 145: 1-25, 1977.

439. **Tangri KK, Saxena AX, Misra N, Mumar A, Bhargava KP.** Nature of receptors in midbrain raphe nuclei concerned in thermoregulation in rabbit, in *Proc 7th Int Cong Pharmacology, Advances in Pharmacology and Therapeutics*. Boissier JR, Lechat P, Fichelle J. Eds., Pergamon Press, Oxford, 1978, 932.

440. **Couch JR.** Responses of neurons in the raphe nuclei to serotonin, norepinephrine and acetylcholine and their correlation with an excitatory synaptic input. *Brain Res* 19: 137-150, 1970.

441. **Aghajanian GK, Vandermaelen CP.** Intracellular recordings from serotonergic dorsal raphe neurons: pacemaker potentials and the effect of LSD. *Brain Res* 238: 463-469, 1982.

442. **Azmitia EC.** The serotonin-producing neurons in the midbrain median and dorsal raphe nuclei, in *Chemical Pathways in the Brain. Handbook of Psychopharmacology*. Vol. 9, Iversen LL, Iversen S, Snyder SH. Eds., Plenum Press, New York, 1978, 223.

443. **Lechin F, Van Der Dijs B.** Slow wave sleep (SWS), REM sleep (REMS) and depression. *Res Commun Psychol Psychiatr Behav* 9: 227-262, 1984.

444. **Mosko SS, Haubrich D, Jacobs BL.** Serotonergic afferents to the dorsal raphe nucleus: evidence from HRP and synaptosomal uptake studies. *Brain Res* 119: 269-290, 1977.

Chapter 6

BIOLOGICAL MARKERS IN THE ASSESSMENT OF CENTRAL AUTONOMIC NERVOUS FUNCTIONING: AN APPROACH TO THE DIAGNOSIS OF SOME PSYCHIATRIC AND PSYCHOSOMATIC SYNDROMES

TABLE OF CONTENTS

INTESTINAL PHARMACOMANOMETRY

Fuad Lechin, Bertha van der Dijs, Francisco Gómez, Alex Lechin,
Emilio Acosta, and Luis Arocha

I. INTRODUCTION

The cholinergic-serotonergic (ACh-5HT), noradrenergic (NE), and dopaminergic (DA) systems are known to influence distal colon motility (DCM) in humans. The administration in usual therapeutic doses of drugs, agonistic or antagonistic to these systems, can enhance or reduce that motor activity.[1-4] Up to the present, we have investigated the effect of drugs such as D-amphetamine (releases catecholamines and inhibits their reuptake by terminals); fenfluramine (releases serotonin); mianserin, chlorprothixene (α_2-blocking agents, which induce NE release from terminals); prazosin (α_1-blocking agent, which antagonizes NE effect at postsynaptic level); phentolamine and dihydroergotamine ($\alpha_1 + \alpha_2$-blocking agents); clonidine (α_2-agonist which inhibits NE release from terminals); propranolol ($\beta_1 + \beta_2$-blocking agent); haloperidol, sulpiride, pimozide, thioproperazine, trifluoperazine, chlorproperazine, and other dopaminergic blocking agents; thioridazine (DA + NE antagonist); domperidone (peripheral DA blocking agent); nomifensine and methylphenidate (inhibitors of DA uptake); metergoline ($5HT_2$-blocking agent); methysergide (peripheral 5HT antagonist); clonazepam (inhibits DA release by acting at presynaptic level); bromocriptine (presynaptic DA agonist); lorazepam, prazepam and other benzodiazepines; biperiden (central acting anti-ACh agent); hioscine (peripheral anti-ACh agent); baclofen (Gabamimetic agent); and other drugs such as diphenyl-hydantoin and cholecystokinin.[5-11]

We found several patterns of DCM which remained constant for each subject when serial motility studies were performed on different days. However, psychotherapy, pharmacotherapy, and change-in-life situations were shown to induce shifts in the motility patterns. Basing ourselves on DCM studies performed in more than 5000 subjects, we found a close relationship between DCM profile and character traits and personality.[12] Further, we have found that psychological and drug-induced motility changes parallel each other. In addition, we have found that these drug-induced DCM and behavior changes vary from one subject to another according to the patient's predrug profile of motility, which suffers no spontaneous oscillations throughout a 5-h DCM study.

II. PROCEDURE

DCM studies were performed according to previously established procedures in sigmoid and rectum, by means of two open-ended polyethylene catheters continuously perfused with saline solution at a rate of 9 ml/h. The speed of paper was 5 mm/min. Catheters (1.6-mm internal diameter) were connected to two differential air pressure transducers (Statham PM131). A Nihon Kohden polygraph was employed to register pressure changes. A control period of 1 to

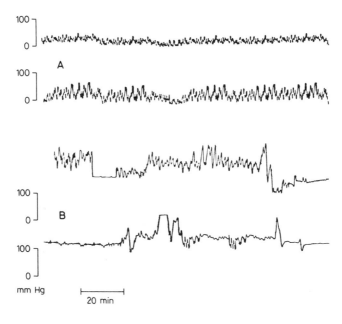

FIGURE 1. Initial segments of DCM records of two normal subjects: one (subject A), low-IT and the other (subject B), high-IT. Motility of the distal colon has two components: (1) IT = the distance between zero-line and base-line pressure, determined on trace segments lacking waves and (2) phasic activity = waves. When very large waves occur frequently, tracing is raised and simulates an elevation of IT (base-line). Consequently, in order to determine the true tone it is necessary to wait for the appearance of a segment without waves, continuing the motility record control period as long as needed. For these reasons, conventional motility indexes are inexact. These motility indexes do not discriminate between IT and phasic activity. Subject A showed low-IT (–20 mm Hg) at both sigmoidal (upper trace) and rectal (lower trace) levels. Subject B showed high-IT (sigmoid 140 mm Hg, rectum 120 mm Hg). The motility record shows segments lacking waves. From *Neuroendocrinology* 40: 253-261, 1985. With permission.

2 h was allowed before administration of the first drug. After the administration of each drug, 90 min of register was obtained.

Motility of the distal colon has two components: (1) intestinal tone (IT) = the distance between zero-line and base-line pressure, determined on segments of trace lacking waves; and (2) phasic activity (PA) = waves. (Figure 1).

We gave up measuring the DCM according to conventional motility indexes more than 11 years ago, because this method was unable to discriminate between the two components of DCM. IT does not vary spontaneously and is influenced little or not at all by emotional factors. PA registers spontaneous variations, within certain limits and for short periods (1 to 5 min), and is influenced by emotions. Regarding these changes, we have devised an index we call reactivity index (RI) = the emotionally induced quantitative changes of PA arbitrarily measured from (+) to (++++). RI is always reduced to zero by the prior administration of NE and DA antagonists. In analyzing a motility record over a period of several hours, we observed that these short-lived (1 to 5 min) oscillations of PA do not affect the general pattern of register. In order to measure a contraction wave (PA) it is first necessary to know the area circumscribed by a triangle limited by the ascending line, the descending line, and the base line. When waves occur frequently, the relaxation phase of the intestinal muscular layer is interrupted by the appearance of a new contraction which is capable of keeping the recording pen raised for a large segment of the tracing, simulating an elevation of IT (base line). Consequently, in order to determine the true tone it is necessary to wait for the appearance of a segment without waves, prolonging the motility record control period as long as needed. For these reasons, conventional motility indexes are inexact. We dealt with this problem by devising a much more precise method arrived at by weighing the total areas of paper limited by the tracing and zero line and, from this, subtracting the paper corresponding to the area below the IT line (previously determined). Thus, we are able to differentiate between PA and IT and to quantify them. A hypotonic intestine can develop intense PA and vice versa, a hypertonic intestine may show segments lacking waves. After the true IT is determined with precision, we prolong the control motility record 1 h, at least. In order to evaluate the drug-induced changes, it is necessary to compare the latter with spontaneous oscillation occuring during the control period.

In order to calculate the spontaneous oscillations the predrug tracing is divided into 20-min segments. The first predrug 20-min segment = 100. Increases or decreases are expressed in ± %. The spontaneous oscillation range we obtained from 5126 motility records = 9%. Motility records obtained in 105 subjects in which speed paper was alternated between 5 and 10 mm/min (20-min segments) showed no significant differences (3% error) between calculations of spontaneous oscillations. Postdrug periods are also divided into 20-min segments. Means of PA and IT obtained from the predrug period (3 or more 20-min segments) are compared with means of PA and IT obtained from the postdrug period (3 or more 20-min segments). Similarly, comparisons are made between the first postdrug period and the second postdrug period and between the second and the third postdrug periods. Drug-induced motility changes are always greater than 50% (usually greater than 80%). In other words, drug-induced changes are present or absent; hence, we prefer to qualify the drug-induced changes.

Extensive experience in DCM studies has taught us that low-IT subjects (less than 20 mm Hg) respond differently to the administration of psychoactive drugs than high-IT subjects (more than 30 mm Hg). IT is lowered by serotonin-releasing agents (fenfluramine[13] mianserin[14]) and catecholamine agonists (D-amphetamine, methylphenidate, nomifensine).[15] IT is increased by catecholamine (NE and DA) antagonists. Chronic administration of clomipramine, a drug which inhibits 5HT uptake and release,[16] also raised IT. Thus, we infer that high IT is associated with increased 5HT-store at distal colon plexa level.

The following generalizations may be made, overlooking some variants which continue to be recorded in detailed published papers: drugs which activate the NE system, besides lowering IT, tend to raise rectal PA; drugs which activate the DA system, besides reducing IT tend to raise sigmoidal PA; and anti-ACh drugs which cross the hemato-brain barrier tend to lower sigmoidal PA and raise rectal PA. The motility changes induced by NE/DA agonists are reverted by NE/DA antagonists. Similarly, antagonistic effects are frequently found between NE/DA antagonists and ACh antagonists, as well as between presynaptic NE/DA and postsynaptic NE/DA blocking agents. Domperidone, a DA-blocking agent lacking central effects, does not affect DCM,[17] whereas hioscine, a peripheral anti-ACh agent lacking central effects, does influence DCM. Furthermore, this drug induces opposite effects to those induced by biperiden, a central acting anti-ACh agent.

Although our findings give no direct evidence as to whether the drug-induced changes in motility stem from central or peripheral effects, nevertheless, a set of facts favors a central action, at least in part: (1) a parallel was found between motility changes and behavior changes, to the point where behavior changes are rarely found without DCM changes; (2) it has been demonstrated that increases of catecholamines in the blood do not modify gastrointestinal motility;[18] (3) it has been shown that the effects of D-amphetamine on respiratory,[19] cardiovascular,[20] and pupil functions[21] are due fundamentally to central stimuli; (4) the structural, functional, pharmacological, and biochemical aspects of the myenteric plexa have been shown to be very similar to those observed at the central level[22] and, furthermore, there is evidence of the existence of a hemato-myenteric barrier very similar to the hemato-brain barrier,[23] which opens speculation that the myenteric plexa may be integrated to the central nervous system; and (5) the fact that nomifensine, a DA agonist lacking peripheral effects,[24] induced DCM changes; whereas, domperidone, a DA-blocking agent lacking central effects, does not induce DCM changes, reinforces the hypothesis that this intestinal segment behaves rather as a part of the central autonomic nervous system.[25] The relationship existing between the central and peripheral effects induced by neurohormones and psychoactive drugs is not yet understood; for instance, we do not know whether the effects of plasmatic serotonin on the metabolism of carbohydrates[26,27] are similar to the effects induced by the release of serotonin at central level.[28] Neither is it known whether changes in the peripheric biochemical parameters, induced by administration of psychoactive drugs, are peripherally or centrally mediated.[29-31]

It is worth mentioning that psychoactive drugs induce different behavior and/or side-effects according to the existence of a low or high IT. For instance, D-amphetamine provokes excitement in low-IT subjects and sedation in high-IT subjects.[9,10] Haloperidol frequently induces acute dyskinesia in low-IT subjects, whereas this undesirable side effect is less frequently observed in high-IT subjects. Biperiden is very badly tolerated by low-IT subjects but leaves high-IT subjects unaffected. Clonidine frequently induces undesirable symptoms in high-IT but not in low-IT subjects. Mianserin provokes deep sleep in high-IT but not low-IT subjects.

Evaluation of the activity of the different components of the autonomic nervous system (ANS) is obtained from the ability displayed by the various ANS drugs to provoke DCM changes. Hyperactivity of the NE system is diagnosed when clonidine and other NE antagonists but not DA and ACh antagonists are able to suppress DCM. Similar inferences derive for DA and ACh systems when DA and ACh antagonists are able to suppress or strongly reduce DCM.

Three basic patterns of DCM are observed: (1) high IT at both sigmoidal and rectal levels; (2) high IT at sigmoidal or rectal level; and (3) low IT at both levels. Each of these three groups may show many degrees of PA at one or both intestinal segments. Sigmoidal and rectal responses to drugs are independent and frequently antagonistic (mirror image response), showing that sigmoidal and rectal motility have different neurological controls. With respect to this, it has been shown that the structural and functional organization of myenteric plexa differs in the two intestinal segments. In addition, there are evidences showing that discrete stimulation at different striatal zones is able to evoke discrete changes in rectal motility in mammals.[32]

A. Manometric-Guided Assessment: Psychotic Syndromes

Two major types of ANS imbalances were found among the 41 psychotic patients included in a published study: those with NE hyperactivity and those with DA hyperactivity. There existed a close correlation between the clinical approach (research diagnostic criteria) and the patterns of DCM in that NE intestinal hyperactivity was found in schizophrenic subjects, while DA intestinal hyperactivity was found in schizoaffective disorder patients. The fact that schizophrenics experienced great improvement with clonidine, an inhibitor of NE release, while schizoaffectives experienced total improvement with clonazepam, an inhibitor of DA release, reinforces the above postulated ANS imbalances amongst psychotic patients.[33-36] These findings have been ratified in 221 schizophrenics and 137 schizoaffectives (Figures 2 to 5).

B. Nonpsychotic Syndromes
1. Psychosomatic Disturbances

Irritable bowel syndrome — An unbalanced neurologic control of the gastrointestinal tract with dominance of α-adrenergic over the cholinergic activity was found in patients affected by nervous diarrhea (those lacking abdominal pain).[1] On the contrary, the opposite DCM pattern was found among spastic colon patients.[2] The DCM profile guided us in designing succesful treatments for both these phases of irritable bowel syndrome. α-Adrenergic agents were

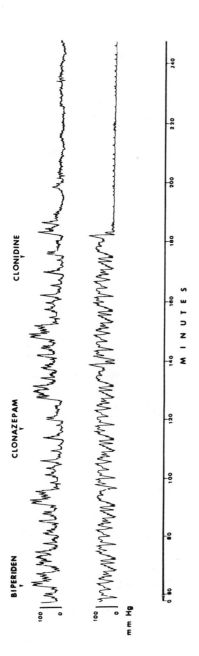

FIGURE 2. Final segment of DCM record obtained from a nontreated schizophrenic patient. Clonidine (0.15 mg intramuscular), a central acting α_2 agonist which inhibits NA release, suppressed DCM. However, both biperiden (2 mg intramuscular), a central acting anti-ACh drug, and clonazepam (2 mg orally), an inhibitor of central dopamine release, failed to modify DCM. According to this intestinal pharmacoma-nometric study, we inferred that the patient had a central NE predominance. Further, this low-IT patient was improved by clonidine therapy.

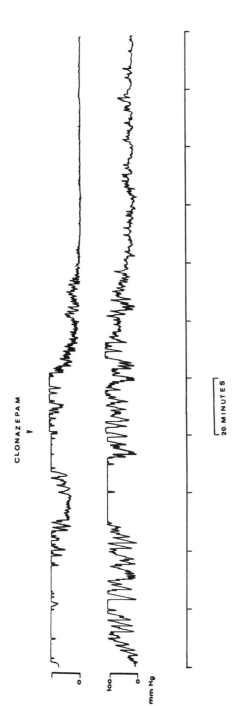

FIGURE 3. Final segment of DCM record obtained from a nontreated schizoaffective disorder patient. Clonazepam (2 mg orally), but neither clonidine nor biperiden previously administered, suppressed DCM. Clonazepam inhibits central DA release from DA neurons by stimulating DA autoreceptors. Moreover, this low-IT patient improved with clonazepam therapy.

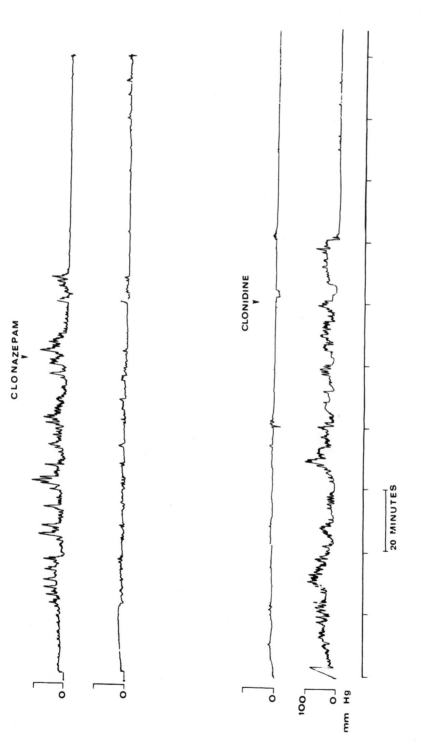

FIGURE 4. Final segments of DCM records from two psychotic patients. One schizoaffective disorder (upper trace) and the other schizophrenic (lower trace). Clonidine but not clonazepam suppressed DCM in the lower. Patients were improved by clonazepam and clonidine, respectively. Further, administration of clonidine to the schizoaffective patient and clonazepam to the schizophrenic patient aggravated their syndromes.

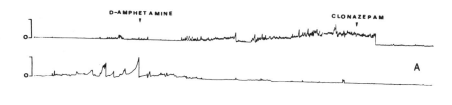

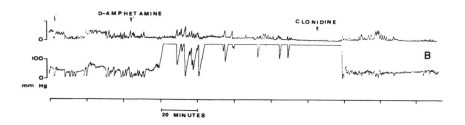

FIGURE 5. Final segments of DCM records obtained from two psychotic patients. One schizoaffective disorder (upper case) and the other, schizophrenic (lower case). The administration of D-amphetamine (20 mg orally) triggered a rebound of DCM in both patients, accompanied by behavioral excitation. Intestinal and emotional rebounds alike were suppressed by clonazepam in the schizoaffective, and clonidine in the schizophrenic patient. We postulate a dopamine release stimulated by D-amphetamine in the schizoaffective patient, and a NA release provoked by D-amphetamine in the schizophrenic subject.

employed in nervous diarrhea patients while pharmacological manipulations tending to enhance the catecholaminergic system were employed to treat spastic colon patients. The latter treatment was also successfully used in patients suffering from biliary dyskinesia. Such patients showed a DCM pattern similar to that found in spastic colon.[37]

Ulcerative colitis — Most of these patients show NE activity, only; thus, they are treated with thioproperazine and other phenothiazine derivates along with clonidine.[38]

Granulomatous colitis (Crohn's disease) — All seven patients examined by pharmacomanometric investigation showed DA hyperactivity. All these patients obtained great improvement with clonazepam, given in very low doses (0.5 to 2 mg, daily).

Headache — Up to the present, we have investigated and treated more than 1800 headache patients. We found several patterns of DCM amongst these patients, thus, we employed different kinds of pharmacological manipulations according to the respective DCM profile.[39-41] For the last 2 years, we have been treating headache patients as depressive patients. Thus, we administer clomipramine to low-IT subjects and fenfluramine or desipramine/imipramine to high-IT patients.[42,43]

Female infertility — We have previously reported that 17 infertile women were successfully treated with small doses of haloperidol.[44] The DCM profile showed DA hypoactivity in all these cases. Only seven of these women showed raised plasma prolactin levels. During the last 3 years, 22 additional cases have been successfully treated with small doses of L-dopa alone or in addition to small doses of D-amphetamine.

Essential hypertension — We have investigated 87 of these patients, and, in addition, we are performing similar investigations in 35 bronchial asthma patients.

Gilles de la Tourette disease — We recently reported the success obtained in a woman with this disease treated with thioproperazine.[45] Pharmacomanometric investigation guided us to the administration of thioproperazine, pimozide (another dopaminergic blocking agent), along with L-tryptophan and biperiden, have also been employed succesfully in these patients.[46] Following this case, we treated four additional GTD patients with thioproperazine and obtained similar success in all four cases.

Trigeminal neuralgia (TN) — Intestinal pharmacomanometric studies have guided us to employ pimozide in the treatment of TN patients. This therapeutical approach provoked dramatical and sustained improvement in these subjects.[47]

2. Affective Disorders

Depressive syndrome — We found two kinds of patients: those showing low IT and those showing high IT. The former improved with clomipramine, whereas, the latter improved with fenfluramine and imipramine.[42,43] Increases in IT paralleled improvement in the low-IT patients, whereas, a reduction of IT paralleled improvement in the high-IT patients.

Hypomanic disorders — (research diagnostic criteria, DSM III). In all these patients, DA hyperactivity was demonstrated during pharmacomanometric investigation. In all of them, clonazepam dramatically eliminated manic symptoms. Prolongation of the clonazepam therapy provoked depressive symptoms in all these cases, thus, clonazepam doses were gradually adjusted according to the attenuation of manic symptoms[47,48] (Figures 6 to 16).

REFERENCES

1. **Lechin F, van der Dijs B, Bentolila A, Peña F.** Antidiarrheal effects of dihydroergotamine. *J Clin Pharmacol* 17: 339-349, 1977.
2. **Lechin F, van der Dijs B, Bentolila A, Peña F.** The spastic colon syndrome. Therapeutic and pathophysiological considerations. *J Clin Pharmacol* 17: 431-440, 1977.
3. **Lechin F, van der Dijs B.** The effects of dopaminergic blocking agents on distal colon motility. *J Clin Pharmacol* 19: 617-624, 1979.
4. **Lechin F, van der Dijs B.** Dopamine and distal colon motility. *Digest Dis Sci* 1979, 24: 86-87.
5. **Lechin F, van der Dijs B.** Effects of diphenylhydantoin on distal colon motility. *Acta Gastroenter Latinoam* 9: 145-152, 1979.
6. **Lechin F, van der Dijs B.** Physiological Effects of Endogenous CCK on Distal Colon Motility. *Acta Gastroenter Latinoam* 9: 195, 1979.
7. **Lechin F, van der Dijs B.** Intestinal pharmacomanometry and glucose tolerance: evidence for two antagonistic mechanisms in the human. *Biol Psychiatry* 16: 969-984, 1981.
8. **Lechin F, van der Dijs B, Lechin E.** The autonomic nervous system. Physiological basis of psychosomatic therapy. Editorial Científico-Médica, Barcelona, 1979.
9. **Lechin F, van der Dijs B, Gómez F, Arocha L, Acosta E.** Effects of D-amphetamine, clonidine and clonazepam on distal colon motility in non-psychotic patients. *Res Commun Psychol Psychiat Behav* 7: 385-410, 1982.
10. **Lechin F, van der Dijs B, Gómez F, Acosta E, Arocha L.** Comparison between the effects of D-amphetamine and fenfluramine on distal colon motility in non-psychotic patients. *Res Commun Psychol Psychiat Behav* 7: 411-430, 1982.
11. **Lechin F, van der Dijs B.** Two postulated alpha$_2$-antagonists (mianserin and chlorprothixene) and one alpha$_2$-agonist (clonidine) induced opposite effects on human distal colon motility. *J Clin Pharmacol* 23: 209-218, 1983.
12. **Lechin F, van der Dijs B.** Colon motility and psychological traits in the irritable bowel syndrome. *Dig Dis Sci* 26: 474-475, 1981.
13. **Clineschmidt BV, Zacchei AG, Totaro JA, Pflueger AB, McGuffin JC, Wishousky TI.** Fenfluramine and brain serotonin. *Ann NY Acad Sci* 305: 222-241, 1978.
14. **Raiteri M, Angellini F, Bertollini A.** Comparative study of the effects of mianserin, a tetracyclic antidepressant, and of imipramine on uptake and release of neurotransmitters in synaptosomes. *J Pharm Pharmacol* 28: 483-488, 1976.
15. **Samanin R, Bernasconi S, Garattini S.** The effect of nomifensine on the depletion of brain serotonin and catecholamines induced respectively by fenfluramine and 6-hydroxidopamine. *Eur J Pharmacol* 34: 377-380, 1975.
16. **Liang-Fu T.** 5-Hydroxytryptamine uptake inhibitors block para-methoxyamphetamine-induced 5HT release. *Br J Pharmacol* 66: 185-190, 1979.
17. **Laduron PM, Leysen JE.** Domperidone, a specific in vitro antagonist, devoid of in vivo central dopaminergic activity. *Biochem Pharmacol* 28: 2161-2165, 1979.
18. **Dubois A, Henry DP, Kopis IJ.** Plasma catecholamines and post-operative gastric emptying and small intestinal propulsion in the rat. *Gastroenterology* 68: 466-469, 1975.

19. **Mediavilla A, Feria M, Fernández JF, Cagigas P, Pazos A, Flórez J.** The stimulatory action of D-amphetamine on the respiratory centre, and its mediation by a central alpha adrenergic mechanism. *Neuropharmacology* 18: 135-142, 1979.

20. **Bolme PK, Fuxe T, Hökfelt T, Goldstein M.** Studies on the role of dopamine in cardiovascular and respiratory control: central vs. peripheral mechanisms, in *Advances in Biochemical Pharmacology*. Costa E, Gessa GL. Eds., Raven Press, New York, 1977.

21. **Koss MC.** Studies on the mechanism of amphetamine mydriasis in the cat. *J Pharmacol Exp Ther* 213: 49-53, 1980.

22. **Weber LJ.** p-Chlorophenylalanine depletion of gastrointestinal 5-hydroxytryptamine. *Biochem Pharmacol* 19: 2169-2172, 1970.

23. **Gershon MD, Bursztajn S.** Properties of the enteric nervous system: limitation of access of intravascular molecules to the myenteric plexus and muscularis externa. *J Comp Neurol* 80: 467-488, 1978.

24. **Braestrup C.** Biochemical differentiation of amphetamine vs. methylphenidate and nomifensene in rats. *J Pharm Pharmacol* 29: 463-470, 1977.

25. **Fox J.** Gut's nervous system: a model for the brain. *J Chem Eng* Dec 1: 32-33, 1980.

26. **Lechin F, Coll-García E, van der Dijs B, Peña F, Bentolila A, Rivas C.** The effect of serotonin (5-HT) on insulin secretion. *Acta Physiol Latinoam* 25: 339-346, 1975.

27. **Lechin F, van der Dijs B.** Glucose tolerance, non-nutrient drink and gastrointestinal hormones. *Gastroenterology* 80: 216, 1981.

28. **Fernstrom JD, Wurtman RJ.** Brain serotonin content: physiological regulation by plasma neutral aminoacids. *Science* 178:414-416, 1972.

29. **Lechin F, Coll-García E, van der Dijs B, Bentolila A, Peña F, Rivas C.** The effects of captivity on the glucose tolerance test in dogs. *Experientia* 35: 876-877, 1979.

30. **Lechin F, Coll-García E, van der Dijs B, Bentolila A, Peña F, Rivas C.** The effects of dopaminergic blocking agents on the glucose tolerance test in six humans and six dogs. *Experientia* 35: 886-887, 1979.

31. **Lechin F, van der Dijs B.** Haloperidol and insulin release. *Diabetologia* 20: 78, 1981.

32. **Pazo JH.** Caudate-putamen and globus pallidus influences on a visceral reflex. *Acta Physiol Lat Am* 26: 260, 1976.

33. **Lechin F, Gómez F, van der Dijs B, Lechín E.** Distal colon motility in schizophrenic patients. *J Clin Pharmacol* 20: 459-464, 1980.

34. **Lechin F, van der Dijs B, Gómez F, Vall JM, Acosta E, Arocha L.** Pharmacomanometric studies of colonic motility as a guide to the chemotherapy of schizophrenia. *J Clin Pharmacol* 20: 664-671, 1980.

35. **Lechin F, van der Dijs B.** Clonidine therapy for psychosis and tardive dyskinesia. *Am J Psychiatry* 138: 390, 1981.

36. **Lechin F, van der Dijs B.** Noradrenergic or dopaminergic activity in chronic schizophrenia? *Br J Psychiatry* 139: 472-473, 1981.

37. **Lechin F, van der Dijs B, Bentolila A, Peña F.** The adrenergic influences on the gallbladder emptying. *Am J Gastroenterol* 69: 662-668, 1978.

38. **Lechin F, van der Dijs B, Insausti CL, Gómez F.** Treatment of ulcerative colitis with thioproperazine. *J Clin Gastroenterol* 4: 445-449, 1982.

39. **Lechin F, van der Dijs B.** A new treatment for headache. Pathophysiological considerations. *Headache* 16: 318-321, 1977.

40. **Lechin F, van der Dijs B, Lechín E, Peña F, Bentolila A.** The noradrenergic and dopaminergic blockades: a new treatment for headache. *Headache* 18: 69-74, 1978.

41. **Lechin F, van der Dijs B.** Physiological, clinical and therapeutical basis of a new hypothesis for headache. *Headache* 20: 77-84, 1980.

42. **Lechin F, van der Dijs B, Acosta E, Gómez F, Lechín E, Arocha L.** Distal colon motility and clinical parameters in depression. *J Affect Dis* 5: 19-26, 1983.

43. **Lechin F, van der Dijs B, Gómez F, Arocha L, Acosta E, Lechín E.** Distal colon motility as a predictor of antidepressant response to fenfluramine, imipramine and clomipramine. *J Affect Dis* 5: 27-35, 1983.

44. **Lechin F, van der Dijs B.** Treatment of infertility with levodopa. *Br Med J* 280: 480, 1980.

45. **Lechin F, van der Dijs B, Gómez F, Acosta E, Arocha L.** On the use of clonidine and thioproperazine in a woman with Gilles de la Tourette's disease. *Biol Psychiatry* 17: 103-108, 1982.

46. **Lechin F, van der Dijs B, Gomez F, Lechin ME, Amat J, Lechin AE, Cabrera A, Rodriguez O.** Effects of tryptophane addition to therapy for Gilles de la Tourette disease: a model of a proposed neurochemical profile. in press.

47. **Lechin F, van der Dijs B, Amat J, Lechin AE, Cabrera A, Lechin ME, Gomez F, Arocha L, Jimenez V.** Definite and sustained improvement with pimozide of two patients with severe trigeminal neuralgia: some neurochemical, neurophysiological and neuroendocrinological findings. in press.

48. **Lechin F, van der Dijs B.** *Intestinal Manometry as a Guide to Psychopharmacological Therapy. Clinical Pharmacology and Therapeutics.* Int Congr Ser No. 604, Excerpta Medica, Amsterdam, 1982.

FIGURE 6.   DCM record obtained from a high-IT depressed patient. Biperiden, a central acting anti-ACh drug did not modify DCM. On the contrary, scopolamine (2.5 mg intramuscular), a peripheral anti-ACh drug which lacks central effects suppressed DCM. Peripheral anti-ACh agents always suppress DCM. For this reason, these agents are not useful in pharmacomanometric investigation.

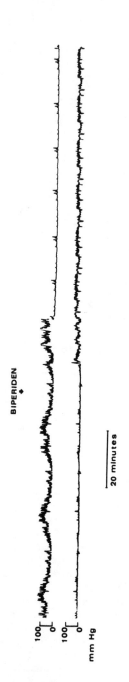

FIGURE 7.   Final segment of DCM record of a high-IT depressed patient. Biperiden, a central anti-ACh drug suppressed sigmoidal phasic activity (waves) but not sigmoidal high tone. The drug increased rectal phasic activity. Furthermore, this patient was improved by biperiden therapy.

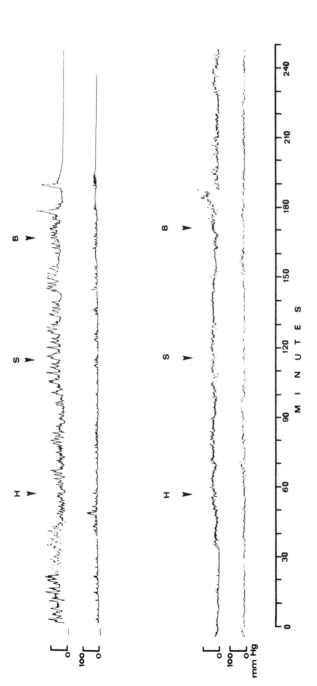

FIGURE 8. Final segments of DCM records of two headache-depressed patients. The upper case shows high-IT whereas the lower case shows low-IT. Biperiden, a central acting anti-ACh drug but not haloperidol (H) or sulpiride (S), suppressed and reduced sigmoidal phasic activity in the upper and lower case, respectively. In the upper case, sigmoidal tone remained elevated after biperiden injection. In both cases, we postulated that central cholinergic activity predominates over DA activity.

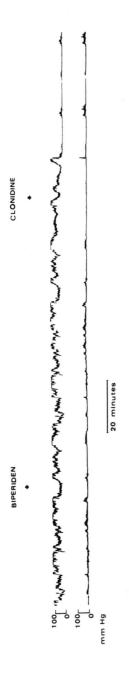

FIGURE 9.   Final segment of DCM record of a headache-depressed patient. Clonidine but not biperiden suppressed sigmoidal motility. This patient showed a profile of NE predominance. He was improved by clonidine therapy.

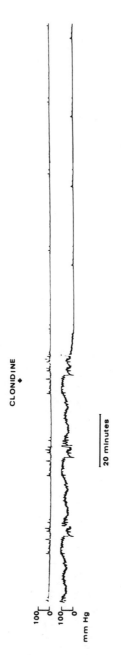

FIGURE 10. Final segment of DCM record of a hypomanic subject. He showed low sigmoidal tone without phasic activity at this level. Clonidine suppressed rectal phasic activity and provoked sleep. Dopamine antagonists and anti-ACh drugs, previously administered, failed to modify DCM and behavior.

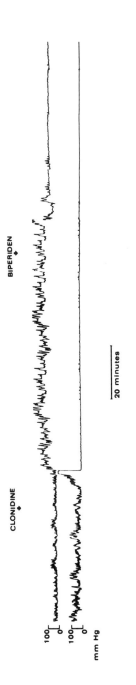

FIGURE 11. Final segment of DCM record of a patient affected by anorexia nervosa. Clonidine abruptly suppressed rectal motility and at the same time increased sigmoidal motility. The rebound of sigmoidal motility was suppressed by biperiden; however, sigmoidal tone remained elevated. In this case, we postulated the existence of a central autonomic system oscillation (see-saw) between NE and cholinergic systems. We also inferred that DA system was inhibited in this patient. She was dramatically improved with clonidine + biperiden therapy.

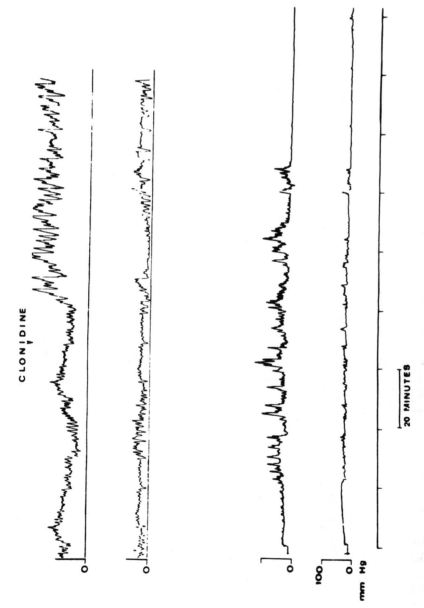

FIGURE 12. Final segments of DCM records of two headache-depressed patients: one high-IT (upper case) and the other low-IT (lower case). Clonidine raised sigmoidal motility in the upper case and triggered headache symptoms. The drug suppressed sigmoidal phasic activity in the lower case and relieved headache symptoms present before start of intestinal pharmacomanometric study.

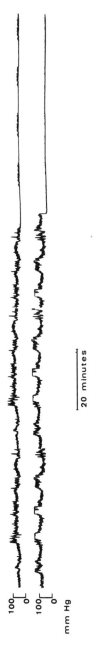

FIGURE 13. Final segment of DCM record of a high-IT depressed patient. Fenfluramine (60 mg orally), a serotonin releaser, but not clonidine, suppressed DCM. We inferred that fenfluramine-induced DCM changes are secondary to an increase of synaptic level. This patient was greatly improved by fenfluramine therapy.

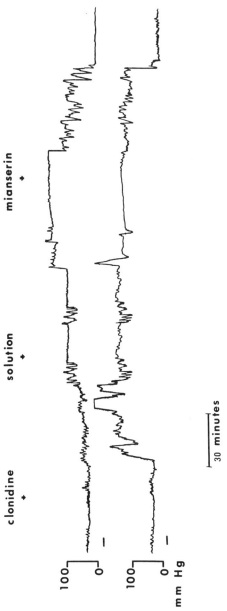

FIGURE 14. Final segment of DCM record of a high-IT depressed patient. Administration of clonidine, an α_2 agonist, increased IT and phasic activity. Mianserin (20 mg orally), an α_2 antagonist suppressed IT and phasic activity, induced by clonidine. This patient was succesfully treated with mianserin. From *J Clin Pharmacol.*

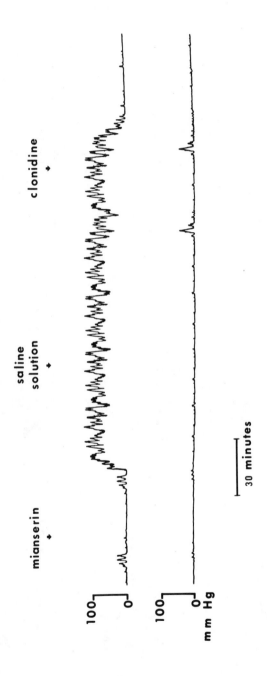

FIGURE 15. Two antagonistic drugs, mianserin (20 mg orally), an α_2 antagonist, and clonidine (0.15 mg intramuscular) displayed opposite effects on DCM in a low-IT hypomanic subject. Mianserin triggered a great rebound of phasic activity (waves) at sigmoidal level. Concommitantly, behavioral excitation was observed. Clonidine suppressed abruptly the rebound of motility and induced sleep. We postulated that both intestinal and behavioral effects were secondary to an acute release of NA at central level which was annulled by clonidine. From *J Clin Pharmacol.*

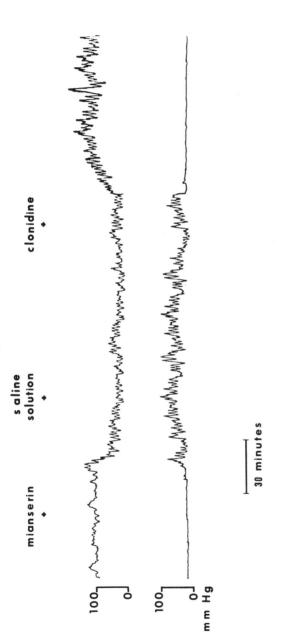

FIGURE 16. Two antagonistic drugs, mianserin (20 mg orally), an α₂ antagonist, and clonidine (0.15 mg intramuscular) displayed opposite effects on DCM in a high-IT depressed patient. Mianserin reduced sigmoidal tone and increased rectal phasic activity. Concommitantly, the subject fell asleep. Clonidine reverted mianserin-induced intestinal effects but not behavioral effects. The patient remained asleep. It is known that mianserin, in addition to its α₂ antagonistic property, is a serotonin releaser because it is able to block serotonin₂ autoreceptors. We speculate that mianserin-induced sleep is due to its serotonin₂ antagonistic effect. We observe that in high intestinal subjects mianserin always provokes intestinal and behavioral effects similar to those registered in the above case. From *J Clin Pharmacol*.

INTESTINAL PHARMACOMANOMETRY: EVIDENCE FOR TWO ANTAGONISTIC DOPAMINERGIC MECHANISMS IN THE HUMAN

Fuad Lechin, Bertha van der Dijs, Francisco Gomez, Marcel Lechin, Luis Arocha, and Emilio Acosta

I. INTRODUCTION

In this section cumulative data are presented dealing with the existence of more than one dopamine receptor in mammals. In addition, evidence is presented for the existence of two antagonistic dopamine-functional expressions in the human. This fact, along with the disclosure of new DA agonistic and antagonistic drugs, should prove useful in the management of diseases in which these drugs have proven to be good therapeutic tools.

Despite the observation that the major tranquilizers or antipsychotic drugs (phenothiazine derivatives) exert prominent actions on central and peripheral ACh[1] and NA[2,3] processes, they are generally accepted as dopaminergic-blocking agents (DBA).[4] However, the fact that many drugs commonly used as anticonvulsants, diphenylhydantoin,[5,6] clonazepam,[7] β-blockers, propranolol,[8] etc. and α-blockers phentolamine, phenoxybenzamine, dihydroergotamine,[9,10] etc. show DA agonistic and DA-antagonistic actions is generally ignored. This is also the case of the ergot alkaloids, which behave as mixed DA agonist-antagonists, depending on the doses administered.[11-15] Other drugs, such as DA-releasing agents: amphetamine-like drugs,[16] L-dopa, and reuptake inhibitors at the DA terminals, methylphenidate,[17,18] nomifensene,[19] etc., enhance DA activity. However, clinicians observe this effect decreases as time elapses. Inhibition of DA synthesis and hyposensitization of postsynaptic DA receptors, secondary to augmented availability of dopamine at the synaptic cleft, are the mechanisms most frequently invoked to explain the attenuation of drug effects.[20,21]

II. A BINARY ANTAGONISTIC DOPAMINERGIC SYSTEM

Two or more receptors have been demonstrated for most neurotransmitters: nicotinic and muscarinic for acetylcholine,[22] α and β for adrenaline (A) and noradrenaline (NA),[23] M and D for serotonin (5HT),[24] and H_1 and H_2 for histamine.[25] These subdivisions allow the understanding of the physiological and pathophysiological mechanisms in which those systems are involved.

Cumulative data dealing with the existence of more than one DA receptor in mammals are fast increasing.[26-30] In addition, evidence for the existence of two antagonistic DA-functional expressions in the human has recently emerged.[31-33] This fact and the development of new DA agonistic and antagonistic drugs together should improve the treatment of schizophrenia and other psychoses,[343,35] parkinsonian,[36,37] affective disorders,[38,39] hyperprolactinemia,[40,41] infertility,[42] epilepsy,[7,43] migraine,[44,45] blood pressure disorder,[39,44] obesity,[46-48] etc., in which DA agonists and DA antagonists are of value.

A. Evidence in Experimental Animals

1. Central Nervous System

Two antagonistic interacting DA receptors or mechanisms have been demonstrated in the brain of the snail *Helix aspersa*, in the cat caudate nuclei, in the striatum-mesolimbic complex of rodents, and in the brain of rhesus monkey.[26,27] These two receptor zones or systems, DA_1 and DA_2 for some authors[5] are better known as inhibition-mediating dopamine receptors (DAi) and excitation-mediating dopamine receptors (DAe).[26,27] There exist well-defined histochemical, anatomical, embryological, electrophysiological, biochemical, pharmacological, and behavioral differences between both systems which can be summarized as follows:

1. DAi zone and/or system is characterized by high density of dotted DA fluorescence homogeneously distributed in striatic and mesolimbic structures. It appears early during postnatal development and has a low DA turnover. These receptors or DAi zone can be stimulated by dopamine, L-dopa, amphetamine, and the imidazoline derivative DPI while they can be blocked by ergometrine, piribedil, and NA. Stimulation of DAi zone in unilateral lesioned animals induces ipsilateral rotation, while studies on the electrical activity of neurons at DAi zone show inhibitory responses to micro-iontophoretically applied dopamine.

2. DAe zone and/or systems is characterized by high density of diffuse DA fluorescence mostly circumscribed to the rostro-medialis part of the caput nuclei caudati (cat) or neostriatum (rat). It appears late during postnatal development and has a high dopamine turnover. DAe zone can be activated by dopamine, L-dopa, amphetamine, and apomorphine, and antagonized by haloperidol and related butyrophenones. Electrical activity of neurons at DAe zone shows excitatory responses to micro-iontophoretically applied DA while stimulation of DAe zone in unilateral lesioned animals shows contralateral turning behavior.

It is noteworthy that carbachol, a cholinemimetic agent, when applied through micro-iontophoresis to the DAi zone counteracts the effects elicited by dopamine, but when applied to the DAe zone it synergizes the effects elicited by dopamine. Hence, central DA and ACh mechanisms are antagonistic at the DAi zone, but cooperate at the DAe zone. In addition, since a decrease in the DAe activity produces symptoms which are to a larger extent identical to those elicited by an increase in the DAi activity and vice versa, a functional antagonism between both systems could be postulated. The finding of the anti-ACh properties of haloperidol and other phenothiazines is coherent with the DAe-ACh synergism.

There are other lines of evidence that support the view of Cools et al.[26] that DAe and DAi receptors are located at presynaptic and postsynaptic levels, respectively. It is now generally recognized that stimulation of DA presynaptic receptors both at central and peripheral levels induces inhibition of DA release and synthesis, along with an inhibition of firing at DA neurons, whereas the blockade of these receptors induces the opposite effects.[20,49-53] The fact that low doses of apomorphine, a DAe agonist, and of haloperidol, a DAe antagonist,[52-54] induce decrease and increase of dopamine release respectively, is consonant with the hypothesis that there are two kinds of DA receptors. Actually, apomorphine has a powerful inhibitory effect on the firing rate of DA neurons when applied directly onto DA cell bodies in substantia nigra by micro-iontophoresis.[20] Other drugs such as clonazepam[7] and bromocriptine[55] (in low doses) modify stereotyped behavior in a fashion similar to apomorphine, by stimulating presynaptic DA receptors.

Biochemical studies show that some kind of DA receptors are associated with adenylate cyclase activity,[56] whereas other DA receptors are not associated with the cAMP generating system.[53] Apomorphine (DAe-agonist) stimulates and haloperidol (DAe-antagonist) inhibits dopamine-sensitive adenyl cyclase activity.[28,53,56,57] The fact that cAMP does not mediate the DAi, but that it probably does mediate the enduring modulatory change in the DAe is in line with the above-mentioned facts.[58]

Since DA-sensitive adenylate cyclase seems to be located at postsynaptic receptor levels[59,60] whereas DAe receptors have been postulated as being present at the presynaptic level,[26,27] some disagreement emerges from these contradictory facts. However, DA-sensitive adenylate cyclase, blockable by phenothiazines and butyrophenones but not by propanolol or phentolamine, has recently been shown at presynaptic level.[61-64] On the other hand, DA receptors not associated with adenylate cyclase have been demonstrated at corpus striatum,[54,65,66] pituitary gland, and other postsynaptic levels.[28,67,68]

The fact that Kebabian and Calne[69] showed that there are multiple classes of brain DA receptors is compatible with the above-mentioned findings. So-called D_1 dopamine receptors were found on neuronal cell bodies intrinsic to the striatum (postsynaptic DA receptors), and were thought to be associated functionally with dopamine-sensitive adenylate cyclase. Another class, D_2 dopamine receptors, was located on axons of the glutaminergic corticostriatic pathway (presynaptic DA receptors). However, pituitary contains predominantly D_2 dopamine receptors (postsynaptic location).[70]

2. Peripheral Tissues

Direct or indirect evidence for dopamine receptors in renal, mesenteric, and coronary vascular beds is now available.[71-74] Dopamine also inhibits low esophagus[75] and gastric motility[76] whereas it contracts[77] or inhibits[31,32] distal colon motility. In addition, the DA system exerts influence on gastric,[76] pancreatic-exocrine,[78-80] pancreatic-endocrine,[33,81-83] pituitary,[84,85] and renin secretions.[86] The action of dopamine is complex: in addition to its action on specific DA receptors, dopamine has been shown to act on β_1- and α-adrenergic receptors and possibly β_2- and serotonin receptors.[72] Finally, inhibitory dopamine receptors have been found on postganglionic sympathetic nerves[51,87] and on sympathetic ganglion.[87-89] These diverse effects may be attributable to the flexibility of the DA molecule. The benzene ring may assume an infinite range of rotation about the side change, and the side chain may be either *cis* or *trans* with respect to the benzene ring.

B. Evidence in Humans

Evidence for a binary antagonistic DA system in the human has emerged from studies showing the effects of two distinct DA blocking agents on distal colon motility,[31,32] on glucose tolerance, and on blood serotonin levels.[33] Both haloperidol and sulpiride were chosen because they are accepted as DA blocking agents despite some differential and opposite characteristics.

1. Haloperidol

This drug is considered a "classical" or "typical" neuroleptic[90] and a DAe blocking agent.[26] At low doses, it blocks presynaptic receptors at both central and peripheral catecholaminergic terminals, thus inducing release of NA and/or dopamine.[52] The drug induces marked catalepsy in experimental animals and sedative effects in humans.[90-92] Haloperidol inhibits the DA-sensitive adenylate cyclase; this effect is shared with other phenothiazines. However the antipsychotic power of the former is greater than that of the latter drugs. This suggests that haloperidol blocks other receptors than those associated with the cAMP-generating system.[28,53,56-59,93] When applied by micro-iontophoresis to the DAe zone in the neostriatum of cats, haloperidol antagonizes the ACh system.[26] It can provoke dyskinesia and extrapyramidalism symptoms in humans and in experimental animals;[91] this fact accords with its ability to increase DA turnover in the corpus striatum.[94] In addition to DA receptors, haloperidol also blocks NA and 5HT receptors at both central and peripheral levels.[2,3]

2. Sulpiride

This drug is known as an "atypical" neuroleptic or antipsychotic agent which has a poor cataleptogenic action and does not induce dyskinetic movements or exprapyramidal manifestations.[90,92,95,96] On the contrary, sulpiride has been proven to quickly eliminate the dyskinesias induced by a single intramuscular injection of 5 mg haloperidol in hypersensitive subjects.[44] The fact that sulpiride does not influence DA turnover at the striatal level is consistent with its poor ability to provoke extrapyramidal symptoms.[68,95] This drug preferentially blocks subcortical and cortical limbic DA receptors,[14] and in addition produces strong endocrine effects[33,96] (hyperprolactinemia, galactorrhea, amenorrhea, hyperserotonemia). Sulpiride is a potent anti-DA drug when tested for antagonism to DA-induced pancreatic juice secretion in dogs[79] and to DA-induced renal vasodilation (haloperidol had a weaker effect).[80] Sulpiride (but not haloperidol) inhibits the vestibular function in experimental studies; this might explain its ability to improve different kinds of vertigo in humans.[97] Similar considerations might be valid in explaining why sulpiride, but not haloperidol, reduces gastric juice and gastrin secretions and is helpful in the cicatrization of peptic ulcers.[98,99] Sulpiride is unable, or shows a poor ability, to block NA receptors[100,101] and has no inhibitory effect on DA-sensitive adenylate cyclase.[95,102]

3. Distal Colon Motility Studies

The effects of haloperidol and sulpiride were investigated in 2729 subjects. The effects of DA-releasing and DA-suppressing agents (D-amphetamine and clonazepam, respectively) were also investigated. Results have been reported in several published papers.[6,32-35,103-105] Motility records were obtained during 3 to 4 h at both sigmoidal and rectal levels, according to previously established procedure.[6,34,38,104-106] Both DA-blocking agents induced deep, long-lasting, and antagonist changes of distal colon motility in 78.7% of the subjects. It is noteworthy that biperiden (a central-acting anti-ACh drug) counteracted the sulpiride- but not the haloperidol-induced changes.

The above-mentioned findings have been largely discussed in other papers, however, two main facts are self-evident: (1) haloperidol and sulpiride block two different and frequently antagonistic DA receptors and/or mechanisms at the doses employed in these studies and (2) the anti-ACh action of biperiden is antagonistic to the action of sulpiride, but synergic with haloperidol (anti-DAe) action. These findings are consonant with those obtained by Cools et al.[26] in the cat caudate nuclei which show that the ACh system cooperates with DAe but antagonizes DAi system. In light of the above it is logical to infer that sulpiride might be a DAi-blocking agent.

4. Serotonin Blood Levels and Glucose Tolerance

Pretreatment periods of 8 d with sulpiride (25 to 50 mg t.i.d.) increased blood 5HT levels and decreased tolerance to glucose in six normal humans and six normal dogs investigated.[33] On the contrary, 8-d pretreatment periods with haloperidol (0.3 to 0.5 mg t.i.d.) provoked effects opposite to those induced by sulpiride in the same humans and dogs.[107] Since impairment of glucose tolerance and elevation of 5HT blood levels were also registered in dogs under captivity stress,[108] it is conceivable that the raised levels of 5HT might act as a diabetogenic factor.[109] This presumption finds support from the fact that serotonin has been shown to inhibit or stimulate insulin secretion, depending on the experimental conditions.[110] On the other hand, diminished tolerance to glucose and impaired insulin secretion were observed in carcinoid patients showing elevated blood-5HT levels.[111] The fact that the inhibitory effects exerted by central DAi-against-DAe mechanisms were shown to be mediated by serotonin[112] is in line with the above-mentioned findings.

III. CONCLUSION

The two DA blocking agents, haloperidol and sulpiride, induce opposite effects on the diverse parameters tested in humans (colon motility, blood serotonin and insulin levels, and glucose tolerance). In addition, experimental and clinical evidence indicate distinct and frequently antagonistic properties of both drugs. All this strongly suggests that these drugs operate through two antagonistic DA systems and/or mechanisms which exist in many animal species and in humans. It is unlikely that these drugs were acting on other than their specific DA receptors because of the low doses employed in our experimental studies. The question of whether the drugs act through blockage of peripheral and/or central DA receptors remains unanswered and must await further research. However, the fact that the doses employed were able to improve a great deal of affective, schizoaffective, and psychotic disorders leads us to infer that, at least in part, the drugs acted at the central level.

IV. MEDICAL IMPLICATIONS

In the light of the two antagonistic DA-system hypothesis, the pathophysiology and treatment of all diseases in which this system is involved should be revised. Clinical and therapeutic advances could emerge from this new point of view. In fact, new succesful therapeutic trials have been made by manipulating jointly or separately the two kinds of DBA, typical (haloperidol and other phenothiazines) and atypical (sulpiride, clozapine, thioridazine). Other drugs exert their therapeutic effects by stimulating different types of DA receptors. Some of them stimulate presynaptic DA receptors, inducing inhibition of DA synthesis and release (clonazepam, low doses of bromocriptine, mazindol, etc.), while others stimulate postsynaptic DA receptors (high doses of bromocriptine, diphenylhidantoin, etc.) Finally, DA-releasing agents and DA reuptake inhibitors (amphetamine, methylphenidate, nomifensine, etc.) would stimulate all kinds of DA receptors.

Clonazepam, an anticonvulsant drug, has been shown to be very useful in controlling affective and schizoaffective disorders.[34] Doses as low as 0.25 mg b.i.d. are effective in treating vertigo, insomnia, and anxiety (unpublished results). Higher doses eliminated psychotic schizoaffective disorder patients.[34] Bromocriptine is a valuable tool in the treatment of hyperprolactinemia and Parkinson's disease, while mazindol is a new anorexic agent. An understanding of the intimate mechanisms of action of all these drugs could explain the failure or the paradoxical effects they can induce. D-Amphetamine, methylphenidate, nomifensine, L-dopa, etc. increase the availability of dopamine at the synaptic cleft, and thus act as stimulating agents. In all these cases, the agonistic action of dopamine decreases as time elapses. According to Martress et al.,[20] inhibitory impulses of DA release and synthesis arising from the stimulation of presynaptic (autoreceptors) and postsynaptic DA receptors by an augmented dopamine at the synapsis are the most generally accepted explanations for the decreasing effects of the drug. In addition, it has been shown recently that after sustained stimulation DA receptors may become hyposensitive to the action of DA agonists; this might result in a decrease of the therapeutic effect in Parkinson patients treated with L-dopa or bromocriptine. However, there is a large body of experimental evidence demonstrating that DA agonists in low doses lose, at least partly, their ability to inhibit DA release. Therefore, their postsynaptic effects are enhanced by the participation of endogenous dopamine, leading to an increased therapeutic response.

Taking into consideration the above-mentioned findings it is logical to assume, from a theoretical point of view, that lowering the dosage of L-dopa and other DA agonists might result in better and more sustained therapeutic benefit in those illnesses in which an enhancement of DA neurotransmission is desired. The fact that very low doses of D-amphetamine, methylphenidate, nomifensene, and L-dopa have proven to be successful in the treatment of headache,[44,45,113] spastic colon,[38] gallbladder hypokinesia,[114] depressive conditions,[115] etc. provides clinical evidence to reinforce the experimental studies cited.

Finally, it has been postulated that haloperidol and other phenotiazines stimulate dopamine release by blocking DA autoreceptors, thus relieving DA neurons from the inhibitory action of released dopamine which would result in an increased postsynaptic DA effect.[20] Since the addition of low doses of phenotiazines to L-dopa and other stimulatory DA agents has proven to enhance the therapeutic effects of the latter in the treatment of depressive patients (unpublished results), this would tend to support the hypothesis.

The concepts put forth in this chapter surely need to be ratified or rectified in the light of new experimental and clinical evidence. There is enough accumulated knowledge on the physiology, pathophysiology, and pharmacology of the DA system to establish its central importance, but at the same time there is an astonishing disparity between the basic information available in these areas and the clinical benefits obtained up to now[116] (Figures 17 to 19).

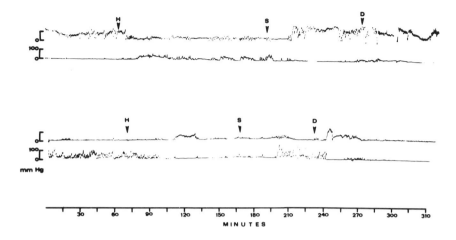

FIGURE 17. Two cases demonstrate the antagonistic effects of two DBAs: haloperidol (H) (5 mm intramuscular), a "typical" and sulpiride (S) (50 mm intramuscular), an "atypical" DBA. In the upper case, a high-IT subject, H reduced sigmoidal tone and increased rectal phasic activity. Concommitantly, the subject showed drowsiness. S increased sigmoidal tone and suppressed rectal activity. An increase of drowsiness was registered and the subject fell asleep. Dihydroergotamine (D) (1 mg intramuscular), a NE blocking agent, did not modify behavior nor DCM. However, biperiden, a central acting anti-ACh drug, reduced sigmoidal activity (not showed in the motility record). In the lower case, showing a low-IT subject, H suppressed rectal phasic activity and increased sigmoidal phasic activity. Concommitantly, H induced sedation. S annulled H-induced rectal effects but not behavioral effects. D suppressed sigmoidal and rectal activity, but not behavioral effects. According to DCM investigation, there exist two opposite dopaminergic receptors and/or mechanisms. Intestinal changes varied in comparison to the previous DCM profile registered before DBA administration. We postulate that different responses of high- and low-IT subjects would depend on which central autonomic system is disinhibited by total blockade of DA system. From Lechin F, van der Dijs. *Biol Psychiatry* 16: 966-986, 1981. With permission.

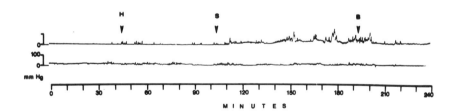

FIGURE 18. Final segments of DCM records of a low-IT subject. Haloperidol (H) (5 mm intramuscular), a "typical" DBA, did not modify intestinal motility but provoked behavioral excitation. Sulpiride (S) (50 mm intramuscular), an "atypical" DBA, triggered a rebound of sigmoidal motility and drowsiness. Biperiden (B) (2 mm intramuscular), a central acting anti-ACh drug, suppressed intestinal and behavioral effects provoked by sulpiride. We inferred that atypical but not typical DBA induces disinhibition of the central ACh parasympathetic system. Further, atypical DBA annulled the excitatory behavior triggered by typical DBA. From Lechin, F, van der Dijs, *Biol Psychiatry* 16: 966-986, 1981. With permission.

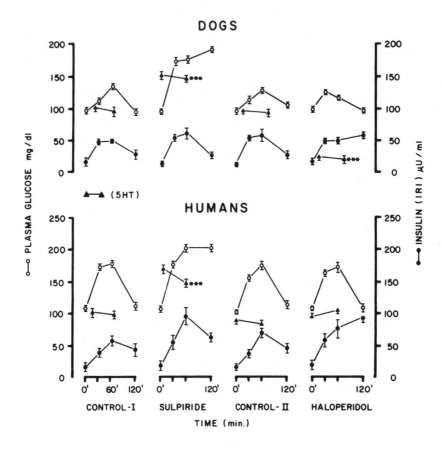

FIGURE 19.   Pretreatment of 8 d of sulpiride, but not haloperidol, impaired oral glucose tolerance and raised blood serotonin (5HT) levels in six normal human subjects and six normal dogs ($p <$ 0.001). Statistical significance against zero-mean values at Control I test = (100%). Sulpiride, but not haloperidol pretreatment induced sedation and depressive symptoms. Raised blood sugar during fasting and raised opalescent serum were registered in four out of the six dogs and in one of the six humans after sulpiride treatment but disappeared following haloperidol treatment. From Lechin F, van der Dijs. *Biol Psychiatry* 16: 966-986, 1981. With permission.

REFERENCES

1. **Sayers AC, Burki HR.** Antiacetylcholine activities of psychoactive drugs: a comparison of the (³H)-quinuclinidinyl benzilate binding assay with conventional methods. *J Pharm Pharmacol* 28: 252-253, 1976.
2. **Anden NE, Butcher SG, Corrodi H, et al.** Receptor activity and turnover of dopamine and noradrenaline after neuroleptics. *J Pharmacol* 11: 303-314, 1970.
3. **Peroutka SJ, U'Prichard DC, Greenberg DA, et al.** Neuroleptic drug interactions with norepinephrine alpha receptors binding sites in rat brain. *Neuropharmacology* 16: 549-556, 1977.
4. **Van Praag HM, Korf J.** Neuroleptics, catecholamines and psychotic disorders. A study of their interrelation. *Am J Psychiatry* 132: 593-597, 1975.
5. **Elliot PNC, Jenner P, Chadwick D, et al.** The effect of diphenylhydantoin on central catecholamine containing neuronal systems. *J Pharm Pharmacol* 29: 41-43, 1977.
6. **Lechin F, van der Dijs B.** Effects of diphenylhydantoin on distal colon motility. *Acta Gastroenterol Lat Am* 9: 145-152, 1979.
7. **Weiner WJ, Goetz C, Nausieda PA, et al.** Clonazepam and dopamine-related stereotyped behavior. *Life Sci* 21: 901-906, 1977.
8. **Harris JE.** Beta adrenergic receptor-mediated adenosine cyclic 3′,5′-monophosphate accumulation in the rat corpus striatum. *Mol Pharmacol* 12: 546-558, 1976.

9. **Walton KG, Liepmann P, Baldessarini RJ.** Inhibition of dopamine-stimulated adenylate cyclase activity by phenoxybenzamine. *Eur J Pharmacol* 52: 231-234, 1978.

10. **Govoni S, Iuliano E, Spano PF, et al.** Effect of ergotamine and dihydroergotamine on dopamine-stimulated adenylate cyclase in rat caudate nucleus. *J Pharm Pharmacol* 29: 45-47, 1977.

11. **Anlezark G, Pycock C, Meldrum B.** Ergot alkaloids as dopamine agonists. Comparison in two rodent models. *Eur J Pharmacol* 37: 295-302, 1976.

12. **Horowski R, Wachtel H.** Direct dopaminergic action of lisuride hydrogen maleate, an ergot derivative, in mice. *Eur J Pharmacol* 36: 373-383, 1976.

13. **Ziegler MG, Lake CR, Williams AC, et al.** Bromocriptine inhibits norepinephrine release. *Clin Pharmacol Ther* 25: 137-142, 1979.

14. **Fuxe K, Fredholm BB, Agnati LF, et al.** Dopamine receptors and ergot drugs. Evidence that an ergolene derivative is a differential agonist at subcortical limbic dopamine receptors. *Brain Res* 146: 295-311, 1978.

15. **Pagnini G, Cammani F, Crispino A, et al.** Effects of bromocriptine on adenylate cyclase and phosphodiesterase activities of rat striatum. *J Pharm Pharmacol* 30: 92-95, 1978.

16. **Raiteri M, Bertollini A, Angellini F, et al.** D-Amphetamine as a releaser or reuptake inhibitor of amines in synaptosomes. *Eur J Pharmacol* 34: 189-195, 1975.

17. **Fuller RW, Snoddy HD.** Inability of methylphenidate or mazindol to prevent the lowering of 3,4-dihydroxyphenylacetic acid in rat brain by amphetamine. *J Pharm Pharmacol* 31: 183-184, 1979.

18. **Ross SB.** The central stimulatory action of inhibitors of the dopamine uptake. *Life Sci* 24: 159-168, 1979.

19. **Hunt P, Kannengiesser MH, Raynaud JP.** Nomifensene: a new potent inhibitor of dopamine uptake into synaptosomes from rat brain corpus striatum. *J Pharm Pharmacol* 26: 370-376, 1974.

20. **Martress MP, Costantin J, Baudry M, et al.** Long-term changes in the sensitivity of pre- and postsynaptic dopamine receptors in mouse striatum evidenced by behavioural and biochemical studies. *Brain Res* 136: 319-337, 1978.

21. **Maggi A, Bruno F, Cattabani F, et al.** Apomorphine-induced inhibition of striatal dopamine release: role of dopaminergic receptors in substantia nigra. *Brain Res* 145: 180-184, 1978.

22. **Curits DR.** Central synaptic transmitters, in *Basic Mechanisms of the Epilepsies.* Jasper HH, Ward AA Jr, Pope A. Eds., Little, Brown, Boston, 1969, 105.

23. **Ahlquist, RP.** A study of adrenotroptic receptors. *Am J Physiol* 152: 586-600, 1948.

24. **Gaddum JH, Piccarelli ZP.** Two kinds of tryptamine receptor. *Br J Pharmacol* 12: 323-328, 1957.

25. **Baudry M, Martress MP, Schwartz JC.** H_1 and H_2 receptors in histamine-induced accumulation of cAMP in guinea pig brain slices. *Nature* 253: 362-363, 1975.

26. **Cools A, Van Rossum JM.** Excitation-mediating and inhibition-mediating dopamine receptors: a new concept towards a better understanding of electrophysiological, biochemical, pharmacological, functional, and clinical data. *Psychopharmacologia* 45: 243-254, 1976.

27. **Cools A, Struyker Boudier HAJ, Van Rossum JM.** Dopamine receptors: selective agonists and antagonists of functionally distinct types within the feline brain. *Eur J Pharmacol* 37: 283-293, 1976.

28. **Seeman P, Tedesco JL, Lee T, et al.** Dopamine receptors in the central nervous system. *Fed Proc Fed Am Soc Exp Biol* 37: 130-136, 1978.

29. **Tye NC, Horsman L, Wright FC, et al.** Two dopamine receptors: supportive evidence with the rat rotational model. *Eur J Pharmacol* 45: 87-90, 1977.

30. **Pycock CJ, Marsden CD.** The rotating rodent: a two component system? *Eur J Pharmacol* 47: 167-175, 1978.

31. **Lechin F, van der Dijs B.** Dopamine and distal colon motility. *Digest Dis Sci* 24: 86-87, 1979.

32. **Lechin F, van der Dijs B.** Effects of dopaminergic blocking agents on distal colon motility. *J Clin Pharmacol* 19: 617-624, 1979.

33. **Lechin F, Coll-García E, van der Dijs B, et al.** The effects of dopaminergic blocking agents on the glucose tolerance test in 6 humans and 6 dogs. *Experientia* 35: 886-887, 1979.

34. **Lechin F, van der Dijs B, Gómez F, et al.** Pharmacomanometric studies of colonic motility as a guide to the chemotherapy of schizophrenia. *J Clin Pharmacol* 20: 664-671, 1980.

35. **Lechin F, van der Dijs B.** Clonidine therapy for psychosis and tardive dyskinesia. *Am J Psychiatry* 138(3), 1981.

36. **Calne DB, Plotkin C, Williams AC, et al.** Long-term treatment of parkinsonianism with bromocriptine. *Lancet* 1: 735-737, 1978.

37. **Kartzinel R, Perlow M, Teychenne PF, et al.** Bromocriptine and levodopa (with or without carbidopa) in parkinsonism. *Lancet* 2: 272-275, 1976.

38. **Lechin F, van der Dijs B, Bentolila A, et al.** The "spastic colon" syndrome: therapeutic and pathophysiologic considerations. *J Clin Pharmacol* 17: 431-440, 1977.

39. **Barbeau A.** Dopamine and diseases. *Can Med Assoc J* 103: 824-832, 1970.

40. **Thorner MD, Besser GM, Jones A, et al.** Bromocriptine treatment of female infertility: report of 13 pregnancies. *Br Med J* 4: 694-697, 1975.

41. **Seppälä M, Hirvonen E, Ranta T.** Bromocriptine treatment of secondary amenorrhea. *Lancet* 1: 1154-1156, 1976.

42. **Lechin F, van der Dijs B.** Treatment of infertility with levodopa. *Br Med J* 280: 480, 1980.
43. **Chadwick D, Harris R, Jenner P, et al.** Manipulation of brain serotonin in the treatment of myoclonus. *Lancet* 2: 434-435, 1975.
44. **Lechin F, van der Dijs B, Lechin E, et al.** The dopaminergic and noradrenergic blockages: a new treatment for headache. *Headache* 18: 69-74, 1978.
45. **Lechin F, van der Dijs B.** A new treatment for headache: pathophysiologic considerations. *Headache* 16: 318-321, 1977.
46. **Burridge SL, Blundell JE.** Amphetamine anorexia: antagonism by typical but not atypical neuroleptics. *Neuropharmacology* 18: 453-457, 1979.
47. **Picotti GB, Carruba MO, Zambotti F, et al.** Effects of mazindol and D-fenfluramine on 5-hydroxytryptamine uptake, storage and metabolism in blood platelets. *Eur J Pharmacol* 42: 217-224, 1977.
48. **Shetty PS, Jung RT, James WPT.** Effect of catecholamine replacement with levodopa on the metabolic response to semistarvation. *Lancet* 1: 77-79, 1979.
49. **Nagy JI, Lee T, Seeman P, et al.** Direct evidence for presynaptic and postsynaptic dopamine receptors in brain. *Nature* 274: 278-282, 1978.
50. **Lokhandwala MF, Buckley JP.** Presynaptic dopamine receptors as mediators of dopamine induced inhibition of neurogenic vasoconstriction. *Eur J Pharmacol* 45: 305-**309, 1977.**
51. **Sharabi FM, Long JP, Cannon JG, et al.** Inhibition of the sympathetic nervous system by a series of heterocyclic congeners of dopamine. *J Pharmacol Exp Ther* 199: 630-638, 1976.
52. **McMillen BA, Shore PA.** The relative functional ability of brain noradrenaline and dopamine storage pools. *J Pharm Pharmacol* 29: 780-786, 1977.
53. **Iversen LL, Rogawski MA, Miller RJ.** Comparison of the effects of neuroleptic drugs on pre- and postsynaptic dopaminergic mechanisms in the rat striatum. *Mol Pharmacol* 12: 251-256, 1976.
54. **Di Chiara G, Porceddu ML, Spano PF, et al.** Haloperidol increases and apomorphine decreases a striatal dopamine metabolism after destruction of striatal dopamine-sensitive adenylate cyclase by kainin acid. *Brain Res* 130: 374-382, 1977.
55. **Snyder SH, Hutt C, Stein B, et al.** Correlation of behavioural inhibition or excitation produced by bromocriptine with changes in brain catecholamine turnover. *J Pharm Pharmacol* 28: 563-566, 1976.
56. **Kebabian JW, Petzold GL, Greengard P.** Dopamine-sensitive adenylate cyclase in caudate nucleus of rat brain and its similarity to the "dopamine receptors". *Proc Natl Acad Sci USA* 69: 2145-2149, 1974.
57. **Snyder SH, Banerjee SP, Yamamura HI, et al.** Drugs, neurotransmitters, and schizophrenia. *Science* 184: 1243-1253, 1974.
58. **Libet B.** Which postsynaptic action of dopamine is mediated by cyclic AMP? *Life Sci* 24: 1043-1058, 1979.
59. **Lew JY, Goldstein M.** Dopamine receptor binding for agonists and antagonists in thermal exposed membranes. *Eur J Pharmacol* 55: 429-430, 1979.
60. **Rosenfeld MR, Seeger TF, Sharpless NS, et al.** Denervation supersensitivity in the mesolimbic system: involvement of dopamine-stimulated adenylate cyclase. *Brain Res* 173: 572-576, 1979.
61. **Traficante LJ, Friedman E, Oleshansky MA, et al.** Dopamine sensitive adenylate cyclase and cAMP phosphodiesterase in substantia nigra and corpus striatum of rat brain. *Life Sci* 19: 1061-1066, 1976.
62. **Kebabian JW, Saavedra JM.** Dopamine-sensitive adenylate cyclase occurs in a region of substantia nigra containing dopaminergic dendrites. *Science* 193: 683-685, 1976.
63. **Phillipson DT, Horn AS.** Substantia nigra of the rat contains a dopamine sensitive adenylate cyclase. *Nature* 261: 418-420, 1976.
64. **Spano PF, Govoni S, Trabucchi M.** Studies on the pharmacological properties of dopamine receptors in various areas of the central nervous system. *Adv Biochem Psychopharmacol* 19: 155-158, 1978.
65. **Waddington JL, Cross AJ, Longden A, et al.** Functional distinction between DA-stimulated adenylate cyclase and ^{3}H-apiperone binding sites in rat striatum. *Eur J Pharmacol* 58: 341-342, 1979.
66. **Quick M, Emson PC, Joyce E.** Dissociation between the presynaptic dopamine-sensitive adenylate cyclase and ^{3}H-apiperone binding sites in rat substantia nigra. *Brain Res* 167: 335-365, 1979.
67. **Briley M, Langer SZ.** Two binding sites for ^{3}H-spiroperidol on rat striatal membrane. *Eur J Pharmacol* 50: 283-284, 1978.
68. **Mishra RK.** Effect of substituted benzamide drugs on rat striatal tyrosine hydroxylase. *Eur J Pharmacol* 51: 189-190, 1978.
69. **Kebabian JW, Calne DW.** Multiple receptors for dopamine. *Nature* 277: 197-200, 1979.
70. **Spano PG, Di Chiara G, Tonon G, et al.** Dopamine-sensitive adenylate cyclase in rat substantia nigra. *J Neurochem* 27: 1565-1568, 1976.
71. **Lokhandwala MF, Buckley JP.** The effect of L-dopa on peripheral sympathetic nerve function: role of presynaptic dopamine receptors. *J Pharmacol Exp Ther* 204: 362-371, 1978.
72. **Goldberg LI, Kohli JD, Kotake AN, et al.** Characteristics of vascular dopamine receptor: comparison with other receptors. *Fed Proc Fed Am Soc Exp Biol* 37: 2396-2402, 1978.
73. **Hope W, McCulloch MW, Story DF, et al.** Effects of pimozied on noradrenergic transmission in rabbit isolated ear arteries. *Eur J Pharmacol* 46: 101-111, 1977.

74. **Murthy VV, Gilbert JC, Goldgerg LI, et al.** Dopamine-sensitive adenylate cyclase in canine renal artery. *J Pharm Pharmacol* 28: 567-571, 1976.

75. **De Carlo DJ, Christensen J.** A dopamine receptor in esophageal smooth muscle of the opposum. *Gastroenterology* 70: 216-219, 1976.

76. **Valenzuela JE.** Dopamine as a possible neurotransmitter in gastric relaxation. *Gastroenterology* 71: 1019-1022, 1978.

77. **Lanfranchi GA, Marzio L, Cortini C, et al.** Motor effect of dopamine on human sigmoid colon. *Am J Dig Dis* 23: 257-263, 1978.

78. **Valenzuela JE, Defilippi C, Diaz G, et al.** Effect of dopamine on human gastric and pancreatic secretion. *Gastroenterology* 76: 323-326, 1979.

79. **Iwatsuki K, Hashimoto K.** Enhancement of dopamine-induced stimulation of pancreatic secretion by 5-dimethyldithio carbamylpicolinic acid (YP-279), a dopamine betahydroxylase inhibitor. *Jpn J Pharmacol* 29: 187-190, 1979.

80. **Caldera R, Ferrari C, Romussi M, et al.** Effect of dopamine infusion on gastric and pancreatic secretion and on gastrin release in man. *Gut* 19: 724-728, 1978.

81. **Ericson LE, Hakanson R, Lundquist I.** Accumulation of dopamine in mouse pancreatic beta cells following injection of L-dopa. Localization to secretory granules and inhibition of insulin secretion. *Diabetologia* 13: 117-124, 1977.

82. **Leblanc H, Lachelin GCL, Abu-Fadil S, et al.** Effect of dopamine infusion on insulin and glucagon secretion in man. *J Clin Endocrinol Metab* 44: 196-198, 1977.

83. **Lorenzi M, Teakilian E, Bohannon NV, et al.** Differential effects of L-dopa and apomorphine on glucagon secretion in man: evidence against central dopaminergic stimulation of glucagon. *J Clin Endocrinol Metab* 45: 1154-1158, 1977.

84. **Leblanc H, Lachelin GCL, Abu-Fadil S, et al.** Effects of dopamine infusion on pituitary hormone secretion in humans. *J Clin Endocrinol Metab* 43: 668-674, 1976.

85. **Masala A, Delitala G, Alagna S, et al.** Effect of dopaminergic blockade on the secretion of growth hormone and prolactin in man. *Metabolism* 27: 921-926, 1978.

86. **Imbs JL, Schmidt M, Velly J, et al.** Effect of apomorphine and of pimozide on renin secretion in the anesthetized dog. *Eur J Pharmacol* 38: 175-178, 1976.

87. **Langer SZ.** Presynaptic receptors and their role in the regulation of transmitter release. *Br J Pharmacol* 60: 481-486, 1977.

88. **Gardier RW, Tsevdos EJ, Jackson DB, et al.** Distinct muscarinic mediation of suspected dopaminergic activity in sympathetic ganglion. *Fed Proc Fed Am Soc Exp Biol* 37: 2422-2428, 1978.

89. **Björklund A, Cegrell L, Falck B, et al.** Dopamine-containing cells in sympathetic ganglia. *Acta Physiol Scand* 78: 334-338, 1970.

90. **Costall B, Naylor RJ.** A comparison of the abilities of typical neuroleptic agents and thioridazine, clozapine, sulpiride, and metoclopramide to antagonize the hyperactivity induced by dopamine applied intracerebrally to areas of the extrapyramidal and mesolimbic systems. *Eur J Pharmacol* 40: 9-19, 1976.

91. **Costall B, Naylor RJ.** Neuroleptic antagonism of dyskinetic phenomena. *Eur J Pharmacol* 33: 301-312, 1975.

92. **Costall B, Funderburk WH, Leonard CA, et al.** Assessment of the neuroleptic potential of some novel benzamide, butyrophenone, phenothiazine and indole derivatives. *J Pharm Pharmacol* 30: 771-778, 1978.

93. **Rosenblatt JE, Shore D, Neckers LM, et al.** Effects of chronic haloperidol on caudate ³H-spiroperidol binding in lesioned rats. *Eur J Pharmacol* 60: 387-388, 1979.

94. **Costall B, Naylor RJ.** Mesolimbic involvement with behavioral effects indicating antipsychotic activity. *Eur J Pharmacol* 27: 46-58, 1974.

95. **Scatton B, Bischoff S, Dedek J, et al.** Regional effects of neuroleptics on dopamine metabolism and dopamine-sensitive adenylate cyclase. *Eur J Pharmacol* 44: 287-292, 1977.

96. **Mielke DH, Gallant DM, Craig K.** An evaluation of a unique new antipsychotic agent, sulpiride: effects on serum prolactin and growth hormone levels. *Am J Psychiatry* 134: 1371-1375, 1977.

97. **Oosterveld WJ.** A comparative study of the effects of cinnarizine sulpiride and thiethylperazine on vestibular nystagmus in rabbits. *Eur J Pharmacol* 50: 91-96, 1978.

98. **Caldera R, Romussi M, Ferrari C.** Inhibition of gastrin secretion by sulpiride treatment in duodenal ulcer patients. *Gastroenterology* 74: 221-221, 1978.

99. **Lam SK, Lam KC, Lai CL, et al.** Treatment of duodenal ulcer with antacid and sulpiride. *Gastroenterology* 76: 315-322, 1979.

100. **Kohli JD, Cripe LD.** Sulpiride: a weak antagonist of norepinephrine and 5-hydroxytryptamine. *Eur J Pharmacol* 56: 283-286, 1979.

101. **Le Fur G, Burgavin MC, Malgouris C, et al.** Differential effects of typical and atypical neuroleptics on alpha-noradrenergic and dopaminergic postsynaptic receptors. *Neuropharmacology* 18: 591-594, 1979.

102. **Jenner P, Elliott PNC, Clow A, et al.** A comparison of in vitro and in vivo dopamine receptor antagonism produced by substituted benzamide drugs. *J Pharm Pharmacol* 30: 46-48, 1978.

103. **Lechin F, van der Dijs B.** Physiological effects of endogenous CCK on distal colon motility. *Acta Gastroenterol Lat Am* 9: 198-204, 1979.

104. **Lechin F, Gómez F, van der Dijs B, et al.** Distal colon motility in schizophrenic patients. *J Clin Pharmacol* 20: 459-464, 1980.

105. **Lechin F, van der Dijs B, Bentolila A, et al.** Antidiarrheal effects of dihydroergotamine. *J Clin Pharmacol* 17: 339-349, 1977.

106. **Lechin F, van der Dijs B.** Colon motility and psychological traits in the irritable bowel syndrome. *Dig Dis Sci* 26(4), 1981.

107. **Lechin F, van der Dijs B.** Haloperidol and insulin release. *Diabetologia* 20: 78, 1981.

108. **Lechin F, Coll-García E, van der Dijs B, et al.** Effects of captivity on glucose tolerance in dogs. *Experientia* 35: 876-877, 1979.

109. **Lechin F, van der Dijs B.** Glucose tolerance, non-nutrient drink, and gastrointestinal hormones. *Gastroenterology* 80: 216, 1981.

110. **Lechin F, Coll-García E, van der Dijs B, et al.** The effect of serotonin on insulin secretion. *Acta Physiol Lat Am* 25: 339-349, 1975.

111. **Feldman JM, Plonk JW, Bivena CH, et al.** Glucose tolerance in the carcinoid syndrome. *Diabetes* 24: 664-671, 1975.

112. **Costall B, Hui SCG, Naylor RJ.** Hyperactivity induced by injection of dopamine into the accumbens nuclei: actions and interactions of neuroleptic, cholinomimetic and cholinolytic agents. *Neuropharmacology* 18: 661-665, 1979.

113. **Lechin F, van der Dijs B.** Physiological, clinical and therapeutical basis of a new hypothesis for headache. *Headache* 20: 77-84, 1980.

114. **Lechin F, van der Dijs B, Bentolila A, et al.** Adrenergic influences on the gallbladder emptying. *Am J Gastroenterol* 69: 662-668, 1978.

115. **Lechin F, van der Dijs B, Lechin E.** *The Autonomic Nervous System, Physiological Basis of Psychosomatic Therapy.* Editorial Científico-Médica, Barcelona, 1979.

116. **Lechin F, van der Dijs B.** Intestinal pharmacomanometry and glucose tolerance: evidence for two antagonistic dopaminergic mechanisms in the human. *Biol Psychiatry* 16: 969-986, 1981.

PHARMACOMANOMETRIC-GUIDED THERAPIES: INTESTINAL PHARMACOMANOMETRY AS A GUIDE TO THE THERAPY OF SCHIZOPHRENIA

Fuad Lechin, Bertha van der Dijs, Francisco Gómez, Alex Lechin, Emilio Acosta, and Luis Arocha

I. INTRODUCTION

Successful therapeutic approaches for physical and psychologic symptoms were outlined for more than 4000 nonpsychotic subjects, on the basis of their pharmacomanometric investigations.[1] In an earlier study carried out in our institute, it was shown that only NE activity was present at the distal colon level in most psychotic patients, while only a small percentage of these patients showed DA activity exclusively.[2] These findings seem to contradict the generally accepted hypothesis that postulates the existence of DA overactivity in schizophrenia.[3] Nevertheless, some pathophysiologic reasoning could conciliate these contradictory facts, i.e., it is well known that presynaptic denervation is always followed by postsynaptic supersensitivity.[4] Hence, it is possible to think that the invoked DA overactivity in schizophrenia could be secondary to a lack of dopamine or to the interference of synaptic dopamine transmission in some brain structures.

On the basis of the two patterns of DCM found in our psychotic patients, we outlined two kinds of therapeutic approaches: (1) in patients showing only NE-induced motility, we chose to reduce the synaptic release of NA by stimulating presynaptic NE receptors with clonidine.[5] (2) In those cases showing only DA-induced motility, we reduced the synaptic release of dopamine by stimulating presynaptic DA receptors with clonazepam.[6] In the former group of patients, a postsynaptic DA receptor blocking agent (sulpiride)[7,8] was added to clonidine in order to counteract the effects of a hypothetical supersensitivity of these receptors, while in the latter group of patients a postsynaptic NE receptor blocking agent (phentolamine or levopromazine)[9] was added to clonazepam in order to counteract the consequence of a hypothetical supersensitivity of these receptors.

II. MATERIALS AND METHODS

Upon receiving phenothiazinic treatment, 153 psychotic inpatients were rated before the study on the Schedule for Affective Disorders and Schizophrenia (SADS).[10,11] The Research Diagnostic Criteria (RDC)[11,12] were determined afterward. Of these patients, 120 were diagnosed as schizophrenics (residual), and the remaining 33 subjects as suffering from schizoaffective disorders (in remission). All the subjects were hospitalized in one of the three following psychiatric centers participating in a collaborative program: Instituto Psiquiátrico de Caracas, Clínica Psiquiátrica Casablanca, and Clínica Psiquiátrica Renacimiento. DCM studies were performed on all the patients 2 d before discharge from the hospital. When patients were discharged, 2 d after the motility studies, phenothiazinic therapy was substituted by placebo. Control of the patients was continued through fortnightly visits to the hospital. A second SADS interview and DCM study were performed 6 weeks after discharge. Out of the 153 patients, 62 relapsed (8 to 20 weeks after phenothiazine withdrawal); hence, they were hospitalized again. The criteria for relapse were the development of new florid symptoms, a worsening of existing positive symptoms, or a change in the pattern of the patient's behavior that could be understood as a part of the illness. In 2 d after hospitalization, a third SADS interview was made (two joint interviews were held with each patient, conducted by two psychiatrists with previous training in the use of the SADS).

Only the 62 relapsed patients were invited to participate in the study. They gave their consent, as did their next of kin who were traceable. Our protocol includes 41 psychotic patients. Of the 62 patients who initially agreed to participate in the investigation, 11 did not complete the study and 10 were dropped because of difficulty in complying with the protocol. All the patients were physically fit and under 55 years of age; one had tardive dyskinesia in both hands, a condition afflicting the patient since 3 years before the beginning of the present study. Of the 41 subjects, 32 were diagnosed as schizophrenics (19 chronic and 13 subchronic). The remaining nine subjects were diagnosed as having schizoaffective disorders (chronic).

III. DISTAL COLON MOTILITY STUDIES (PHARMACOMANOMETRIC)

Pharmacomanometric studies were carried out twice in all the subjects according to prior established procedure.[2,13] In those patients who showed disappearance of DCM after the injection of clonidine, hyperactivity of the NE system was diagnosed. In the remaining cases whose DCM was unchanged by clonidine, the effects of clonazepam (a dopamine release inhibitor), haloperidol, and sulpiride (DA blocking agents) were investigated; in all of these cases hyperactivity of the dopaminergic system was established. Technical details of pharmacomanometric procedure have been previously described.[2,13] The first motility study was performed during phenothiazinic therapy, while the second study was performed 6 weeks after phenotiazine withdrawal (placebo period).

A. Medication

The choice of drug to be prescribed varied according to the result obtained from the second pharmacomanometric study. Clonidine (an NA release inhibitor) (0.075 to 0.30 mg t.i.d.) was administered orally to those patients in whom colon motility was shown to be generated by NA activity, while clonazepam (a dopamine release inhibitor) (0.25 to 4 mg t.i.d.) was prescribed orally for those patients showing a DA pattern of colon motility. Magnitude of the drug dosage was decided according to intensity of symptoms and in relation to therapeutic effect. Oral biperiden (an anti-ACh drug) (0.5 to 1 mg t.i.d.) was added to clonidine and clonazepam. Blood pressure, pulse rate, physical condition, and mental state were monitored hourly during the first 72 h, after which visits were spread out. At 4 weeks after the beginning of clonidine and clonazepam therapies, a postsynaptic DA blocking agent, was added to clonazepam. Once again, hourly physical and mental conditions were monitored during the first 72-h period following the addition of these drugs.

B. Assessment

Five SADS joint interviews were conducted with every patient, in which two psychiatrists independently rated the patient's mental state. Two studies were carried out during each interview: study A directed by one physician and study B directed by the other physician. The first SADS interview was held before phenothiazine withdrawal. The second interview was held 6 weeks after phenothiazine withdrawal (placebo period). The third interview was held during the current relapse. The fourth interview took place 4 weeks after clonidine or clonazepam prescription, and the fifth interview was carried out 4 weeks after the addition of sulpiride or phentolamine. We did not use the nurses observation scales because we have serious objections to these evaluation procedures.

After the 8-week trial period the patient's condition was followed for a further 3 months in an open evaluation carried out by the psychiatric staff. Many patients were checked for longer periods of time.

C. Statistical Evaluation

One-way and three-way analyses of variance[14] were employed interpreting the time series data yielded by this investigation. For one value, the two-way interactions — patient X scale, patient X treatment, and scale X treatment — were investigated. The mean squares (MS) were calculated by dividing the sums of squares (SS) by the corresponding degrees of freedom (df), and the F-ratios by dividing the M.S. between groups by the MS residual. A Compucorp 327 Scientist was used for the calculation.

D. Results

Of the 41 psychotic patients, 32 showed only NE activity in their pharmacomanometric study, and all of them satisfied the RDC for chronic or subchronic schizophrenia (12 paranoid and 20 undifferentiated subtypes); the remaining 9 subjects showed only DA activity in their pharmacomanometric study and satisfied the RDC for chronic schizoaffective disorders (manic subtype). Hence, the NE (schizophrenic) patients were treated with clonidine (a NA release inhibitor), while the DA (schizoaffective disorders) patients were treated with clonazepam (a dopamine release inhibitor).

1. Noradrenergic (Schizophrenic) Patients

These include 12 female and 20 male, age range 19 to 34 years, mean 28.7 years. All of them received clonidine, which almost completely eliminated their characteristic schizophrenic symptoms as defined by the Diagnostic and Statistical Manual of Mental Disorders,[12] such as delusions of being controlled, thought broadcasting, thought insertion, thought withdrawal, absurd-somatic-grandiose-religious delusions, auditory hallucinations, incoherence, loosening of associations, marked illogicality, and marked poverty of content of speech if accompanied by blunted effect. Abrupt disappearance of low diagnostic significance for schizophrenia was also observed: insomnia, psychomotor agitation, suicidal tendencies, difficulty in concentrating, loss of interest or pleasure, panic attacks, anxiety, phobias, elevated mood, grandiosity, accelerated speech, financial indiscretion, hostile suspiciousness, and so on.

First and maximal clinical improvement occurred at the 1st and the 15th to 16th days, respectively. All the clinical improvements were sustained in the individuals who continued to take clonidine. Three patients, however, relapsed severely within 2 d of stopping clonidine on their own initiative during weekends at home, although the symptoms remitted quickly when the drug was again introduced.

a. Side Effects

Depression, dry mouth, and drowsiness not associated with hypotension were the main side effects observed. The same tended to occur when the dose exceeded 0.6 mg/d, although these effects lessened with time. Slight decrease of blood pressure occurred in all the cases. However, a return to premedication blood pressure levels was observed after 2 to 3 weeks. Supine blood pressure remained over 105/65 mm Hg in all cases but one (80/50 mm Hg); nevertheless, in this subject standing blood pressure did not decrease below 75/45 mm Hg and recuperation occurred spontaneously after 3 h. It is noteworthy that there were no hypertensive subjects among our patients.

b. Schizophrenia Ratings

The scores presented in Table 1 were obtained by rating the scales of the current section (2nd ed.) of the SADS as follows: 5 = anxiety, 6 = manic syndrome, 7 = delusion-hallucinations, and 8 = formal thought disorders. The values were expressed in percentage of the maximal scale score (100%); these values are the mean of those obtained by two raters during two joint interviews. The interrater correlation was high; the majority of the κ coefficients lie between 0.70 and 0.85. The values presented in Table 1 are the mean of all the individual values. Table 1 shows that the greatest difference levels were registered between SADS-III (relapsing period) and the other SADS interviews. Clonidine (SADS-IV) provoked great reduction in all the scale scores; all but scale 7 (delusion-hallucinations) fell below SADS-I (phenothiazine period) rating scores. The addition of sulpiride to clonidine provoked further significant decrease in all scale scores, including scale 7. However, the latter did not fall below the SADS-I rating level (Table 1).

Significance of global improvement ratings in schizophrenic patients during placebo and treatment periods (ANOVAR) was as follows: between groups, $F = 12563.82$; placebo period vs. clonidine period, $F = 13031.74$; clonidine period vs. clonidine + sulpiride period, $F = 23124.42$; and in all the cases $p < 0.0005$.

Triple interaction F-ratios (ANOVAR) were total, $F = 15.911$ ($p < 0.0005$); between patients, $F = 0.910$ (p nonsignificant); between scales, $F = 73.132$ ($p < 0.0005$); between treatments, $F = 523.508$ ($p < 0.0005$); patient X scale, $F = 0.917$ (p nonsignificant); patient X treatment, $F = 1518$ (p nonsignificant); and scale X treatment, $F = 30.064$ ($p < 0.0005$).

No dyskinesias were observed during the present study. On the contrary, one patient suffering from phenothiazine-induced tardive dyskinesia in his hands, a condition of 3-year standing, improved progressively until total remission was reached 4 months after the starting of the clonidine trial.

c. Followup

The patients were discharged from the hospitals at the end of the 8-week trial period. However, clonidine + sulpiride + biperiden therapy was maintained in all the patients, and their controls were continued by means of fortnightly visits to the hospitals.

Although the improvements were maintained in all the subjects, most of them tended to be depressed and suffered insomnia plus some catatonic traits (mutism, negativism, withdrawal, etc.). Reduction of these symptoms was obtained by decreasing the dosages of clonidine and sulpiride. Some patients showed excitation when they returned to their homes. These symptoms were accentuated in those subjects lacking family affection. Attenuation of these symptoms was

Table 1
CLINICAL RATINGS OF SCHIZOAFFECTIVE DISORDERS AND
SCHIZOPHRENICS

SADS interview[a]	N	Scale 5	Scale 6	Scale 7	Scale 8	Global score
			Schizoaffective Disorders			
Ia	9	45.9 ± 3.0	27.3 ± 1.5	5.8 ± 0.6	13.0 ± 1.1	92.0 ± 3.9
IIa	9	47.0 ± 3.3	25.0 ± 0.9	8.6 ± 0.4	12.3 ± 0.9	93.0 ± 4.0
IIIa	9	95.1 ± 1.6	78.3 ± 0.9	40.8 ± 1.0	79.7 ± 1.5	293.9 ± 2.7
IVa	9	0.0 ± 0.0	0.0 ± 0.0	10.7 ± 0.8	20.3 ± 1.6	31.2 ± 2.2
Va	9	0.0 ± 0.0	0.0 ± 0.0	0.0 ± 0.0	0.0 ± 0.0	0.0 ± 0.0
			Schizophrenic[b] Disorders			
Ib	32	10.7 ± 1.0	4.8 ± 0.5	9.6 ± 0.2	19.4 ± 0.6	44.5 ± 1.6
IIb	32	12.7 ± 1.1	12.5 ± 0.4	12.2 ± 0.2	22.8 ± 1.2	60.2 ± 1.9
IIIb	32	62.0 ± 0.5	38.7 ± 0.4	46.8 ± 0.6	74.1 ± 0.8	221.6 ± 1.2
IVb	32	5.2 ± 0.3	1.1 ± 0.1	18.1 ± 0.4	9.8 ± 0.8	34.0 ± 1.1
Vb	32	2.3 ± 0.3	0.6 ± 0.1	10.9 ± 0.2	9.9 ± 0.5	23.7 ± 0.5

Note: % mean SE, maximal scale score = 100%.

[a] Schedule for affective disorders and schizophrenia: Ia, Ib = phenothiazine; IIa, IIb = placebo; IIIa, IIIb = relapse during placebo; IVa = clonazepam; IVb = clonidine; Va = clonazepam + phentolamine; and Vb = cloniodine + sulpiride.

[b] RDC: scale 5 = anxiety; scale 6 = manic syndrome; scale 7 = delusion-hallucinations; and scale 8 = formal thought disorder.

From *J Clin Pharmacol.*

obtained by increasing the dosage of clonidine and/or sulpiride. Psychotherapy (three to five weekly visits) was begun by 17 patients, and all of these patients kept up their improvement taking minimal doses of clonidine and/or sulpiride (0.45 and/or 150 mg, respectively).

2. Dopaminergic (Schizoaffective Disorder) Patients

They include six female and three male, age range 28 to 52 years, mean 39.5 years. Clonazepam administered to these patients strongly reduced all the above-mentioned psychotic characteristic symptoms and other less characteristic ones. Clonazepam patients showed the first and maximal clinical improvements at the 1st and 5th to 6th d, respectively. All clinical improvements were sustained in the individuals who continued to take clonazepam. Two patients relapsed severely within 2 d of stopping the drug on their own initiative during a short-term stay at their homes, but symptoms remitted quickly when clonazepam was taken again.

a. Side Effects

Drowsiness and depression were frequently observed during clonazepam treatment; however, these tended to lessen upon reduction of the dosage.

b. Phentolamine

The addition of this drug to clonazepam potentiated the therapeutic effects of the latter (Table 1). However, phentolamine may slightly increase drowsiness and depression not associated with hypotension.

c. Schizoaffective Disorder Rating

Table 1 shows that the greatest differences were registered between SADS-III score (relapsed period) and the scores of the other SADS interviews. Scores fell below SADS-I (phenothiazine period) and SADS-II (placebo period) levels during SADS-IV interview (clonazepam period). However, scale 7 (delusion-hallucinations) and scale 8 (formal thought disorders) did not differ from those of SADS-I and SADS-II interviews. These two scales fell to zero during SADS-V interview (clonazepam + phentolamine period).

Significance of global improvement ratings in schizoaffective disorder patients during placebo and treatment periods (ANOVAR) was as follows: between groups, F = 6623.47; placebo period vs. clonazepam period, F = 5847.28; clonazepam period vs. clonazepam + phentolamine period, F = 209.92; in all the cases $p < 0.0005$.

Triple interactions F-ratios (ANOVAR) were total, F = 19.329 ($p < 0.0005$); between patients, F = 1.348 (p nonsignificant); between scales, F = 311.412 ($p < 0.0005$); between treatments, F = 2485.482 ($p < 0.0005$); patient X scale, F = 0.747 (p nonsignificant); patient X treatment, F = 1.128 (p nonsignificant); scale X treatment, F = 74.387 ($p < 0.0005$).

d. Followup

The patients discharged from the hospitals at the end of the 8-week trial period; however, clonazepam + phentolamine + biperiden therapy was maintained in all the patients, and their controls were continued by means of fortnightly visits to the hospitals.

Despite a gradual reduction in the dosage of drugs, some patients tended to show insomnia, depression, and some catatonic traits (mutism or delay on verbal answers). No relapses were registered up now. Moreover, it is noteworthy that the schizoaffective disorder patients achieved overall improvement with the clonazepam + phentolamine treatment; in addition they showed less difficulty in readapting to life outside the hospital than schizophrenic patients.

IV. DISCUSSION

Despite the fact that the present study was not "double-blind", if our results are accepted, several conclusions spring from them. (1) Two major types of autonomic nervous system imbalances were found among the 41 psychotic patients included in the protocol: those with NE hyperactivity and those with DA hyperactivity. (2) There exists a close correlation between the clinical approach (RDC) and the patterns of DCM (pharmacomanometric study), in that noradrenergic intestinal hyperactivity is found in schizophrenic subjects, while DA intestinal hyperactivity is found in schizoaffective disorder subjects. (3) The fact that schizophrenics improved with clonidine, which inhibits NA release from terminal ends, and experienced a further improvement with sulpiride, a drug which blocks postsynaptic dopamine receptors, allows us to speculate that NA was acting on dopamine receptors. (4) The fact that clonazepam, which inhibits dopamine release from terminal ends, impairs schizophrenics (unpublished results) could reflect a deficiency of dopamine or some interference on dopamine neurotransmission. With regard to this it is possible to think that the often postulated supersensitivity of dopamine receptors supposedly present in schizophrenia[15] could be secondary to a lack of dopamine at the synaptic level (supersensitivity of deafferentation). (5) The fact that schizoaffective disorder subjects improved with clonazepam, which inhibits dopamine release from terminal ends, and experienced a further improvement with phentolamine, a drug which blocks noradrenaline receptors, allows us to think that dopamine was acting on NA receptors.

This hypothesis is reinforced by the fact that clonidine impairs these patients (unpublished results). A lack of NA or some interference in NA neurotransmission could be invoked in these patients. In support of the above hypothesis is the fact that although the three monoamines (NA, dopamine, and serotonin) display their major action on their own receptors, they also bind and probably stimulate the other receptors.[16-20]

Also worth noting is the disappearance of the phenothiazine-induced tardive dyskinesia observed in the patients who showed this complication before clonidine treatment. This finding allows us to assign an important role to NA in the origin of this syndrome. The fact that five other subjects suffering phenothiazine-induced tardive dyskinesia (not included in the present study) were successfully treated with clonidine gives support to this hypothesis.

In the light of the physiologic and therapeutic results obtained from the present study, we postulate the existence of two main psychotic mechanisms, one showing hyperactivity of the NA system and the other showing hyperactivity of the DA system. Supersensitivity of postsynaptic dopamine and NA receptors, respectively, would accompany the ANS imbalances. A supersensitivity of these receptors would be secondary to deficiencies of dopamine neurotransmission in the former and of NA neurotransmission in the latter syndrome.[21]

V. SUMMARY

DCM studies performed in 41 psychotic subjects demonstrated that 32 of them had hyperactivity of the NE system at this peripheral level, while the remaining nine cases showed hyperactivity of the DA system. The NE-hyperactive patients fulfilled the RDC of schizophrenia, whereas the DA-hyperactive patients were diagnosed as having schizoaffective disorders. NE-hyperactive subjects were successfully treated with clonidine, a drug which inhibits release of NA, while DA-hyperactive subjects were successfully treated with clonazepam, a drug which inhibits release of dopamine. The addition of sulpiride (a postsynaptic DBA) and of phentolamine (a postsynaptic NE blocking agent) to clonidine and clonazepam, respectively, induced further significant improvements in both types of psychotic patients.

REFERENCES

1. **Lechin F, van der Dijs B, Lechin E.** *Autonomic Nervous System: Physiological Basis of Psychosomatic Therapy.* Editorial Científico-Médica, Barcelona, 1979.
2. **Lechin F, Gómez F, van der Dijs B, Lechin E.** Distal colon motility in schizophrenic patients. *J Clin Pharmacol* 20: 459, 1980.
3. **Snyder SH.** The dopamine hypothesis of schizophrenia: focus of the dopamine receptor. *Am J Psychiatry* 133: 197, 1976.
4. **Trulson ME, Eubanks EE, Jacobs BL.** Behavioral evidence for supersensitivity following destruction of central serotonergic nerve terminals by 5,7-dihydroxytryptamine. *J Pharmacol Exp Ther* 198: 23, 1976.
5. **Starke K, Altmann KP.** Inhibition of adrenergic neurotransmission by clonidine: an action on prejunctional alpha-receptors. *Neuropharmacology* 12: 339, 1973.
6. **Weiner WJ, Goetz C, Nausieda PA, Klawans HL.** Clonazepam and dopamine-related stereotyped behavior. *Life Sci* 21: 901, 1977.
7. **Honda F, Satch Y, Shimomura K, Satch H, Noguchi H, Uchida S, Kato R.** Dopamine receptor blocking activity of sulpiride in the central nervous system. *Jpn J Pharmacol* 27: 397, 1977.
8. **Lechin F, Coll-García E, van der Dijs B, Bentolila A, Peña F, Rivas C.** The effects of dopaminergic blocking agents on the glucose tolerance test in six humans and six dogs. *Experientia* 35: 886, 1979.
9. **Drew GM.** Effects of alpha-adrenoceptor agonists and antagonists on pre- and postsynaptically located alpha-adrenoceptors. *Eur J Pharmacol* 36: 313, 1976.
10. **Endicott J, Spitzer RL.** A diagnostic interview. The schedule for affective disorder and schizophrenia. *Arch Gen Psychiatry* 35: 837, 1978.
11. **Endicott J, Spitzer RL.** Use of the Research Diagnostic Criteria and the Schedule for Affective Disorders and Schizophrenia to study affective disorders. *Am J Psychiatry* 136: 52, 1979.
12. American Psychiatric Association. Diagnostic and Statistical Manual of Mental Disorders, 3rd ed. DSM-III Draft, APA, 1978.
13. **Lechin F, van der Dijs B.** Effects of dopaminergic blocking agents on distal colon motility. *J Clin Pharmacol* 19: 617, 1979.
14. **Campbell RC.** *Statistics for Biologists.* Cambridge University Press, Cambridge, 1967.
15. **Owen F, Crow TJ, Poulter M, Cross AJ, Longden A, Riley GJ.** Increased dopamine-receptor sensitivity in schizophrenia. *Lancet* ii: 223, 1978.
16. **Hope W, McCulloch MW, Story DF, Rand MJ.** Effects of pimozide on noradrenergic transmission in rabbit isolated ear arteries. *Eur J Pharmacol* 46: 101, 1977.

17. **Bianchine JR, Shaw GM, Greenwala JE, Dandalides SM.** Clinical aspects of dopamine agonists and antagonists. *Fed Proc Fed Am Soc Exp Biol* 37: 2434, 1978.

18. **Goldberg LI, Kohli JD, Kotake AN, Volkman PH.** Characteristics of the vascular dopamine receptor: comparison with other receptors. *Fed Proc Fed Am Soc Exp Biol* 37: 2396, 1978.

19. **Christoph GR, Kuhn DM, Jacobs BL.** Electrophysiological evidence for a dopaminergic action of LSD: depression of unit activity in the substantia nigra of the rat. *Life Sci* 21: 1585, 1977.

20. **Bonkowski L, Dryden WF.** Effects of iontophoretically applied neurotransmitters on mouse brain neurones in culture. *Neuropharmacology* 16: 89, 1977.

21. **Lechin F, van der Dijs B, Gómez F, Valls JM, Acosta E, Arocha L.** Pharmacomanometric studies of colonic motility as a guide to the chemotherapy of schizrenia. *J Clin Pharmacol* 20: 664-671, 1980.

INTESTINAL PHARMACOMANOMETRY AS A GUIDE TO ANTIDEPRESSANT THERAPY

Fuad Lechin, Bertha van der Dijs, Francisco Gómez, Luis Arocha, Emilio Acosta, and Alex E. Lechin

I. INTRODUCTION

D-Fenfluramine, a trifluoromethyl N-ethyl substituted analogue of amphetamine, an anorexigen drug, is a serotonin releasing agent which increases the availability of this neurotransmitter at synaptic level.[1] However, the prolonged administration of this amphetamine induces serotonin depletion.[2] The drug has been employed in the treatment of obesity despite the failure in a percentage of patients and despite the paradoxical effects observed in some others. The collateral antidepressant effects observed in many of our obese patients induced us to attempt a trial with this drug. Further, taking into account that in our experience some correlation exists between DCM profile and character traits,[3] we decided to investigate DCM in every depressed patient in order to differentiate subjects according to their distal colon tone. Previously, we have observed that distal colon tone is more stable and constant than phasic activity (waves) as a component of distal colon motility.[4-6] Thus, we designed two separate double-blind trials, one for low intestinal tone (IT) and another for high-IT depressed patients. For comparative purposes, we divided each group into three subgroups receiving fenfluramine, clomipramine, and imipramine, respectively.

II. PATIENTS AND METHODS

This study was carried out as a double-blind investigation. The patients were randomly allocated to receive FENF, IMI, or CMI and were treated for 45 d.

A. Assessment

The therapeutic effect was assessed fortnightly by means of changes in an 18-item (0 to 58 score) slightly modified HRS for depression[7] and the 21-item (0 to 63 score) self-rating BDI,[8] starting 15 d before active treatment. Both scales were administered on the same day and the pretreatment assessments were made twice: (1) at the time of the patient's first visit (visit 00) and (2) at the time of the patient's second visit (visit 0), 2 weeks after the first; during this 15-d period all patients received placebo. Three fortnightly successive assessments were made during the in-treatment period (visits I, II, and III). The HRS was completed by two psychiatrists (with previous experience in this procedure) who assessed the patients simultaneously. The raters were not aware of the BDI score at the time of the HRS rating. The HRS values are the mean of those obtained by the two interviewers during the joint interviews. The interrater correlation was high; the majority of the κ coefficients lay between 0.75 and 0.90. Psychiatrists were unaware of the DCM study, while physiologists were unaware of the HRS and BDI scores.

There was no significant difference between the two groups, (1) low-IT and (2) high-IT, or the subgroups of FENF, IMI, and CMI, insofar as mean ages, mean episodes, mean duration of illness, or gender composition. Informed consent was obtained from all patients.

B. Statistical Methods

For comparison purposes, the results from the three treatment subgroups (FENF, IMI, and CMI) = HRS-scores (mean ± SE) were obtained during the three in-treatment visits (I, II, and III). Differences between subgroup-means were analyzed using Student *t* test. The criterion for significance was $p < 0.05$. A minimum HRS score of 17 was required for a patient to enter the study. Maximal improvement was obtained when final HRS score was <10.

Correlations between HRS and BDI scores are presented in this study. The BDI values, however, are not presented; these are available from the authors on request. Correlations between IT and total HRS values are also presented in this study.

C. Distal Colon Motility

Two DCM studies were performed immediately after visit 00 and before visit 0. No major variations in IT were found between the two motility records; the values are the mean of those obtained during the two DCM records. More details dealing with this procedure are given in the preceding paper.[9] We found 40 low-IT (<20 mm Hg) patients: 7.3 ± 1.15 (mean ± SE) and 46 high-IT (<30 mm Hg) patients: 64.7 ± 5.1 mm Hg. Values between 20 and 25 mm Hg = 20 mm Hg, and values between 25 and 30 mm Hg = 30 mm Hg. A third DCM record was performed in 29 low-IT and in 38 high-IT patients, 1 to 2 weeks after visit III.

III. RESULTS

Two separated double-blind trials were carried out in this study with the two groups included in our population: 40 low-IT and 46 high-IT depressed patients. The subjects were randomly assigned to treatment with FENF, IMI, or CMI and were systematically assessed over the first 6 weeks of treatment. All patients received 40 mg a day of fenfluramine or 150 to 300 mg a day of tryciclic antidepressant. When the data for all 86 patients were examined we found the following results.

A. High Intestinal Tone Group

This group was composed of 46 subjects (23 male and 23 female), age range 15 to 60 (mean 37.7). Episode numbers ranged from 1 to 15 (mean 6.6), whereas duration of illness ranged from <1 to 36 years (mean 16.2). The mean HRS and BDI scores for all patients entering the group were 27.0 ± 2.0 and 33.3 ± 2.1 respectively (mean ± SE); ranges were 18 to 40 and 22 to 39, respectively. Correlation between both scales at visit 0 was r = 0.81, $p < 0.001$. Correlation between IT values and HRS scores: r = 0.56, $p < 0.01$.

1. Clomipramine Subgroup

This group was composed of 14 cases (7 male and 7 female), age range 15 to 57 years (mean 37.2). Episode numbers ranged from 1 to 15 (mean 7.2), while duration of illness ranged from <1 to 36 years (mean 14.6 years) These patients showed little or no improvement as indicated by change scores on the total HRS and BDI. HRS scores at visit 0, I, II, and III (mean ± SE) were 25.1 ± 1.5; 19.5 ± 1.4; 17.2 ± 1.3; and 16.2 ± 0.9. The lowest HRS mean score recorded (16.2) was clearly higher than the normal HRS score = 10. However, it is obvious that some degree of improvement was obtained. Correlations between HRS and BDI scores (r) were

1.   Visit I — r = 0.73, $p < 0.001$
2.   Visit II — r = 0.51, $p < 0.01$
3.   Visit III — r = 0.56, $p < 0.01$

IT in this subgroup before CMI trial was 62.4 ± 8.5 (mean $\pm$ SE), whereas after visit III it was 52.9 ± 10.6.

2. Imipramine Subgroup

This group was composed of 15 cases (7 male and 8 female), age range 14 to 58 years (mean 38.1). Episode numbers ranged from 1 to 15 (mean 6.9), whereas duration of illness ranged from <1 to 36 years (mean 14.3). All these patients showed considerable improvement at the first fortnightly visit (visit I) and further improvements were registered at visits II and III. Mean $\pm$SE of HRS scores at visits 0, I, II, and III, were 27.9 ± 1.7; 16.1 ± 1.6; 8.2 ± 0.8; and 6.8 ± 0.7. It is observed that IMI-HRS scores fell below 10 at visit II, $p < 0.001$ when compared with CMI-HRS score. Correlations (r) between HRS and BDI values were

1. Visit I — r = 0.93, $p < 0.001$
2. Visit II — r = 0.67, $p < 0.001$
3. Visit III — r = 0.78, $p < 0.001$

IT in this subgroup before IMI-trial was 66.5 ± 6.3; whereas after visit III it was 23.9 ± 5.8.

3. Fenfluramine Subgroup

This group was composed of 17 cases (9 male and 8 female), age range 15 to 60 years (mean 37.8). Episode numbers ranged from 1 to 14 (mean 5.7); whereas duration of illness ranged from <1 to 41 years (mean 16.9). This subgroup showed normalization of mean HRS score since the first fortnightly visit (visit I): 10.1 ± 1.0. Comparison with mean HRS scores from IMI and CMI subgroups showed highly significant differences, $p < 0.001$ and $p < 0.001$, respectively. Further reductions in HRS scores were registered at visits II and III; 6.9 ± 0.3 and 5.1 ± 0.2, respectively. These last two mean scores differ significantly from CMI mean scores $p < 0.001$, but not from IMI mean score, p not significant. Correlations (r) between HRS and BDI values were

1. Visit I — r = 0.59, $p < 0.001$
2. Visit II — r = 0.60, $p < 0.001$
3. Visit III — r = 0.41, $p < 0.001$

IT in this subgroup before FENF-trial was 67.1 ± 7.4, whereas after visit III it was 7.3 ± 4.6.

Summarizing, IMI and FENF but not CMI markedly improved high-IT depressed patients and strongly reduced IT. The FENF-induced effects on HRS scores since the first fortnightly visit (visit I) were registered, whereas the IMI-induced normalization of HRS scores was obtained at visit II (the second fortnightly visit). Although both drugs strongly reduced IT, this effect was more pronounced with FENF than with IMI.

It is worth mentioning that 20 out of the 46 high-IT depressed patients (43.5%) were unable to complete the self-rating BDI scale (7 from CMI, 6 from IMI, and 7 from FENF subgroups). All the IMI- and FENF-treated patients were able to complete BDI scale during visit I; however, only two from the CMI subgroup completed BDI scale on visit I, and the remaining five patients on visit II. HRS scores and BDI scores correlated well during the in-treatment periods.

B. Low Intestinal Tone Group

This group was made up of 40 cases (21 male and 19 female), age range 15 to 63 years (mean 37.8). Episode numbers ranged from 1 to 13 (mean 5.7). Duration of illness ranged from <1 to 39 years (mean 15.5 ± 2.6). Correlation between IT values and HRS scores: r = 0.07, p not significant.

1. Clomipramine Subgroup

This group was composed of 12 cases (7 male and 5 female), age range 17 to 59 years (mean 38.2). Episode numbers from 1 to 11 (mean 5.9). Duration of illness from <1 to 36 years (mean

16.3). These patients showed marked improvement since the first visit (visit I), as indicated by change scores on the total HRS and BDI; HRS scores at visits 0, I, II, and III (mean ± SE) were 24.8 ± 1.6; 12.1 ± 1.3; 6.4 ± 0.7; and 4.8 ± 0.4. Correlations (r) between HRS and BDI scores were

1. Visit I — r = 0.87, *p* <0.001
2. Visit II — r = 0.15. *p* not significant
3. Visit III — r = 0.40, *p* not significant

IT in this subgroup before CMI-trial was 6.8 ± 1.0 (mean ± SE), whereas after visit III it was 35.7 ± 8.9.

2. Imipramine Subgroup

This group was made up of 13 cases (6 male and 7 female), age range 15 to 61 years (mean 39.1). Episode numbers ranged from 1 to 13 (mean 5.8). Duration of illness was <1 to 39 years (mean 14.6). These patients showed little or no improvement as indicated by mean-HRS change scores at visits 0, I, II, and III: 24.3 ± 1.4; 20.6 ± 1.3; 19.6 ± 1.0; 19.6 and ± 0.9. Correlations (r) between HRS and BDI scores were

1. Visit I — r = 0.82, *p* <0.001
2. Visit II — r = 0.65, *p* <0.001
3. Visit III — r = 0.66, *p* <0.001

Comparisons between CMI and IMI mean scores showed highly significant differences at visits I, II, and III: *p* <0.001 in all cases. IT in this subgroup before IMI-trial was 9.1 ± 2.1, whereas after visit III it was 7.3 ± 1.4.

3. Fenfluramine Subgroup

This group was composed of 15 cases (8 male and 7 female), age range 16 to 63 years (mean 36.3). Episode numbers ranged from 1 to 13 (mean 5.3); duration of illness was <1 to 33 years (mean 15.3). These patients did not tolerate the drug. All of them interrupted the trial during the 1st week because of undesirable side effects (nausea, tachycardia, insomnia, anxiety, headache, diarrhea, irritability, etc.).

Summarizing, CMI but not IMI markedly improved low-IT depressed patients. Normalization of HRS scores was observed following the first visit (visit I) in the CMI subgroup. FENF was not tolerated by these patients. CMI but not IMI strongly increased IT. HRS and BDI correlated well during the in-treatment period.

IV. RESULTS

The results presented in our study show that FENF and IMI, but not CMI, markedly improved high-IT depressed patients whereas CMI, but neither IMI nor FENF, markedly improved low-IT depressed patients. Furthermore, improvements in high-IT depressed patients were paralleled by deep reduction of IT, while improvements in low-IT depressed patients were accompanied by elevation of IT. These findings suggest that two different pathophysiological mechanisms underlie the two types of depression and that IT at distal colon level could be a good predictor for establishing this difference. On the other hand, the fact that FENF was well tolerated by high-IT patients and very badly tolerated by low-IT subjects could be used as a guide to differentiate and treat depressed patients.

The fact that the two groups of patients showed significantly different preponderances when tested on many of the individual HRS-items reinforces the hypothesis of two different types of depressive disorders (Figure 20).

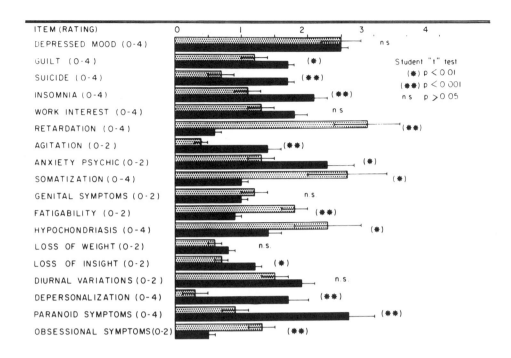

FIGURE 20. The 18-item Hamilton rating depression scores (mean ± SE) in 40 low-IT (▨) and 46 hi- IT (■) patients. From *J Affect Dis.*

We have no satisfactory explanation for understanding the pathophysiologic mechanisms underlying these kinds of depression; we can only offer indirect evidence. However, the present findings show once again that the DCM profile is a good guide to the pharmacotherapy of many psychiatric and psychosomatic disorders.[10-17]

Our speculations are addressed to the supposition that some kind of relationship exists between the central autonomic nervous system (CANS) and the peripheral one at distal colon level. With regard to this, many authors have postulated that the nervous system of the gut could be a good model to approach the brain model.[18] In addition, it has been shown that the hemato-enteric barrier possesses some characteristics similar to those of the hemato-brain barrier.[19]

The different responses to FENF, a serotonin releasing agent, seem to involve this neurotransmitter in both groups of depressive disorders found in this study. On the other hand, the finding that high-IT patients responded to treatment with FENF and IMI, whereas the low-IT patients responded to CMI, seems paradoxical since both FENF and CMI supposedly increase serotonin neurotransmission. The therapeutic effects of FENF could be more related to its ability to induce brain serotonin depletion than to the early increase of this neurotransmitter at synaptic level. In this respect, it would be interesting to investigate the changes in CSF-5HIAA levels in FENF in-treatment patients.

We cannot explain why inversion of motility pattern paralleled improvements in depressive disorders, whereas no changes in distal colon tone were registered in those cases not improved by treatments.

It has been shown that central serotonin releasing agents mianserin,[20] amphetamines,[21] and fenfluramine[1] abruptly reduce distal colon tone whereas drugs which block ACh receptors or inhibit catecholamine release, induce decreased distal colon tone.[22,23]

According to the above-mentioned findings, high- and low-IT could parallel increased and reduced 5HT storage capacity at both intestinal and brain levels. Since 5HIAA increases in brain and CSF levels when 5HT neurons are activated,[24] low-IT depressed patients should have greater

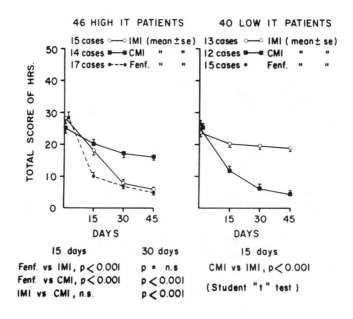

FIGURE 21.   Antidepressive effects of imipramine, clomipramine, and fenfluramine. From *J Affect Dis.*

CSF-5HIAA levels than high-IT depressed patients. With respect to this, Banki et al.[25,26] found a negative correlation between CSF-5HIAA and the symptoms of anxiety, suicide, and agitation (predominating in high-IT depressed patients); whereas the correlation was positive between CSF-5HIAA and the symptoms somatization, obsession-compulsion, hypochondria, and retardation (predominating in low-IT depressed patients). On the other hand, severe obsessive-compulsive disorder patients who improved with CMI treatment[27] showed significantly higher CSF levels of 5HIAA before treatment. The amelioration of their obsessive-compulsive symptoms was positively correlated with the reduction of CSF concentrations of 5HIAA during CMI treatment. In fact, CMI reduced 5HIAA levels in CSF, whereas CMI and other 5HT-uptake inhibitors block the induced release of $[^3H]5HT$ into the ventricle brain preloaded with $[^3H]5HT$.[28] Finally, it has been shown that the depletion of brain 5HT induced by FENF is antagonized by CMI but not by IMI or desipramine[29,30] (Figure 21).

In conclusion, our findings suggest that distal colon tone can be useful as a guide to the pharmacotherapy of depressive disorders, and that FENF is a good antidepressant drug in one type of patient, those showing high distal colon tone.

V. SUMMARY

The present study gathers results obtained in the treatment of 86 depressed outpatients. The patients were grouped according to their distal colon tone, low-IT, and high-IT. All the patients fulfilled the RDC for nonpsychotic unipolar major depressive disorders and were rated on a slightly modified 18-item Hamilton Rating Scale for Depression (HRS). They also completed the 21-item self-rating Beck Depression Inventory (BDI). The 46 subjects from the high-IT group and the 40 subjects from the low-IT group were divided into 3 subgroups each for double-blind trials performed with CMI, IMI, and FENF. Normalization of HRS and BDI scores and reduction of IT were obtained with IMI and FENF but not CMI in high-IT patients; whereas normalization of both scores and increase in IT were obtained with CMI but not with IMI or

FENF in low-IT patients. FENF was not tolerated in this last group. HRS and BDI scores correlated well during in-treatment periods. Our results suggest that distal colon tone is a guide to determining the pharmacotherapy in the depressive disorders. Furthermore, it was shown that FENF is a useful antidepressant drug in one type of patient.

REFERENCES

1. **Trulson ME, Jacobs BL.** Behavioural evidence for the rapid release of CNS serotonin by PCA and fenfluramine. *Eur J Pharmacol* 36: 149-154, 1976.
2. **Reuter CJ.** A review of the CNS effects of fenfluramine, 780SE and norfenfluramine on animals and man. *Postgrad Med* 51: 18-27, 1975.
3. **Lechin F, van der Dijs B.** Colon motility and psychological traits in the irritable bowel syndrome. *Dig Dis Sci* 26: 474-475, 1981.
4. **Lechin F, van der Dijs B, Bentolila A, Peña F.** Antidiarrheal effects of dihydroergotamine. *J Clin Pharmacol* 17: 339-349, 1977.
5. **Lechin F, van der Dijs B, Bentolila A, Peña F.** The "spastic colon" syndrome — therapeutic and pathophysiologic considerations. *J Clin Pharmacol* 17: 431-440, 1977.
6. **Lechin F, van der Dijs B.** Effects of dopaminergic blocking agents on distal colon motility. *J Clin Pharmacol* 19: 617-625, 1979.
7. **Hamilton M.** A rating scale for depression. *J Neurol Neurosurg Psychiatry* 23: 56-62, 1960.
8. **Beck AT, Ward CH, Mendelson M, Mock J, Erbaugh J.** An inventory for measuring depression. *Arch Gen Psychiatry* 4: 561-571, 1961.
9. **Lechin F, van der Dijs B, Gómez F, Lechin E, Arocha L.** Distal colon motility and clinical parameters in depression. *J Affect Dis* 5: 19-26, 1983.
10. **Lechin F, van der Dijs B, Lechin E.** *The Autonomic Nervous System — Physiological Basis of Psychosomatic Therapy.* Editorial Científico-Médica, Barcelona, 1979.
11. **Lechin F, Gómez F, van der Dijs B, Lechin E.** Distal colon motility in schizophrenic patients. *J Clin Pharmacol* 20: 459-464, 1980.
12. **Lechin F, van der Dijs B, Gómez F, Valls JM, Acosta E, Arocha L.** Pharmacomanometric studies of colonic motility as a guide to the chemotherapy of schizophrenia. *J Clin Pharmacol* 20: 664-671, 1980.
13. **Lechin F, van der Dijs B, Gómez F, Acosta E, Arocha L.** On the use of clonidine and thioproperazine in a woman with Gilles de la Tourette's disease. *Biol Psychiary* 17: 103-108, 1982.
14. **Lechin F, van der Dijs B.** Treatment of infertility with levodopa. *Br Med J* 280: 480, 1980.
15. **Lechin F, van der Dijs B.** Physiological, clinical and therapeutical basis of a new treatment for headache. *Headache* 20: 77-84, 1980.
16. **Lechin F, van der Dijs B.** Clonidine therapy for psychosis and tardive dyskinesia. *Am J Psychiatry* 138: 390, 1981.
17. **Lechin F, van der Dijs B.** Intestinal pharmacomanometry and glucose tolerance — evidences of two antagonistic dopaminergic mechanisms in the human. *Biol Psychiatry* 16: 969-986, 1981.
18. **Fox J.** Gut's nervous system. A model for the brain. *J Chem Eng* 1: 32-33 (Dec), 1980.
19. **Gershon MD, Bursztajn S.** Properties of the enteric nervous system — limitation of access of intravascular macromolecules to the myenteric plexus and muscularis externa. *J Comp Neurol* 80: 467-488, 1978.
20. **Raiteri M, Angellini F, Bertollini A.** Comparative study of the effects of mianserin, a tetracyclic antidepressant, and of imipramine on uptake and release of neurotransmitters in synaptosomes. *J Pharm Pharmacol* 28: 483-488, 1976.
21. **Breese GR, Cooper BR, Mueller RA.** Evidence for involvement of 5HT in the actions of amphetamine. *Br J Pharmacol* 52: 307-311, 1974.
22. **Lechin F, van der Dijs B, Gómez F, Acosta E, Arocha L.** Effects of D-amphetamine, clonidine, and clonazepam on distal colon motility in non-psychotic patients. *Res Commun Psychol Psychiatr Behav* 7: 385-410, 1982.
23. **Lechin F, van der Dijs B, Gómez F, Acosta E, Arocha L.** Comparison between the effects of D-amphetamine and fenfluramine on distal colon motility in non-psychotic patients. *Res Commun Psychol Psychiatr Behav* 7: 411-430, 1982.
24. **Reinhardd JF, Wurtman RJ.** Relation between 5HIAA levels and the release of serotonin into brain synapses. *Life Sci* 21: 1741-1746, 1977.
25. **Banki CM, Vojnik M, Molnar G.** Cerebrospinal fluid amine metabolites, tryptophan and clinical parameters in depression. I. Background variables. *J Affect Dis* 3: 81-89, 1981.
26. **Banki CM, Molnar G, Vojnik M.** Cerebrospinal fluid amine metabolites, tryptophan and clinical parameters in depression. II. Psychopathological symptoms. *J Affect Dis* 3: 91-109, 1981.

27. **Thorén P, Asberg M, Bertilsson L, Mellström B, Sjöqvist F, Traskman L.** Clomipramine treatment of obsessive-compulsive disorder. II. Biochemical aspects. *Arch Gen Psychiatry* 37: 1289-1294, 1980.

28. **Liang-Fu T.** 5-Hydroxytryptamine uptake inhibitors block para-methoxyamphetamine-induced 5HT release. *Br J Pharmacol* 66: 185-190, 1979.

29. **Ghezzi D, Samanin R, Bernasconi S, Tognoni G, Gerna M, Garattini S.** Effect of thymoleptics on fenfluramine-induced depletion of brain serotonin in rats. *Eur J Pharmacol* 24: 205-210, 1973.

30. **Lechin F, van der Dijs B, Gómez F, Arocha L, Acosta E, Lechin E.** Distal colon motility as a predictor of antidepressant response to fenfluramine, imipramine and clomipramine. *J Affect Dis* 5: 27-35, 1983.

INTRAMUSCULAR CLONIDINE TEST IN HIGH AND LOW INTESTINAL TONE SUBJECTS

Fuad Lechin, Bertha van der Dijs, Marcel Lechin, Rheyna Camero, Simon Villa, and Francisco Gómez

I. INTRODUCTION

Systolic blood pressure (SBP), diastolic blood pressure (DBP), norepinephrine, cortisol (CRT), growth hormone (GH), and prolactin (PRL) plasma levels were investigated in 46 normal subjects, 28 high-IT and 18 low-IT, before and after the administration of a single intramuscular dose of clonidine (2.5 µg/kg). High-IT subjects had lower mean values of DBP than low-IT subjects, and basal norepinephrine was significantly greater in low-IT than in high-IT subjects. A negative correlation between norepinephrine and IT values was found for the high-IT, but not for the low-IT group, during the preclonidine periods. The drug reduced SBP in high-IT, whereas it reduced SBP plus DBP and norepinephrine in low-T subjects. Clonidine induced significant reductions of CRT and increases of GH in both groups; furthermore, a slight but significant reduction of PRL was registered in high-IT group. The drug also induced increase of distal colon tone in high-IT subjects and suppressed phasic activity (waves) in low-IT subjects. While a significant positive correlation was found between norepinephrine and DBP in low-IT subjects during postclonidine periods, no correlation was found between the two parameters in high-IT subjects. Other significant positive (+) and negative (-) correlations during postclonidine periods were CRT/GH (-), CRT/PRL (+), and GH/PRL (-) in high-IT subjects and NE/CRT (+), NE/GH (-), CRT/GH (-), CRT/DBP (+), and GH/DBP (-) in low IT subjects. Finally, significant negative correlation was found between norepinephrine and distal colon tone during postclonidine periods in high-IT subjects. Our results suggest that there is a sympathetic (NE) hypoactivity in high IT and a sympathetic (NE) hyperactivity in low-IT subjects. Postsynaptic and presynaptic pharmacological mechanisms are postulated in order to explain these two different patterns of responses to the challenge of clonidine.

Clonidine, an α_2-agonist drug, is an antihypertensive agent which possesses other pharmacologic effects. It induces changes in DCM,[1-3] in behavior,[3,4] and in GH secretory patterns.[5-7] In addition, clonidine is able to modify norepinephrine plasma levels.[8] Clonidine-induced effects are associated with its ability to inhibit sympathetic nervous discharge through centrally mediated mechanisms[9,10] and in addition with its ability to stimulate α_2-postsynaptic central receptors.[11,12] Up to the present we have tested the effects of clonidine on the DCM of more than 3000 subjects, normal and diseased (somatic, psychosomatic, and psychotic).[4,13,14] We have demonstrated that the changes induced in DCM and behavior vary according to the DCM profile obtained during the preclonidine motility record.[2,3] Furthermore, we have also shown that a close relationship exists between DCM, behavior, and character traits.[15-17]

We have distinguished two types of subjects according to the changes induced by clonidine on their DCM: (1) those showing a high-IT who respond to the drug with an increase of tone and (2) those showing a low-IT and frequently high phasic activity. In the latter, clonidine suppresses the phasic activity. We postulated that group 1 is composed of subjects showing hypoactivity of the sympathetic or NE system, whereas group 2 subjects show sympathetic (NE) hyperactivity.[2,3,18] In order to test this presumption we carried out the present study in which we investigated 46 normal subjects. Investigations included the following measurements: DCM, SBP, DBP, and plasma levels of norepinephrine, CRT, GH, and PRL. All these parameters were determined before and after the administration of clonidine.

II. MATERIALS AND METHODS

A. Subjects

Our protocol included 46 volunteer subjects free of physical and mental disorders, most of them medical students (28 women and 18 men), aged 19 to 31 years (mean age 24.3 years). All these drug-free subjects gave written consent before admission to the study, and the protocol was approved by the Ethical Committee of Fundaime. None had a recent history of weight loss or any noteworthy disease, and none exceeded ideal body weight by more than 10%. All were clinically, radiologically, and biochemically investigated to exclude infectious, metabolic, or organic diseases. The subjects were also screened for psychiatric illnesses, using the RDC and the independent agreement of two psychiatrists. Only those subjects on whom there was a full diagnostic agreement were included in this study. Studies of women were performed only outside premenstrual (1 week prior) and menstrual periods. The study was designed in a double-blind crossover manner. Each volunteer was tested with clonidine intramuscularly and with saline placebo injection on two separate occasions 2 weeks apart. The sequence in which drug and placebo were given was randomized.

B. Test Procedure

All experiments were started on recumbent subjects at 07.00 h after an overnight fast. A venous catheter was inserted in a forearm vein at least 30 min prior to beginning the experiments, and patency was achieved by a slow infusion of normal saline. Previously, two open-ended polyethylene catheters were placed into sigmoid and rectum, in order to perform the DCM record. After that, the subjects lay comfortably on a bed. Blood samples were collected at 08.00, 08.30, 09.00, 09.30, 10.00, and 10.30 h. Clonidine (2.5 μg/kg; Catapres, Boehringer Ingelheim, Ridgefield, CT) was intramuscularly injected immediately after the second blood sample (08.30 h). The blood samples were immediately placed on ice and spun at 3000 rpm for 20 min in a refrigerated centrifuge and serum aliquot specimens stored at –80°C until required for hormonal assays. Blood for catecholamine assay was transferred to tubes containing 4 mg dithiothreitol and 20 μl EGTA solution (100 mg/ml) and was inverted carefully several times. Blood specimens were centrifuged immediately at –4°C, and plasma was separated and stored at –80°C until assayed. Blood pressure was monitored every 30 min by two different doctors by duplicate (auscultatory and ultrasonic sphygmomanometer) methods.

C. Analytical Methods

Plasma norepinephrine, epinephrine, and dopamine levels were determined simultaneously by the radioenzymatic method of Sole and Hussein[19] with the modifications suggested by Carey et al.[20] The sensitivity of this method was 2 pg. for norepinephrine, epinephrine, and dopamine. For plasma catecholamine levels, the intraassay coefficients of variation were 12, 14, and 10%, respectively.

For hormone plasma levels the samples were measured in duplicate, and all samples belonging to the same experimental set were measured in the same assay. CRT was assayed by a competitive protein-binding radioimmunoassay[21] using a Cortisol Diagnostic Kit (Corti-Shure, NML). The concentrations are expressed as micrograms per deciliter in terms of the standards supplied with the kits. The sensitivity of the assay was 0.5 µg/dl. The intraassay and interassay coefficients of variations were 2.5 and 5.1%, respectively. Human GH was assayed by specific radioimmunoassay[22] using the Phadebas Human GH Prist Kit with a sensitivity of 0.25 ng/ml. Intra- and interassay coefficients of variation were 9 and 11.2%, respectively. PRL was assayed by specific radioimmunoassay[23] using the Prolactin Diagnostic Kit (Lacti-Quant, NML) with a sensitivity of 0.5 ng/ml. Intra- and interassay coefficients of variation were 5 and 12%, respectively.

D. Distal Colon Motility Studies

The DCM study procedure was described in detail in our previous paper.[3] In short, two open-ended polyethylene catheters were placed in sigmoid and rectum (25 and 10 cm from the anal opening, respectively) and continuously perfused with saline solution at a rate of 9 ml/n. The catheters were connected to two differential air pressure transducers. A Nihon Kohden polygraph was employed to register pressure changes. Motility of the distal colon has two components: (1) IT = distance between zero line and baseline pressure, determined by segments of trace-lacking waves and (2) phasic activity = waves. IT was measured on a 10 mm of Hg graded scale.

E. Statistical Methods

Results are expressed as the mean values ± SE. The hypothesis to be tested was that intramuscularly injected clonidine brings about an alteration in the mean values of parameters recorded for the same subjects given placebo and that, in addition, clonidine-induced changes vary according to two distinct DCM patterns found among the 46 subjects integrating our protocol. On the other hand, our research work looked for possible correlations between DCM and the other parameters altered by clonidine. The null hypotheses were examined using a two-way analysis of variance (Anova), a Pearson Product Moment Test, and an unpaired t test. p Values less than 0.05 were considered significant. A Hewlett Packard 41C Statistics PAC 00041-15002 was used for calculations.

III. RESULTS

A. Preclonidine Periods

Of the 46 subjects (17 women and 11 men), 28 showed high-IT (>20 mm Hg), whereas the remaining 18 subjects (11 women and 7 men) showed low-IT (<20 mm Hg, most of them 0 mm Hg). When comparing both groups during the preclonidine period, it was found that high-IT subjects showed significantly lower DBP and norepinephrine plasma levels than low-IT subjects (p <0.01 and <0.001, respectively). Taking into consideration that high-IT subjects showed comparatively lower NE plasma levels, we will refer to them as hyposympathetic subjects. On the contrary, since low-IT subjects showed comparatively higher norepinephrine plasma levels, we will refer to them as hypersympathetic or hypernoradrenergic subjects. A negative correlation between norepinephrine plasma levels and IT values was found for the high-IT group (r = –0.23; p not significant). The lack of correlation in this group probably reflects low variability of IT.

B. Postclonidine Periods
1. High Intestinal Tone (Hyposympathetic) Subjects
a. Blood Pressure

Clonidine induced a small but significant reduction in SBP at 30, 60, and 90 min (F = 2.79, 3.21, and 2.86, respectively; p <0.01 in all cases). On the contrary, the drug did not modify DBP at any time.

b. Norepinephrine Plasma Levels

Clonidine also reduced slightly but significantly norepinephrine levels at 30 and 60 min (F = 2.99 and 2.78, respectively; $p < 0.01$ in both cases). However, NE values returned to control levels during 90- and 120-min periods. No significant correlation was found between SBP and norepinephrine plasma levels.

c. Distal Colon Motility

Clonidine induced an increase of sigmoidal and/or rectal tone in all the 28 cases. The percent mean $\pm$ SE increase was 308.4 ± 36.6 over the mean preclonidine value (100%). The increases in IT were registered during the first 30-min postclonidine period. The same remained constant through the following postclonidine periods. The correlation between norepinephrine plasma levels and IT values during the four postclonidine periods was $r = -0.77, -0.79, -0.71,$ and $-0.55,$ respectively; $p < 0.001$ in all cases. No significant correlation was found between IT and the other parameters during postclonidine periods (DBP, CRT, GH, and PRL).

d. Cortisol

Small but significant reductions of the mean values were registered at 60, 90, and 120 min (F = 3.68, 3.77, 3.81, respectively; $p < 0.005$ in all cases).

e. Growth Hormone

Significant and important increases of GH were registered at 60, 90, and 120 min (F = 12.2, 15.63, and 15.34, respectively; $p < 0.001$ in all cases).

f. Prolactin

Although some tendency to PRL lowering was observed, the decrease was only significant at the 90-min period (F = 1.98; $p < 0.05$).

Highly negative correlations were found between CRT and GH individual values: $r = -0.69,$ $-0.78, -0.89,$ and $-0.83,$ respectively; $p < 0.001$ in all cases. Slight but significant positive correlations were found between CRT and PRL values during the four postclonidine periods: $r = 0.43, 0.52, 0.57,$ and $0.55;$ $p < 0.02$ in all cases. Finally, slight but significant negative correlations were found between GH and PRL values at 60, 90, and 120 min: $r = -0.41, -0.49,$ and $-0.43,$ respectively; $p < 0.02$ in all cases. No correlation was found between hormonal values and other parameters (SBP, DBP, IT, and NE) (Figures 22 and 23).

2. Low Intestinal Tone (Hypersympathetic) Subjects

a. Blood Pressure and Norepinephrine

Clonidine induced significant reduction of SBP only in the 60- and 90-min periods (F = 3.48 and 3.29, respectively; $p < 0.01$ in both cases). However, no difference between clonidine and placebo was observed during the 120-min period. On the contrary, a sustained and very significant reduction in DBP was observed during the four postclonidine periods (F = 3.56, 5.22, 5.15, and 4.98, respectively; $p < 0.001$ in all cases). A sustained progressive and strong reduction in norepinephrine plasma levels paralleled DBP reductions (F = 6.91, 10.85, 15.32, and 17.46, respectively; p 0.001 in all cases).

Although there was a poor correlation between norepinephrine and DBP during preclonidine periods, these two parameters correlated significantly during the four postclonidine periods: $r = 0.58, 0.63, 0.67,$ and $0.57,$ respectively; $p < 0.02, <0.01, <0.01,$ and $<0.02,$ respectively. These findings contrast strikingly with the group of high-IT subjects, in which norepinephrine and DBP showed no significant correlation among levels.

b. Distal Colon Motility

Clonidine suppressed waves (phasic activity) in all low-IT subjects. No increase of IT was registered in these subjects.

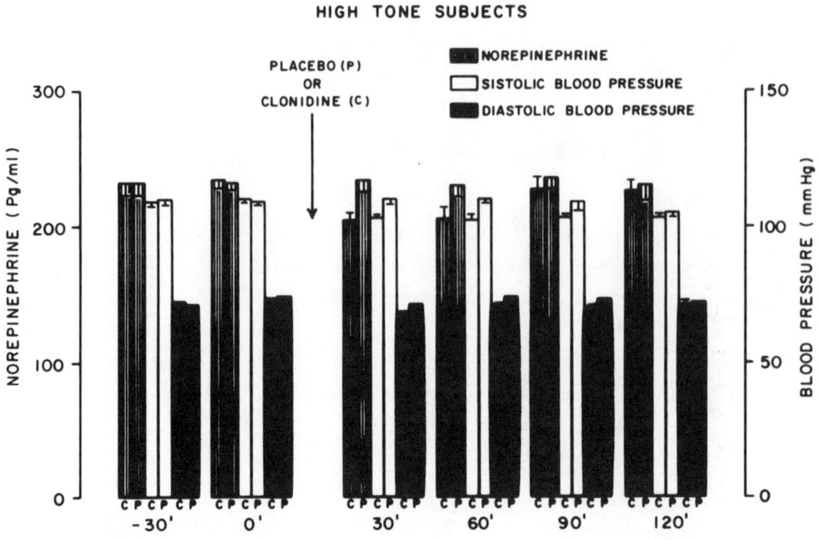

HIGH TONE SUBJECTS

FIGURE 22.   Effects of clonidine (2.5 µg/kg i.m.) and placebo on norepinephrine plasma level, SBP, and DBP in 28 high-IT subjects. Each bar represents the mean value ± SEM. SBP was significantly reduced at 30, 60, and 90 min. Norepinephrine was significantly reduced at 30, 60 and 90 min only. DBP was not reduced at any period of time. No correlations were found between SBP, DBP, and norepinephrine. From *Neuroendocrinology* 40: 253-261, 1985. With permission.

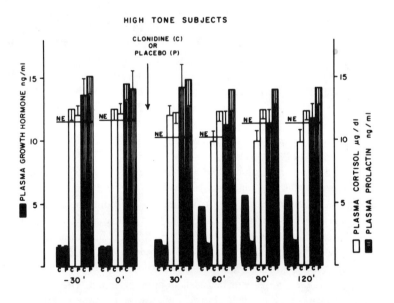

FIGURE 23.   Effects of clonidine (2.5 µg/kg i.m.) and placebo on plasmatic levels of GH, CRT, and PRL in 28 high-IT subjects. Each bar represents the mean value ± SEM. CRT was significantly reduced at 60, 90, and 120 min, while GH was significantly increased at the same time periods. Norepinephrine values (horizontal bars; mean ± SEM: pg/ml) were: -30 min: 232.62 ± 8.07; 0 min: 295.36 ± 6.07; 30 min: 206.35 ± 5.06; 60 min: 207.83 ± 8.09; 90 min: 228.60 ± 9.31; and 120 min: 227.59 ± 8.09. From *Neuroendocrinology* 40: 253-261, 18985. With permission.

c. Cortisol

Sustained and progressive reduction of cortisol mean values were registered in this group at 60, 90, and 120 min (F = 2.02, 6.76, and 5.89, respectively; p <0.05, <0.001, and <0.001, respectively).

d. Growth Hormone

Important and significant increases of this hormone were registered simultaneously with cortisol decreases during the 60-, 90-, and 120-min periods (F = 10.84, 11.38, and 11.06, respectively; p <0.001 in all cases).

e. Prolactin

No changes of PRL mean values were registered at any postclonidine period.

Significant negative correlations were found between CRT and GH individual values during the four postclonidine periods: r = –0.33, –0.56, –0.63, and –0.69, respectively; p = not significant <0.01, <0.001, and <0.001. In addition, significant positive correlations were found between CRT, DBP, and norepinephrine at 60, 90, and 120 min. CRT-norepinephrine: r = 0.48, 0.76, and 0.79, respectively; p <0.01, <0.001, and <0.001. CRT-DBP: r = 0.57, 0.61, and 0.65, respectively; p <0.01, <0.001, and <0.001.

Significant negative correlations were found between GH and NE and between GH and DBP at 60, 90, and 120 min. GH-NE: r = –0.87, –0.91, and –0.83, respectively; p <0.001 in all cases. GH-DBP: r = –0.49, –0.52, and –0.58, respectively; p <0.01 in all cases (Figures 24 and 25).

IV. DISCUSSION

The results of this study strongly suggest an inhibitory action of clonidine on basal norepinephrine release at NE terminals in the central nervous system and/or periphery. Indeed, it is widely accepted that clonidine exerts its pharmacological effects through central mechanisms, preferentially;[24] thus, the reduction of blood pressure and plasma norepinephrine levels would be secondary to the proven inhibition of sympathetic nervous discharge evoked by clonidine, centrally.[9,10,25] The well-known fact that the sympathetic nerves are the main origin of circulating norepinephrine during the resting state[26] reinforces this presumption.

The different colonic responses to clonidine registered in low- and high-IT subjects are coherent with the biochemical and physiological differences registered during pre- and postclonidine periods among both groups and, in addition, ratify the presumption that the two types of DCM profiles correspond to two prototypes of ANS subjects: (1) high IT (hyposympathetic) and (2) low-IT (hypersympathetic).

Two main reasons argue against the possibility that clonidine-induced changes were peripherally induced: (1) the raised norepinephrine plasma levels registered in low-IT (hypersympathetic) subjects would favor peripheral pre- and postsynaptic adrenoceptors subsensitivity and (2) in peripheral tissues, α_2-adrenoceptors are predominantly presynaptically located on norepinephrine terminals.[27] Thus, in high-IT-low NE plasma level subjects, it is difficult to attribute the pronounced biochemical, physiological, and endocrinological changes induced by clonidine to the small decrease in already low plasma norepinephrine level registered in the present study. On the other hand, evidence suggests that cardiovascular,[28] pancreatic,[29] and gastrointestinal[30-32] effects are triggered by the drug through central mechanisms.

Although the magnitude of central NE activity cannot be assessed directly in these subjects, peripheral norepinephrine levels in plasma of normal subjects correlate highly with central norepinephrine levels derived from cerebrospinal fluid of these same subjects.[33] If the raised peripheral norepinephrine levels registered in low-IT (hypersympathetic) subjects reflect — at least partially — a central norepinephrine hyperactivity, one may speculate that an increased presynaptic activity and decreased postsynaptic α-adrenergic responsiveness could occur

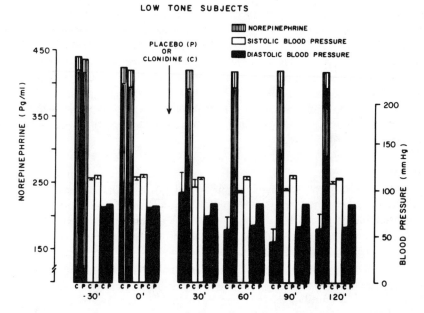

FIGURE 24.   Effects of clonidine (2.5 μg/kg im) and placebo on NE plasma level, SBP, and DBP in 18 low-IT subjects. Each bar represents the mean value ± SEM. DBP and NE were significantly reduced during the four postclonidine periods. SBP was significantly reduced during 60 and 90 min. NE and DBP correlated positively during the four postclonidine periods. No correlation was found between NE and SBP. From *Neuroendocrinology* 40: 253-261, 1985. With permission.

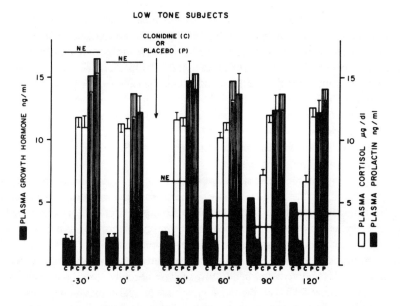

FIGURE 25.   Effects of clonidine (2.5 μg/kg i.m.) and placebo on plasmatic levels of GH, CRT, and PRL in 18 low-IT subjects. Each bar represents the mean value ± SEM. CRT was significantly reduced at 60, 90, and 120 min, while GH was significantly increased at the same time periods. NE values (horizontal bars; mean ± SEM: pg/ml) were -30 min: 441 ± 21.90; 0 min: 425.16 ± 24.26; 30 min: 236.84 ± 31.12; 60 min: 180.95 ± 19.20; 90 min: 167.31 ± 21.90; and 120 min: 183.12 ± 22.85. From *Neuroendocrinology* 40: 253-261, 1985. With permission.

centrally in these subjects. Conversely, the low norepinephrine plasma levels registered in high-IT (hyposympathetic) subjects would reflect decreased presynaptic activity and increased postsynaptic α-adrenergic responsiveness. With respect to this, although α_2-presynaptic receptors are densely located on the somatodendrites of norepinephrine mesencephalic neurones,[34] most of hypothalamic α_2-receptors are postsynaptically located.[35] Thus, according to the above reasons, our speculations are addressed to the presumption that clonidine-induced changes are centrally presynaptically mediated in low-IT (hypersympathetic; high norepinephrine plasma levels) and centrally postsynaptically mediated in high-IT (hyposympathetic; low norepinephrine plasma levels) subjects.

With regard to DCM changes, although our findings give no direct evidence as to whether the drug-induced changes stem from central or peripheral effects, nevertheless a series of facts favors a central action for DCM changes: (1) A parallel was found between motility changes and behavior changes.[3] (2) It has been demonstrated that increases of catecholamines in the blood do not modify gastrointestinal motility.[36] (3) It has been shown that the structural,[37] functional,[38] pharmacological,[39] and biochemical[40] aspects of the myenteric plexa are very similar to those observed at the central level, and, furthermore, there is evidence of the existence of a hemato-myenteric barrier,[41] very similar to the hemato-brain barrier. (4) Drugs lacking peripheral effects (nomifensene) induce DCM changes, whereas drugs lacking central effects (domperidone)[42] do not.[14]

According to our hypothesis, the close correlation found between NE, CRT, GH, and DBP registered in low-IT (hypersympathetic) subjects might imply that all these clonidine-induced changes share a common presynaptic mechanism which evokes a reduction of central and peripheral NE activity. This reduction of sympathetic activity would also explain the disappearance of distal colon waves (phasic activity) registered in these subjects. On the other hand, the less pronounced SBP reduction registered in these subjects after clonidine administration, when compared with high-IT subjects, as well as the lack of correlation between this parameter and the other parameters suggest the involving of a mechanism other than the presynaptic mechanism. The weak SBP reduction registered in low-IT subjects might reflect a clonidine effect at subsensitive central α_2-postsynaptic receptors. Conversely, the strong reduction of SBP registered in high IT (hyposympathetic) subjects might reflect a clonidine effect at supersensitive central α_2-postsynaptic receptors. With respect to this, the lack of correlation registered between SBP and the other parameters in high-IT (hyposympathetic) subjects is consistent with the existence of different types of central postsynaptic α_2-receptors: blood pressure lowering,[43] sedation,[44,45] iris dilation,[5] GH secretion,[35,46] ACTH secretion,[46] etc. Conversely, the close correlation found between clonidine-induced hormonal changes in these subjects would favor the presumption that these changes are exerted — at least partially — through common postsynaptic α_2-receptors. This predominance of postsynaptic events seems to be more probable for cortisol[47,48] and GH,[46,49,50] the values of which were closely correlated. Moreover, the slight although significant correlations found between PRL and CRT-GH values in these high IT subjects are consistent with the possibility that the former hormone shares some common postsynaptic receptors or mechanisms with the latter hormones.[51,52]

The differences in the response to clonidine registered among both groups are consistent with the multiplicity of factors involved in the control of secretion of these hormones. With respect to this, previous evidence for the inhibiting influence of norepinephrine on adrenocorticol regulation has been derived from animal studies.[53] Norepinephrine administered in physiologic doses to rat hypothalamus *in vitro* abolished the ACh stimulation of corticotropin-releasing factor;[47] evidence has been presented *in vivo* and *in vitro* that norepinephrine inhibits corticotropin-releasing factor secretion.[54] In general, systemic and central administration of norepinephrine, its precursors, and pharmacologic agonists, has been found to block the stress-induced release of ACTH and the increase of plasma corticosteroids. This inhibitory effect of norepinephrine has been shown to be dose dependent and is itself blocked by the α_2-antagonist phentolamine but not by the β-blocker propanolol.[48,55] All this accumulated evidence may explain the ability of clonidine to lower corticosteroid plasma levels, registered in this study. With respect to the positive correlation found between norepinephrine and CRT in low-IT subjects, excitatory effects of the norepinephrine system on CRT secretion may be based on peripheral mechanisms mediated by β-adrenergic[56]and α_1-adrenoceptors[57]both located outside the BBB.[58]

REFERENCES

1. **Lechin F, van der Dijs B.** Opposite effects on human distal colon motility of two postulated alpha$_2$-antagonists (mianserin and chlorprothixene) and one alpha$_2$-agonist (clonidine). *J Clin Pharmacol* 23: 209-218, 1983.

2. **Lechin F, van der Dijs B, Gómez F, Acosta E, Arocha L.** Comparison between the effects of D-amphetamine and fenfluramine on distal colon motility in non-psychotic patients. *Res Commun Psychol Psychiatr Behav* 7: 411-430, 1982.

3. **Lechin F, van der Dijs B, Gómez F, Arocha L, Acosta E.** Effects of D-amphetamine, clonidine and clonazepam on distal colon motility in non-psychotic patients. *Res Commun Psychol Psychiatr Behav* 7: 385-410, 1982.

4. **Lechin F, van der Dijs B, Gómez F, Valls JM, Acosta E, Arocha L.** Pharmacomanometric studies of colonic motility as a guide to the chemotherapy of schizophrenia. *J Clin Pharm* 20: 664-671, 1980.

5. **Berridge TL, Gadie B, Roach AG, Tulloch IF.** Alpha$_2$-adrenoceptor agonists induce mydriasis in the rat by an action within the central nervous system. *Br J Pharmacol* 78: 507-515, 1983.

6. **Lal S, Tolis G, Martin JB, Brown GM, Guyda H.** Effect of clonidine on growth hormone, prolactin, luteinising hormone, follicle-stimulating hormone and thyroid stimulating hormone in the serum of normal men. *J Clin Endocrinol Metab* 41: 827-832, 1975.

7. **Lancranjan I, Marbach P.** New evidence for growth hormone modulation by the alpha-adrenergic system in man. *Metabolism* 26: 1225-1230, 1977.

8. **Hoefke W.** Clonidine, in *Pharmacology of Antihypertensive Drugs.* Scriabine A. Ed., Raven Press, New York, 1980, 55.

9. **Starke K, Montel H.** Involvement of alpha-receptors in clonidine-induced inhibition of transmitter release from central monoamine neurons. *Neuropharmacology* 12: 1073-1080, 1973.

10. **Wolf DL, Mohrland JS.** Lateral reticular formation as a site for morphine- and clonidine-induced hypotension. *Eur J Pharmacol* 98: 93-98, 1984.

11. **Anden NE, Corrodi H, Fuxe K, Hökfelt B, Hökfelt T, Rydin C, Svensson T.** Evidence for a central noradrenaline receptor stimulation by clonidine. *Life Sci* 9: 513-523, 1970.

12. **Koss MC, Christensen HD.** Evidence for a central postsynaptic action of clonidine. *Arch Pharmacol* 307: 45-50, 1979.

13. **Lechin F, van der Dijs B.** Intestinal pharmacomanometry and glucose tolerance: evidence for two antagonistic dopaminergic mechanisms in the human. *Biol Psychiatry* 16: 969-986, 1981.

14. **Lechin F, van der Dijs B.** Intestinal pharmacomanometry as a guide to psychopharmacological therapy. *Excerpta Med Int Congr* Ser No. 604, 1983, 166.

15. **Lechin F, van der Dijs B.** Colon motility and psychological traits in the irritable bowel syndrome. *Dig Dis Sci* 26: 474-475, 1981.

16. **Lechin F, van der Dijs B, Acosta E, Gómez F, Lechin E, Arocha L.** Distal colon motility and clinical parameters in depresssion. *J Affect Dis* 5: 19-26, 1983.

17. **Lechin F, van der Dijs B, Gómez F, Arocha L, Acosta E, Lechin E.** Distal colon motility as a predictor of antidepressant response to fenfluramine, imipramine and clomipramine. *J Affect Dis* 5: 27-35, 1983.

18. **Lechin F, Gómez F, van der Dijs B, Lechín E.** Distal colon motility in schizophrenic patients. *J Clin Pharmacol* 20: 459-464, 1980.

19. **Sole MJ, Hussein MN.** A simple, specific radioenzymatic assay for the measurement of picogram quantities of norepinephrine, epinephrine and dopamine in plasma and tissues. *Biochem Med* 18: 301-307, 1977.

20. **Carey RM, Van Loon GR, Baines AD, Kaiser DL.** Suppression of basal and stimulated noradrenergic activities by the dopamine agonist bromocriptine in man. *J Clin Endocrinol Metab* 56: 595-602, 1983.

21. **Murphy BEP.** Some studies of the protein-binding of steroids and their application to the routine micro and ultramicro measurement of various steroids in body fluids by competitive protein-binding radioassay. *J Clin Endocrinol Metab* 27: 973-990, 1967.

22. **Wide L.** Radioimmunoassays employing immunoabsorbents. *Acta Endocrinol* 63: suppl 142, 207-210, 1969.

23. **Sinha YN, Selby FW, Lewis UJ, Vanderlaan WP.** A homologous radioimmunoassay for human prolactin. *J Clin Endocrinol Metab* 36: 509-512, 1973.

24. **Lowenstein J.** Clonidine. *Ann Intern Med* 92: 74-77, 1980.

25. **Hamilton TC, Hunt AAE, Poyser RH.** Involvement of central alpha$_2$-adrenoceptors in the mediation of clonidine-induced hypertension in the cat. *J Pharm Pharmacol* 32: 788-789, 1980.

26. **Lake CR, Ziegler MG, Kopin IJ.** Use of plasma norepinephrine for evaluation of sympathetic neuronal function in man. *Life Sci* 18: 1315-1326, 1976.

27. **Langer SZ.** Presynaptic receptors and the regulation of transmitter release in the peripheral and central nervous system: physiological and pharmacological significance, in *Catecholamines, Basic and Clinical Frontiers.* Usdin E, Kopin IJ, Barchas J. Eds., Pergamon Press, Elmsford, NY, 1979, 387.

28. **Ganong WF, Wise BL, Reid IA, Holland J, Kaplan S, Schackleford R, Boryczka AT.** Effect of spinal cord transection on the endocrine and blood pressure responses to intravenous clonidine. *Neuroendocrinology* 25: 105-110, 1978.

29. **Rozé C, Chariot J, Appia F, Pascaud X, Vaille Ch.** Clonidine inhibition of pancreatic secretion in rats: a possible central site of action. *Eur J Pharmacol* 76: 381-390, 1981.

30. **Lal H, Shearman GT, Ursillo RC.** Non narcotic antidiarrheal action of clonidine and lofexidine in the rat. *J Clin Pharmacol* 21: 16-22, 1981.

31. **Osumi Y, Aibara S, Sakae K, Fujiwara M.** Central noradrenergic inhibition of gastric mucosal blood flow and acid secretion in rats. *Life Sci* 20: 1407-1415, 1977.

32. **Pascaud X, Roger A, Genton M, Rozé C.** Further support for the central origin of the gastric antisecretory properties of clonidine in conscious rats. *Eur J Pharmacol* 86: 247-257, 1983.

33. **Ziegler MG, Lake CR, Wood JH.** Relationship between norepinephrine in blood and cerebrospinal fluid in the presence of a blood cerebrospinal fluid barrier for NE. *J Neurochem* 28: 677-679, 1977.

34. **Weinreich P, Seeman P.** Binding of adrenergic ligands, (^{3}H)-clonidine and (^{3}H)-WB-4101, to multiple sites in human brain. *Biochem Pharmacol* 30: 3115-3120, 1981.

35. **Eriksson E, Eden S, Modigh K.** Up- and down- regulation of central postsynaptic alpha$_2$-receptors reflected in the growth hormone response to clonidine in reserpine pre-treated rats. *Psychopharmacology* 77: 327-331, 1982.

36. **Dubois A, Henry DP, Kopis IJ.** Plasma catecholamines and postoperative gastric emptying and small intestinal propulsion in the rat. *Gastroenterology* 68: 466-469, 1975.

37. **Cook R, Burnstock G.** The ultrastructure of Auerbach's plexus in the guinea pig. I. Neuronal elements. *J Neurocytol* 5: 171-194, 1976.

38. **Gershon MD, Robinson RG, Ross LL.** Serotonin accumulation in the guinea pig myenteric plexus: ion dependence, structure activity relationship and the effect of drugs. *J Pharmacol Exp Ther* 198: 548-561, 1976.

39. **Weber LJ.** *p*-Chlorophenylalanine depletion of gastrointestinal 5-hydroxytryptamine. *Biochem Pharmacol* 19: 2169-2172, 1970.

40. **Gershon MD, Jonakait GM.** Uptake and release of 5HT by enteric 5HT neurons: effects of fluoxetine (Lilly 110140) and clorimipramine. *Br J Pharmacol* 66: 7-9, 1979.

41. **Gershon MD, Bursztajn S.** Properties of the enteric nervous system: limitation of access of intravascular macromolecules to the myenteric plexus and muscularis externa. *J Comp Neurol* 80: 467-488, 1978.

42. **Laduron PM, Leysen JE.** Domperidone, a specific in vitro antagonist, devoid of in vivo central dopaminergic activity. *Biochem Pharmacol* 28: 2161-2165, 1979.

43. **Struyker-Boudier HAJ, van Rossum JM.** Clonidine-induced cardiovascular effects after stereotaxic application in the hypothalamus of rats. *J Pharm Pharmacol* 24: 410-411, 1972.

44. **Drew GM, Marriot AS.** Alpha$_2$-adrenoceptors mediate clonidine-induced sedation in the rat. *Br J Pharmacol* 67: 133-141, 1979.

45. **Spkyracki C, Fibiger HC.** Clonidine-induced sedation in rats: evidence for mediation by postsynaptic alpha$_2$-adrenoceptors. *J Neural Transm* 54: 153-163, 1982.

46. **Weiner RI, Ganong WF.** Role of brain monoamines and histamine in regulation of anterior pituitary secretion. *Physiol Rev* 53: 905-976, 1978.

47. **Hillhouse EW, Burden JL, Jones MT.** The effect of various putative neurotransmitters on the release of CRF hormone from the hypothalamus of the rat brain in vitro. *Neuroendocrinology* 17: 1-7, 1975.

48. **Jones MT, Hillhouse E, Burden J.** Secretion of CRF hormone in vivo, in *Frontiers in Neuroendocrinology.* Ganong WF, Martini L. Eds., Raven Press, New York, 1976.

49. **Martin JB, Reichlin S, Brown GM.** *Clinical Neuroendocrinology.* F.A. Davis, Philadelphia, 1977.

50. **Aloi JA, Post RM, Murphy DL.** Growth hormone response to clonidine as a probe of noradrenergic receptor responsiveness in affective disorder patients and controls. *Psychiatr Res* 6: 171-183, 1982.

51. **Meltzer HY, Simonovic M, Gudelsky GA.** Effect of yohimbine on rat prolactin secretion. *J Pharmacol Exp Ther* 224: 21-27, 1983.

52. **Neill JD.** Neuroendocrine regulation of prolactin secretion, in *Frontiers in Neuroendocrinology,* Vol. 6. Martini L, Ganong WF. Eds., Raven Press, New York, 1980, 125.

53. **Ganong WF.** Neurotransmitters and pituitary function. Regulation of ACTH secretion. *Fed Proc Fed Am Soc Exp Biol* 39: 2923-2930, 1980.

54. **Ganong WF.** Evidence for a central noradrenergic system that inhibits ACTH secretion in brain-endocrine interaction median eminence, in *Structure and Function*. S. Karger, Basel, 1972, 239.
55. **Scapagnini V, Preziosi P.** Receptor involvement in the control of ACTH secretion. *Neuropharmacology* 12: 32-38, 1973.
56. **Anisman H.** *Psychopharmacology of Aversively Motivated Behavior*. Anisman H, Bignami G. Eds. ,Plenum Press, New York, 1978, 119.
57. **Balestreri R, Bertolini S, Castello C.** The neural regulation of ACTH secretion in man, in *Neuroendocrinology: Biological and Clinical Aspects*. Polleri A, MacLeod RM. Eds., Academic Press, London, 1979, 155.
58. **Lechin F, van der Dijs B, Jakubowicz D, Camero R, Villa S, Lechin E, Gomez F.** Effects of clonidine on blood pressure, noradrenaline, cortisol, growth hormone, and prolactin plasma levels in high and low intestinal tone subjects. *Neuroendocrinology* 40: 253-261, 1985.

INTRAMUSCULAR CLONIDINE TEST IN HIGH AND LOW INTESTINAL TONE DEPRESSED PATIENTS

Fuad Lechin, Bertha van der Dijs, Marcel Lechin, Rheyna Camero, Simon Villa, Luis Arocha, and Alex Lechin

I. INTRODUCTION

SBP, DBP, norepinephrine plasma levels, CRT, GH, and PRL plasma levels were investigated in 26 high-IT and 24 low-IT depressed patients, before and after the intramuscular injection of clonidine (2.5 µg/kg). A positive correlation was found between norepinephrine, DBP, and HRS values in low-IT depressed patients, while a negative correlation was found between HRS/IT and norepinephrine in high-IT depressed patients. Although clonidine induced significant reduction of SBP in both groups, the drug reduced DBP and NE in the low-IT group, only. CRT mean level was greater in the high-IT than in the low-IT depressed group. However, clonidine was unable to induce changes in CRT, GH, and PRL mean levels in any depressed group. Our results suggest that the clonidine-induced DBP reduction is a reliable index of sympathetic activity in depressed patients and that both parameters (DBP and IT) are useful physiological markers to differentiate two types of depressive syndrome.

The existence of two prototypes of normal and depressed subjects has been postulated according to their DCM profiles: low-IT and high-IT subjects.[1,2] Because low-IT normal subjects proved to have a higher mean noradrenaline plasma level than high-IT normals, we labeled the first group hypersympathetic (NE) and the latter group hyposympathetic (hyponoradrenergic).[3] It has also been shown that while clonidine induces great reductions of norepinephrine, SBP, and DBP in low-IT subjects, the induced norepinephrine and SBP reductions were significantly less pronounced in high-IT subjects. Moreover, whereas norepinephrine values correlated positively with DBP in low-IT subjects, a negative correlation was registered between norepinephrine and IT in high-IT normals. Finally, although clonidine induced significant GH increases and CRT decreases in both groups, the negative correlation observed between these hormonal changes was greater in high-IT than in low-IT normals. In accordance with the above findings and those of other studies,[4-6] we have suggested that clonidine effects are best understood as a presynaptic action in low-IT (hypersympathetic) subjects and as a postsynaptic action in high-IT (hyposympathetic) subjects.[3]

In view of clinical differences between low- and high-IT depressed patients demonstrated by other studies[2] using the HRS,[7] and the fact that the two types of depressed subjects respond differently to NE potentiating and 5HT potentiating antidepressants,[8] we decided to investigate the effects of clonidine on SBP, DBP, DCM, NE, GH, CRT, and PRL in both high-IT and low-IT depressed subjects.

II. MATERIALS AND METHODS

A. Subjects

Our protocol included 50 depressed patients (26 women and 24 men) who met the RDC[9] for nonpsychotic unipolar major depressive disorder. Two psychiatrists, independently diagnosing each patient, agreed in their diagnoses in 92% of the cases. Patients were rated on a modified HRS.[7] The 18 items had a maximum score of 58. Each patient was interviewed by two psychiatrists with previous experience in the procedure who assessed the patient simultaneously. The mean of the values obtained during the joint interviews constitutes the HRS value. The interrater correlation was high: the majority of the κ coefficients lay between 0.73 and 0.89. In our group of subjects the highest HRS score obtained was 41 and the lowest 18. All patients were clinically, radiologically, and biochemically investigated to exclude infectious, metabolic, or organic diseases. All patients with a primary anxiety disorder, schizophrenia, organic brain syndromes, epilepsy, or mental retardation were excluded from this study, as were patients who had been treated in the previous month with antidepressants, neuroleptics, stimulants, anxiolytics, anticholinergic agents, or any psychoactive drug. Studies of women were only performed outside of premenstrual (1 week prior) and menstrual periods. The study was designed in a double-blind crossover manner. Each patient was tested with intramuscular clonidine and with saline placebo injection on two separate occasions 2 weeks apart. The sequence in which drug and placebo were given was randomized. All the patients gave written consent before being admitted to the study and the protocol was approved by the Ethical Committee of FUNDAIME. Neither the mean ages of low-IT patients, nor the gender composition of the same, were significantly different from high-IT patients (see Table 1).

B. Test Procedure

All experiments were started on recumbent subjects at 07.00 a.m. after an overnight fast. A venous catheter was inserted in a forearm vein at least 30 min prior to beginning the experiment, and patency was achieved by a slow infusion of normal saline. Previously, two open-ended polyethylene catheters were placed into sigmoid and rectum in order to perform the DCM record. After this, the subjects lay comfortably on a bed. Blood samples were collected at 08.00, 08.30, 09.00, 09.30, 10.00, and 10.30 a.m. Clonidine (2.5 µg/kg) (Catapres, Boehringer, Ingelheim, Ridgefield, CT) was intramuscularly injected immediately after the second blood sample (08.30 a.m.). The blood samples were immediately placed on ice and spun at 3000 rpm for 20 min in a refrigerated centrifuge and serum aliquot specimens stored at –80°C until required for hormonal assays. Blood for catecholamines assay was transferred to tubes containing 4 mg dithiothreitol and 20 µl EGTA solution (100 mg/ml) and was inverted carefully several times. Blood specimens were centrifuged immediately at –4°C and plasma was separated and stored at –80°C until assayed. Blood pressure was monitored every 30 min by two different doctors by duplicate (auscultatory and ultrasonic sphygmomanometer) methods.

C. Analytical Methods

Plasma norepinephrine, epinephrine (E), and DA levels were determined simultaneously by the radioenzymatic method of Sole and Hussein[10] with the modifications suggested by Carey et al.[11] The sensitivity of this method was 2 pg for norepinephrine, E and DA. For plasma catecholamine level, the intra-assay coefficients of variation were 12, 14, and 14%, respectively.

For hormone plasma levels, the samples were measured in duplicate and all samples belonging to the same experimental set were measured in the same assay. CRT was assayed by a competitive protein-binding radioimmunoassay (RIA)[12] using a Cortisol Diagnostic Kit (Corti-shure, NML); the concentrations are expressed as micrograms per deciliter in terms of the standards supplied with the kits. The sensitivity of the assay was 0.5 µg/dl. The intra-assay and interasaay coefficients of variations were 2.5 and 5.1%, respectively. Human GH was assayed by specific RIA[13] using the Phadebas hGH Prist Kit. Sensitivity = 0.25 ng/ml. Intra- and interassay coefficients of variation were 9 and 11.2 %, respectively. PRL was assayed by specific RIA[14] using the Prolactin Diagnostic Kit (Lacti-quant, NML). Sensitivity = 0.5 ng/ml. Intra- and interassay coefficients of variation were 5 and 12%, respectively.

D. Distal Colon Motility Studies

DCM study procedure, described in detail in our previous paper,[15] employs two open-ended polyethylene catheters placed in sigmoid and rectum (25 and 10 cm from the anal opening, respectively) and continuously perfused with saline solution at a rate of 9 ml/h. The catheters were connected to two differential air pressure transducers. A Nihon Kohden polygraph was employed to register pressure changes. Motility of the distal colon has two components: (1) IT = distance between zero- and base-line pressure, determined by segments of trace lacking waves and (2) phasic activity = waves. IT was measured on a 10-mm Hg graded scale.

E. Statistical Method

Results are expressed as the mean ± SE. The hypothesis to be tested was that intramuscularly injected clonidine brings about an alteration in the mean values of parameters recorded for the same subjects given placebo and that, in addition, clonidine-induced changes vary according to two distinct DCM patterns found among the 50 subjects integrating our protocol. On the other hand, our research looked for possible correlations between the DCM and the other parameters altered by clonidine. The null hypotheses were (1) values for the clonidine experiments have the same mean value as the values for placebo experiments; (2) there were no different responses to clonidine among high-IT and low-IT depressed patients; and (3) there were no correlations between the different parameters investigated during preclonidine and postclonidine periods. The null hypotheses were examined using a two-way analysis of variance (Anova), a multiple comparison test, a Pearson product moment test, and an unpaired t test. p values less than 0.05 were considered significant. A Hewlett Packard HP-41C Statistics PAC 00041-15002 was used for calculations.

III. RESULTS

A. Preclonidine Periods

Of the 50 patients (14 women and 12 men), 26 showed high-IT (>20 mm Hg), whereas the remaining 24 patients (12 women and 12 men) showed low-IT (<20 mm Hg, most of them = 0 mm Hg). The HRS scores in high-IT depressed patients ranged between 18 and 41 (mean ± SE = 29.3 ± 5.2); whereas in low-IT depressed patients, the HRS score ranged between 17 and 36 (mean ± SE = 26.7 ± 4.3). There was no difference between mean values. When comparing both groups during preclonidine periods, it was found that high-IT depressed subjects showed significantly higher SBP and lower DBP and NE values than low-IT patients: SBP = 111.3 ± 1.3 vs. 102.5 ± 2.2 mm Hg, p <0.01; DBP = 71.0 ± 0.2 vs. 80.3 ± 1.1 mm Hg, p <0.01; norepinephrine = 227.3 ± 15.4 vs. 421.6 ± 20.8 mm Hg , p <0.001, respectively.

IT ranged between 30 and 200 mm Hg (mean ± SE = 80.6 ± 19.4 mm Hg) in the high-IT patients. IT individual values correlated positively with HRS values in high-IT depressed patients, r = 0.51, p <0.001, and negatively with norepinephrine values, r = –0.63, p <0.001. Correlation between HRS and norepinephrine in high-IT depressed patients showed an r = –0.54, p <0.01. No significant correlation was found between SBP-HRS-norepinephrine and the other parameters (DBP and hormonal values) in high-IT patients.

In low-IT depressed patients, HRS values correlated positively with DBP and norepinephrine values: r = 0.47 and 0.55, respectively, p <0.01 in both cases. Correlation between DBP and norepinephrine showed r = 0.63, p <0.001. Although there was no quantitatively significant difference between high- and low-IT HRS mean scores, both groups did differ significantly in some of the mean HRS items. In effect, the low-IT group showed a greater degree of "agitation", "psychic anxiety", "somatization", and "obsessional symptoms" than the high-IT group (p <0.01 in all cases); while the high-IT group showed a greater degree of "suicide" and "paranoid symptoms" (p <0.01 in both cases).

Taking into consideration that low-IT subjects showed a comparatively higher mean norepinephrine plasma value, we will refer to them as hypersympathetic depressed subjects, while the high-IT group will be referred to as hyposympathetic depressed subjects.

Hormonal parameters showed that CRT and PRL mean values were significantly higher in the high-IT than in the low-IT group: CRT = 15.9 ± 0.8 vs. 6.1 ± 0.9 µg/dl and PRL = 17.1 ± 1.9 vs. 10.1 ± 0.9 ng/ml, p <0.001 in both cases. No significant difference was found between GH mean values registered in both groups. No significant correlations were found between hormonal values and the other parameters in either group.

B. Postclonidine Periods
1. High Intestinal Tone Hyposympathetic Depressed Patients
a. Blood Pressure
Clonidine induced a significant reduction in SBP at 30, 60, 90, and 120 min (F = 3.01, 3.56, 3.77, and 3.82, respectively; p <0.05 in all cases) in high-IT depressed patients. Clonidine also induced slight reductions of DBP in high-IT normals,[3] but did not modify DBP at any time in high-IT depressed subjects.

b. Norepinephrine Plasma Levels
Although clonidine induced slight reductions of norepinephrine in low-IT normals,[3] the drug did not change norepinephrine mean value at any time in high-IT depressed patients.

c. Distal Colon Motility
Clonidine induced an increase of sigmoidal and/or rectal tone in all 26 high-IT depressed subjects. The mean increase was 467.7 ± 51.1% mm Hg over the mean preclonidine period. The IT remained constant through the following postclonidine periods. The correlations between norepinephrine and IT during the four postclonidine periods were higher than those registered during the preclonidine periods: r = –0.82, –0.76, –0.69, and –0.58, respectively. In all cases, p <0.001. No significant correlation was found between IT and the other parameters during postclonidine periods.

d. Hormonal Parameters
Although clonidine induced significant mean CRT decrease and mean GH increase in high-IT normals,[3] the drug did not modify any mean hormonal values at any time in high-IT depressed patients (Figures 26 and 27).

2. Low Intestinal Tone Hypersympathetic Depressed Patients
a. Blood Pressure
Clonidine induced a small but significant reduction of SBP at 30, 60, and 90 min, only (F = 2.98, 3.02, and 3.07, respectively; p <0.05 in all cases). A sustained and very significant reduction in DBP was observed during the four post-clonidine periods (F = 9.35, 9.79, 9.98, and 8.35, respectively; p <0.001 in all cases).

b. Norepinephrine Plasma Levels
A sustained, progressive, and strong reduction in norepinephrine mean values parallelled DBP reductions (F = 15.7, 9.6, 14.5, and 10.1, respectively; p <0.001 in all cases). The correlation between DBP and norepinephrine values during the four postclonidine periods was higher than during the preclonidine periods; r = 0.73, 0.77, 0.81, and 0.72, respectively; p <0.001 in all cases.

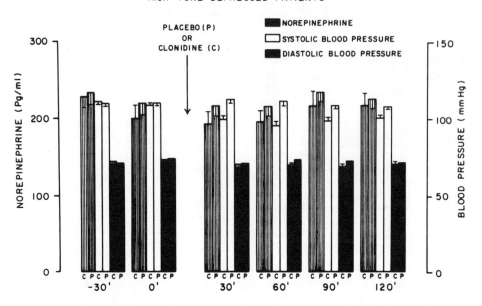

FIGURE 26.   Effects of intramuscular clonidine (2.5 µg/kg) and placebo on NE plasma level, SBP and DBP in 26 high-IT depressed patients. SBP was significantly reduced at 30, 60, 90 and 120 min. Neither DBP nor NE were reduced at any period. Each bar represents the mean value ± SEM here and in all the figures. *p <0.05. From *Neuroendocrinology* 40: 253-261, 1985. With permission.

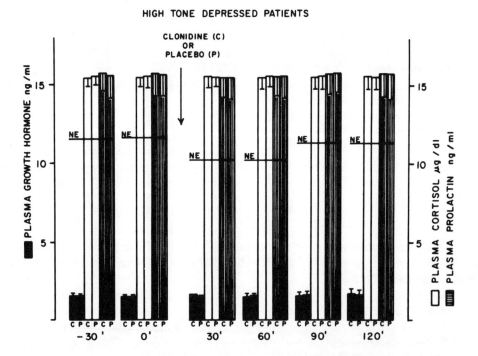

FIGURE 27.   Effects of intramuscular clonidine (2.5 µg/kg) and placebo on plasmatic levels of GH, CRT, and PRL in 26 high-IT depressed patients. The drug did not modify any mean hormonal values at any time. From *Neuroendocrinology* 40: 253-261, 1985. With permission.

LOW TONE DEPRESSED PATIENTS

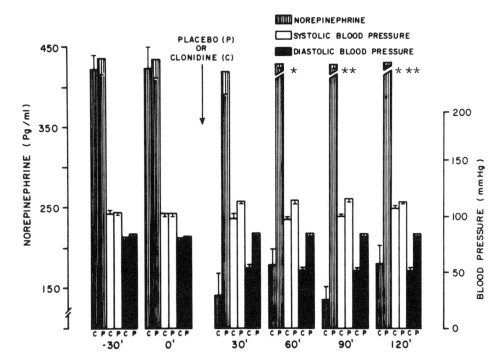

FIGURE 28. Effects of intramuscular clonidine (2.5 µg/kg) and placebo on NE plasma level, SBP, and DBP in 24 low-IT depressed patients. SBP showed a small but significant reduction at 30, 60, and 90 min. A sustained and very significant reduction in DBP was observed in all postclonidine periods, paralleled by a strong reduction in NE mean values. *p <0.05; **p <0.001. From *Neuroendocrinology* 40: 253-261, 1985. With permission.

c. Distal Colon Motility

Clonidine suppressed waves (phasic activity) in all the low-IT patients. No increases of IT were registered in these patients.

d. Hormonal Parameters

Although clonidine induced significant mean CRT decrease and mean GH increase in low-IT normals,[3] the drug did not induce any change of hormonal values, at any time (Figures 28 and 29).

IV. DISCUSSION

In a prior study[3] performed in 46 normal subjects (28 high-IT and 18 low-IT) we likewise investigated the same parameters (SBP, DBP, NE, DCM, CRT, GH, and PRL) before and after the administration of clonidine.

Although the results obtained from the present study in high- and low-IT depressed patients showed many similarities with those obtained in normals, some important differences were observed. The main difference proved to be the inability of clonidine to modify CRT-GH plasma levels. Indeed, it is a known fact that GH plasma levels rise to the challenge of clonidine in normals but not in depressives.[16,17] Nevertheless, reports referring to clonidine-induced CRT changes have been controversial.[18-20]

Up to the present, most arguments invoked to explain the failure of clonidine to induce plasma GH increase in depressed subjects postulate a subsensitivity of central postsynaptic α_2-

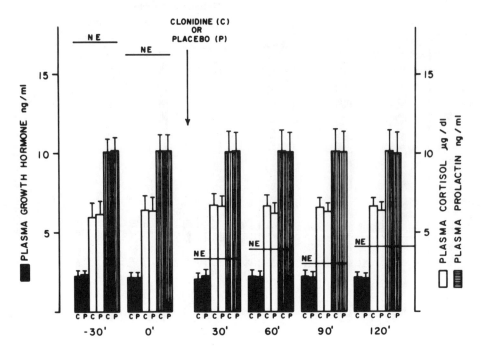

FIGURE 29.   Effects of intramuscular clonidine (2.5 µg/kg) and placebo on plasmatic levels of GH, CRT, and PRL in 24 low-IT depressed patients. The drug did not modify any hormonal values. From *Neuroendocrinology* 40: 253-261, 1985. With permission.

receptors,[4,21,22] secondary to an NE presynaptic overactivity. This postulation might be logical for low-IT hypersympathetic depressed patients but does not fit well with high-IT hyposympathetic depressed patients, in whom a decreased NE turnover with a central α_2-postsynaptic supersensitivity has been postulated.[3] In this respect, the fact that both CRT and PRL mean plasma levels were greater in high-IT depressed patients than in low-IT depressed patients, during preclonidine periods, suggests the stimulating role that the 5HT system might play in releasing these hypothalamic factors in the former group. In effect, it is a well-known fact that the central 5HT system is involved in the releasing of corticotropin-releasing hormone and triggers PRL-releasing mechanisms.[23-29] Furthermore, the proven fact that FENF (a serotonin depletor agent)[10] is able to induce a quick improvement of high-IT depressed patients[8] provides additional reinforcement to this presumption.

One of the most interesting findings emerging from our research is the demonstration that DBP is reduced by clonidine in low-IT hypersympathetic but not in high-IT hyposympathetic depressed patients. This finding provides a valid physiological marker for differentiating between the two types of depressive syndromes. The fact that DBP and NE plasma levels were so closely correlated in hypersympathetic but not in hyposympathetic patients, transforms the clonidine-induced DBP reduction into a useful tool to asses sympathetic activity. In addition, the correlation registered between DBP, NE, and HRS score values in the hypersympathetic group is coherent with this hypothesis. The fact that other authors failed to find a correlation between norepinephrine plasma levels and clinical rating of depression might be due to the fact that they investigated mixed subpopulations of depressed patients.[2,31]

A great deal of research has been carried out in order to establish adequate biological markers of depression.[30,32] With regard to this, many attempts have been made to find neuroendocrine parameters as a guide to diagnosis and therapy of depression.[33-35] Now, with results emerging from the present study, it has been demonstrated that high-IT and low-IT depressed patients can be recognized by the ability of clonidine to reduce DBP in the latter group, only. Taking into account that high-IT depressed patients are best improved by IMI, desipramine, and other antidepressants[36] which potentiate NE activity, whereas low-IT depressed patients are best improved by clomipramine[36] which potentiates 5HT activity, the DBP response to clonidine can be used as a practical and simple physiological marker to differentiate depressive syndromes and to guide antidepressant therapy. This statement derives additional support from the comparison of mean hormonal values obtained from normal and depressed subjects during preclonidine periods. In effect, it was found that the CRT preclonidine mean value of the high-IT normals (10.1 ± 0.3 μg/dl) was significantly higher than that of high-IT depressed group (14.8 ± 0.7 μg/dl, $p < 0.001$). On the contrary, CRT preclonidine mean value of low-IT normal subjects (10.4 ± 1.8 μg/dl) was significantly higher than that of low-IT depressed patients (5.3 ± 1.1 μg/dl, $p < 0.001$). For comparison purposes, we chose 13 high-IT depressed and 12 low-IT normals, drawn from the population presented in our prior study. These findings are consistent with the hypothesis that hyper- and hypo-CRT secretion accompany high-IT hyposympathetic and low-IT hypersympathetic depressive syndromes, respectively.

The positive and negative correlations between IT/HRS and IT/NE, respectively, found in the high-IT depressed group are coherent with the hypothesis that DCM is a good physiological marker to differentiate two prototypes of human ANS profiles and that these, when depressed, display distinct physiopathologic mechanisms. Further, the correlation existing between psychological symptoms and peripherally measured parameters (neuroendocrine, intestinal, and cardiovascular) suggests that the latter parameters share, at leat partially, common central mechanisms[37] (Figures 26 to 29).

REFERENCES

1. **Lechin F, van der Dijs B.** Opposite effects on human distal colon motility of two postulated alpha$_2$-antagonists (mianserin and chlorprothixene) and one alpha$_2$-agonist (clonidine). *J Clin Pharmacol* 23: 209-218, 1983.
2. **Lechin F, van der Dijs B, Acosta E, Gómez F, Lechin E, Arocha L.** Distal colon motility and clinical parameters in depression. *J Affect Dis* 5: 19-26, 1983.
3. **Lechin F, van der Dijs B, Jakubowicz D, Camero RE, Villa S, Lechin E, Gómez F.** Effects of clonidine on blood pressure, noradrenaline, cortisol, growth hormone, and prolactin plasma levels in high and low intestinal tone subjects. *Neuroendocrinology* 40: 253-261, 1985.
4. **Eriksson E, Eden B, Modigh K.** Up- and down-regulation of central postsynaptic alpha$_2$ receptors reflected in the growth hormone response to clonidine in reserpine pre-treated rats. *Psychopharmacology* 77: 327-331, 1982.
5. **Weinreich P, Seeman P.** Binding of adrenergic ligands, [^{3}H]-clonidine and [^{3}H]-WB-4101, to multiple sites in human brain. *Biochem Pharmacol* 30: 3115-3120, 1981.
6. **Ziegler MG, Lake CR, Wood JH.** Relationship between norepinephrine in blood and cerebrospinal fluid in the presence of a blood cerebrospinal fluid barrier for NE. *J Neurochem* 28: 677-679, 1977.
7. **Hamilton M.** A rating scale for depression. *J Neurol Neurosurg Psychiatry* 23: 41-43, 1960.
8. **Lechin F, van der Dijs B, Gómez F, Arocha L, Acosta E, Lechin E.** Distal colon motility as a predictor of antidepressant response to fenfluramine, imipramine, and clomipramine. *J Affect Dis* 5: 27-35, 1983.
9. American Psychiatric Association. *Diagnostic and Statistical Manual of Mental Disorders*. DSM III; 3rd ed. American Psychiatric Association, Washington D.C., 1980.
10. **Sole MJ, Hussein MN.** A simple, specific radioenzymatic assay for the measurement of picogram quantities of norepinephrine, epinephrine and dopamine in plasma and tissues. *Biochem Med* 18: 301-307, 1977.
11. **Carey RM, Van Loon GR, Baines AD, Kaiser DL.** Suppression of basal and stimulated noradrenergic activities by the dopamine agonist bromocriptine in man. *J Clin Endocrinol Metab* 56: 595-602, 1983.
12. **Murphy BEP.** Some studies of the protein-binding of steroids and their application to the routine micro and ultramicro measurement of various steroids in body fluids by competitive protein-binding radio-assay. *J Clin Endocrinol Metab* 27: 973-990, 1967.
13. **Wide L.** Radioimmunoassays employing immunoabsorbents. *Acta Endocrinol* 63: Suppl 142, pp 207-210, 1969.

14. **Sinha YN, Selby FW, Lewis UJ, Vanderlaan WP.** A homologous radioimmunoassay for human prolactin. *J Clin Endocrinol Metab* 36: 509-512, 1973.

15. **Lechin F, van der Dijs B, Gómez F, Arocha L, Acosta E.** Effects of D-amphetamine, clonidine and clonazepam on distal colon motility in non-psychotic patients. *Res Commun Psychol Psychiatr Behav* 7: 385-410, 1982.

16. **Checkley SA, Slade AP, Shur E.** Growth hormone and other responses to clonidine in patients with endogenous depression. *Br J Psychiatry* 138: 51-55, 1981.

17. **Matussek N, Ackenheil M, Hippius H, Mueller F, Schroeder HT, Schultes H, Wasilewski B.** Effect of clonidine on growth hormone release in psychiatric patients and controls. *Psychiatr Res* 2: 25-36, 1980.

18. **Chambers JW, Brown GM.** Neurotransmitter regulation of growth hormone and ACTH in the rhesus monkey: effects of biogenic amines. *Endocrinology* 98: 420-426, 1976.

19. **Lal S, Tolis G, Martin JB, Brown GM, Guyda H.** Effect of clonidine on growth hormone, prolactin, luteinizing hormone, follicle-stimulating hormone, and thyroid-stimulating hormone in the serum of normal men. *J Clin Endocrinol Metab* 41: 827-831, 1975.

20. **Siever LJ, Uhde TW, Jimerson DC, Post RM, Lake CR, Murphy DL.** Plasma cortisol responses to clonidine in depressed patients and controls. *Arch Gen Psychiatry* 41: 63-68, 1984.

21. **Charney DS, Heninger GR, Sternberg DE, Hafstad KM, Giddings S, Lancis DH.** Adrenergic receptor sensitivity in depression. Effects of clonidine in depressed patients and healthy subjects. *Arch Gen Psychiatry* 39: 290-294, 1982.

22. **Siever LJ, Uhde TW, Silberman EK, Jimerson DG, Aloi JA, Post RM, Murphy DL.** Growth hormone response to clonidine as a probe of noradrenergic receptor responsiveness in affective disorder patients and controls. *Psychiatr Res* 6: 171-183, 1982.

23. **Casanueva FF, Villanueva L, Peñalva A, Cabezas-Cerrato J.** Depending on the stimulus, central serotonergic activation by fenfluramine blocks or does not after growth hormone secretion in man. *Neuroendocrinology* 38: 302-308, 1984.

24. **Fuller RW.** Stimulation of pituitary-adrenocortical function in rats. *Neuroendocrinology* 32: 118-127, 1981.

25. **MacLeod RM.** Regulation of prolactin secretion, in *Frontiers in Neuroendocrinology*, Vol. 4. Martini L, Ganong WF. Eds., Raven Press, New York, 1976.

26. **Mendelson WB, Jacobs LS, Reichman JD, Othmer E, Cryer PE, Trivedi B, Danghaday WH.** Methysergide suppression of sleep-related prolactin secretion, an enhancement of sleep-related growth hormone secretion. *J Clin Invest* 56: 690-717, 1975.

27. **Richards GE, Holland FJ, Aubert ML, Ganong WF, Kaplan SL, Grumbach MM.** Regulation of prolactin and growth hormone secretion. *Neuroendocrinology* 30: 139-143, 1980.

28. **Scapagnini U, Moberg GP, Van Loon GR, De Groot J, Ganong WF.** Relation of brain 5-hydroxytryptamine content to the diurnal variation of plasma corticosterone in the rat. *Neuroendocrinology* 7: 90-96, 1971.

29. **Van de Kar LD, Bethea CL.** Pharmacological evidence that serotonergic stimulation of prolactin secretion is mediated via dorsal raphe nucleus. *Neuroendocrinology* 35: 225-230, 1982.

30. **Garattini S, Buczko W, Jory A, Samanin A.** The mechanisms of action of fenfluramine. *Postgrad Med J* 51: Suppl 1, pp 27-35, 1975.

31. **Lake CR, Picker D, Ziegler D, Lipper S, Slater S, Murphy DL.** High plasma norepinephrine levels in patients with major affective disorders. *Am J Psychiatry* 139: 1315-1321, 1982.

32. **Wyatt RJ, Portnoy B, Kupfer DJ.** Resting plasma catecholamine concentrations in patients with depression and anxiety. *Arch Gen Psychiatry* 24: 65-70, 1971.

33. **Checkley SA.** Neuroendocrine studies of monoamine function in man; a review of basic theory and its application to the study of depressive illness. *Psychol Med* 10: 35-53, 1980.

34. **Targum SD.** Persistent neuroendocrine dysregulation in major depressive disorder: a marker for early relapse. *Biol Psychiatry* 19: 305-318, 1984.

35. **Von Zerssen D, Berger M, Doerr P.** Neuroendocrine dysfunction in subtypes of depression, in *Psychoneuroendocrine Dysfunction in Psychiatric and Neurological Illness: Influence of Psychopharmacological Agents*. Shah NS, Donald AG. Eds. ,Plenum Press, New York, 1984.

36. **Lechin F, van der Dijs B.** Slow wave sleep (SWS), REM sleep (REMS), and depression. *Res Commun Psychol Psychiatr Behav* 9: 227-262, 1984.

37. **Lechin F, van der Dijs B, Jakubowicz D, Camero R, Villa S, Arocha L, Lechin A.** Effects of clonidine on blood pressure, noradrenaline, cortisol, growth hormone, and prolactin plasma levels in high and low intestinal tone depressed patients. *Neuroendocrinology* 41: 156-162, 1985.

INTRAMUSCULAR CLONIDINE TEST IN SEVERELY DISEASED PATIENTS

Fuad Lechin, Bertha van der Dijs, Marcel Lechin, Rheyna Camero, Alex Lechin, Simon Villa, and B. Reinfeld

I. INTRODUCTION

Two prototypes of ANS profiles have been identified according to their DCM and their plasma norepinephrine concentrations: the high-IT-low norepinephrine profile and the low-IT-high norepinephrine profile.[1] The former group is termed hyposympathetic, while the latter is considered to be hypersympathetic. In hypersympathetic subjects, a positive correlation has been found between norepinephrine and DBP but not SBP, whereas no correlation at all was found between norepinephrine and blood pressure in hyposympathetic subjects. The two groups respond differently to the challenge of clonidine: whereas the drug provokes SBP + DBP decrease in hypersympathetics, only SBP reduction is observed in hyposympathetics. In addition, although clonidine induces a major reduction of norepinephrine in hypersympathetics, only slight decreases occur in hyposympathetics. Finally, in hypersympathetic subjects a close positive correlation has been found between norepinephrine and DBP, both before and after clonidine administration. The two ANS profiles also can be differentiated by the effects of clonidine on CRT, GH, and PRL plasma levels and their correlations.

The most striking finding obtained from studies of depressed subjects, both hypersympathetic and hyposympathetic, is the observation that clonidine, besides its well-known ability to modify plasma GH concentrations in depressed subjects, is likewise unable to modify these subject's plasma CRT concentrations,[2] although the drug did alter plasma CRT in normals.[1]

Considerable evidence suggests that experimentally induced stress provokes a strong increase in the turnover rate of central NA, the prolonging of which leads to norepinephrine depletion.[3-10] Moreover, pharmacological manipulations capable of provoking interference with NE neurotransmission and capable of provoking subsequent α_2-postsynaptic supersensitivity at the hypothalamic level have produced an exaggerated plasma GH response to clonidine in experimental animals.[11-14] Further, a reduction of glucocorticoid binding in the hypothalamus has been observed after experimentally induced norepinephrine depletion.[15,16] This deficit of steroid hypothalamic binding is assumed to interfere with feedback inhibition mechanisms, thus contributing to the raised plasma cortisol levels found in noncoping animals.[17]

Our preliminary studies showed that severely chronically ill patients had raised plasma norepinephrine and CRT levels during exacerbation but not during nonexacerbation periods (unpublished data). This finding led us to speculate that depressive or stress mechanisms might play some role during exacerbation periods. A hyperresponsiveness of plasma GH levels to clonidine, and not a hyporesponsiveness (as found in depressed subjects), should be obtained during exacerbation periods, if we are to support a hypothetical role of stress mechanisms. We therefore decided to investigate the effects of clonidine administration in severely chronically ill subjects during both exacerbation and nonexacerbation periods.

II. METHODS

A. Subjects

Our study included 193 severely chronically ill outpatients (101 women and 92 men) and 193 normal subjects matched by age and sex. All the severely ill patients were investigated during exacerbation periods, and 92 of them (60 women and 32 men) were also investigated during nonexacerbation periods. Exacerbation periods were considered as the positive manifestations of impairment and progressive disease. Nonexacerbation periods were considered to be those lacking manifestations of disease (symptom free). In some cases, the nonexacerbation period preceded the exacerbation period, because the disease was diagnosed during a clinical check-up. In other cases, nonexacerbation periods occurred during remission of symptoms. All the patients were exhaustively investigated (clinically, radiologically, biochemically, endoscopically, and immunologically). Histological diagnosis was carried out on tissue samples obtained through endoscopy, laparoscopy, or surgery. Definite diagnosis in our 193 chronic patients were 108 cancer patients (breast, 15; lung, 12; stomach, 11; prostate gland, 16; melanoma, 8; sigmoid colon, 6; right colon, 5; rectum, 4; kidney, 2; bladder, 2; pancreas, 5; esophagus, 2; primary hepatoma, 3; testis, 1; uterus, 8; larynx, 1; and lymphoma, 7) and 51 with ulcerative colitis, 7 with Crohn enteritis, 13 hepatic with cirrhosis, and 4 with systemic lupus erythematosus.

All patients were investigated by two psychiatrists, independently, to exclude psychotic disorders and epilepsy. Only 13 (6.7%) of the 193 patients met criteria for nonpsychotic major depressive disorder (DSM-III, American Psychiatric Association, 1980).[18] No patient had received antidepressants, neuroleptics, stimulants, anxiolytics, anticholinergics, or psychoactive drugs in the month prior to study. Patients requiring analgesic drugs because of pain were not included in our protocol. Studies of women were performed only outside of premenstrual (1 week prior) and menstrual periods. Each patient was tested with intramuscular clonidine and with saline placebo injection on two separate occasions, 2 weeks apart. The sequence in which drug and placebo were given was randomized. All the patients gave written consent being admitted to the study, and the protocol was approved by the Ethical Committee of FUNDAIME.

B. Test Procedure

All experiments were started on recumbent subjects at 0700 h after an overnight fast. A venous catheter was inserted in a forearm vein at least 30 min prior to the experiments, and patency was achieved by a slow infusion of normal saline solution. Patients lay comfortably on a bed. Blood samples were collected at 0800, 0830, 0900, 0930, 1000, and 1030 h. Clonidine (2.5 µg/kg) (Catapres, Boehringer Ingelheim, Ridgefield, CT) was injected intramuscularly immediately after the second blood sample (0830 hr). The blood samples were immediately placed on ice and spun at 3000 rpm for 20 min in a refrigerated centrifuge; serum aliquots were stored at −80°C until required for hormone assays. Blood for catecholamine assay was transferred to tubes containing 4 mg dithiothreitol and 20 µl EDTA solution (100 mg/ml) which were inverted carefully several times. Blood specimens were centrifuged immediately at −4°C, and plasma was separated and stored at −80°C until assayed. Blood pressue was monitored every 30 min by two different doctors with two methods (auscultatory and sphygmomanometer).

C. Analytical Methods

Plasma norepinephrine, E, and DA concentrations were determined by the radioenzymatic method of Sole and Hussein[19] with the modifications suggested by Carey et al.[20] The sensitivity of this method was 2 pg for norepinephrine, E, and DA. For plasma catecholamine levels, the intraassay coefficients of variation were 12, 14, and 10%, respectively.

For plasma hormone levels the samples were assayed in duplicate, and all samples belonging to the same experimental set were measured in the same assay. Cortisol was assayed by a competitive protein-binding RIA[21] using a Cortisol Diagnostic Kit (Corti-Shure, NML); the concentrations are expressed as µg/dl in terms of of the standards supplied with the kits. The sensitivity of the assay was 0.5 µg/dl. The intraassay and interassay coefficients of variation were 2.5 and 5.1%, respectively. Human GH was assayed by a specific RIA[22] with the Phadebas hGH Prist Kit. Sensitivity was 0.25 ng/ml. Intra- and interassay coefficients of variation were 9 and 11.2%, respectively. PRL was assayed by a specific RIA[23] with the Prolactin Diagnostic Kit (Lacti-Quant, NML). Sensitivity was 0.5 ng/ml. Intra- and interassay coefficients of variation were 5 and 12%, respectively.

D. Statistical Methods

Results are expressed as mean ± SEM. The hypotheses to be tested were (1) intramuscularly injected clonidine brings about an alteration in the mean values of parameters recorded for the same subjects given placebo; (2) there are differences in the same subjects, during exacerbation vs. nonexacerbation periods; and (3) there are differences between severely-chronically ill subjects (exacerbation and nonexacerbation periods) vs. normals. Furthermore, our research considered possible correlations between the different parameters during pre- and postclonidine periods. The null hypotheses were (1) values for the clonidine experiments have the same mean as the values for placebo experiments; (2) values during pre- and postclonidine periods were similar for exacerbation and nonexacerbation periods; (3) there were no differences between nonexacerbation vs. normals; (4) there were no differences between exacerbation periods vs. normals; and (5) there were no correlations between parameters investigated during pre- and postclonidine periods. The hypotheses were examined by two-way analysis of variance (ANOVA), multiple comparison tests, Pearson product moment correlations, and paired t test, p values less than 0.05 were considered significant.

III. RESULTS

A. Normal Subjects
1. Preclonidine Periods

The mean (±SEM) values of the measures in the 193 normal subjects were similar to those previously obtained in normal mixed (hypersympathetic + hyposympathetic) subjects.[1]

2. Postclonidine Periods
a. Blood pressure

Clonidine induced significant reductions in SBP but not in DBP at 30, 60, 90, and 120 min (F = 4.89, 4.91, 4.85, and 3.83, respectively); $p < 0.01$ in all cases.

b. Norepinephrine

Norepinephrine plasma levels showed some tendency to be lower after clonidine administration; however, the changes were not significant at any time.

c. Cortisol

Similar to norepinephrine, plasma cortisol levels showed some decrease after clonidine injection, but these were not significant at any time.

d. Growth Hormone

Significant increases in plasma GH levels occurred at 60, 90, and 120 min but not at 30 min (F = 4.38, 4.42, 4.35, and 2.25, respectively; $p < 0.001$ at 60, 90, and 120 min).

e. Prolactin

Nonsignificant changes in plasma PRL levels occurred during the postclonidine period.

B. Chronically Ill Patients
1. Preclonidine Periods

The mean (±SEM) plasma norepinephrine value in the 193 severely ill patients during their exacerbation periods was 440.0 ± 20.1 pg/ml. Compared to the mean value obtained in normals (302.2 ± 56.7 pg/ml), this value was significantly higher (F = 6.84; $p < 0.001$). This value is similar to that for normal hypersympathetic subjects (421.6 ± 20.8 pg/ml);[1] thus, these patients showed an increased norepinephrine activity similar to that observed in normal hypersympathetic subjects. The mean (±SEM) DBP value obtained in these patients was 94.6 ± 6.8 mm Hg, which was significantly higher than that in normals (72.1 ± 4.5 mm Hg) (Table 1). The NE and DBP values obtained in these severely chronically ill patients showed a strong positive correlation ($r = 0.67$, $p < 0.001$).

Raised mean CRT and PRL levels also occurred in these patients during their exacerbation periods (29.3 ± 1.1 µg/dl and 13.9 ± 1.3 ng/ml, respectively). These values were significantly higher than those obtained in normals ($F = 6.71$ and 3.85; $p < 0.001$ and $p < 0.001$, respectively), whereas the mean GH level was normal (1.5 ± 0.2 ng/ml) (Table 1). No correlation was found between these hormonal values and norepinephrine, SBP, and DBP.

2. Postclonidine Periods

a. Blood Pressure

Clonidine induced a significant reduction in SBP at 30, 60, 90, and 120 min in this group of patients ($F = 3.07, 3.52, 3.49$, and 3.05, respectively, $p < 0.05$ in all cases). A sustained and very significant reduction in DBP was observed at the four postclonidine times ($F = 7.64, 9.78, 9.81$, and 9.37, respectively; $p < 0.001$ in all cases).

b. Norepinephrine

A sustained, progressive and strong reduction in NE mean values parallelled DBP reduction ($F = 16.3, 19.9, 22.4$, and 20.7, respectively; $p < 0.001$ in all cases). The correlation between DBP and NE values at the four postclonidine times was higher than during the preclonidine period ($r = 0.69, 0.75, 0.82$, and 0.74, respectively; $p < 0.001$ in all cases).

c. Cortisol

Our patients showed a sustained and progressive reduction of mean CRT at all postclonidine times ($F = 11.2, 15.6, 15.4$, and 15.6, respectively; $p < 0.001$ in all cases).

d. Growth Hormone

Large increases of this hormone occurred simultaneously with CRT decreases at all postclonidine times ($F = 19.4, 24.6, 25.1$, and 21.8, respectively; $p < 0.001$ in all cases). Significant negative correlations were found between CRT and GH values at all postclonidine times ($r = -0.56, -0.67, -0.71$, and -0.69, respectively). Conversely, no correlation was found between CRT or GH and NE or blood pressure.

e. Prolactin

Although some tendency to PRL lowering was observed, the decrease was only significant at 90 min ($F = 3.5$; $p < 0.05$). No correlation was found between PRL and the other parameters.

3. Nonexacerbation Chronically Ill Patients vs. Their Matched Paired Normals

The 92 patients (54 women and 38 men) in the nonexacerbation subgroup of chronically ill patients showed mean values of plasma norepinephrine, CRT, GH, and PRL, SBP, and DBP which were not significantly different from those in their normal matched subjects, for both the preclonidine and the postclonidine times.

4. Nonexacerbation vs. Exacerbation Periods

Comparison between the mean values obtained in the 92 severely ill patients during nonexacerbation periods vs. the mean values obtained during exacerbation in the same subjects showed highly significant differences in CRT, GH, norepinephrine, and DBP at both the preclonidine and postclonidine times. No differences were observed for PRL and SBP. Although clonidine induced GH increases in both the exacerbation and the nonexacerbation subgroups, the increases observed during exacerbation were significantly thigher than those observed during nonexacerbation. Nonsignificant differences in preclonidine GH was observed between exacerbation and nonexacerbation.

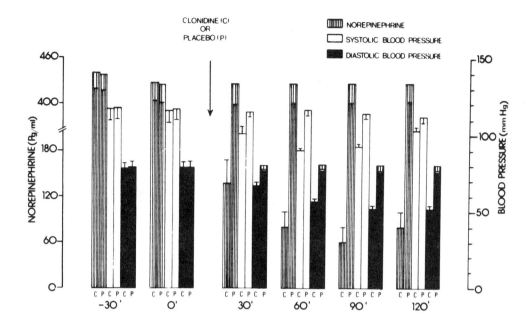

FIGURE 30. Clonidine, but not placebo, induced sudden and sustained reduction of the raised NE plasma levels occurring in severely chronically ill patients during exacerbation periods. Significant SBP and DBP reductions also occurred after clonidine. Each bar represents the mean value ± SEM. From *Psychoneuroendocrinology.*

5. Exacerbation vs. Normal Groups

The mean values (± SEM) of norepinephrine, CRT, and DBP obtained in the 193 severely ill patients during their exacerbation periods were significantly greater than those observed in the normals before clonidine administration (p <0.001, p <0.001 and p < .05, respectively). Clonidine induced marked decreases in these parameters in the patients during their exacerbation periods (Figures 1 and 2), but not in the normals (Table 1). Although preclonidine mean GH did not differ between the exacerbation and the normal groups, clonidine induced significantly greater GH increases in the exacerbation group than in the normals (F = 8.93, 7.81, 7.26, and 5.43; p <0.001 in all cases) (Figures 30 to 32 and Table 2).

IV. DISCUSSION

The results of this study show that the severely chronically ill subjects included in our protocol, investigated during their exacerbation periods, had higher DBP and plasma norepinephrine, CRT, PRL, and CRT than during their nonexacerbation remission periods, and also compared to normal subjects. The administration of clonidine provoked a strikingly greater reduction of DBP, NE, and CRT during the exacerbation periods than during the nonexacerbation periods, and also compared to normals. Moreover, the plasma GH increase following clonidine administration during the exacerbation periods was considerably greater than that during the nonexacerbation periods, and compared to normals.

The raised mean DBP and norepinephrine values, as well as the great reduction of these values provoked by clonidine during illness exacerbation, are similar to those observed in

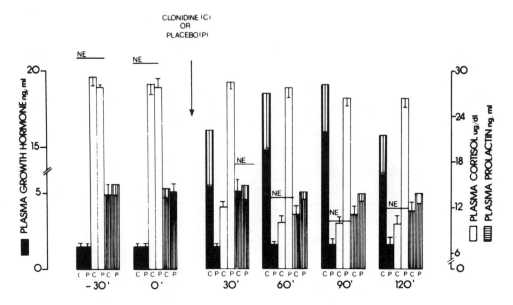

FIGURE 31.   Clonidine, but not placebo, provoked an exaggerated GH increase in severely chronically ill patients during exacerbation periods. Conversely, the drug induced large reductions in raised plasma cortisol levels, occurring during exacerbation periods. At 90 min only, clonidine induced a slight but significant reduction of the raised prolactin levels in these patients. Each bar represents the mean value ± SEM. NE values are shown from Figure 30. From *Psychoneuroendocrinology.*

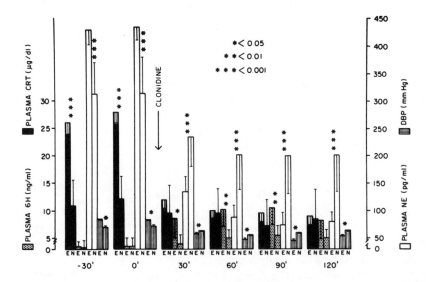

FIGURE 32.   Mean (± SEM) values of CRT, GH, norepinephrine, and DBP in 92 severely chronically ill patients during exacerbation (E) and nonexacerbation (N) periods, before and after administration of clonidine. *p* values are shown for comparison of (E) to (N) mean values at each time (paired t test for preclonidine times and ANOVA for postclonidine times). From *Psychoneuroendocrinology.*

Table 2
MEAN (±SEM) PLASMA CORTISOL, GROWTH HORMONE, PROLACTIN, NOREPINEPHRINE, AND SYSTOLIC AND DIASTOLIC BLOOD PRESSURE IN 193 NORMAL SUBJECTS BEFORE AND AFTER THE INTRAMUSCULAR INJECTION OF CLONIDINE
(2.5 µg/kg)

	Preclonidine		Postclonidine			
	−30 min	0 min	30 min	60 min	90 min	120 min
CRT (mg/dl)	12.65 ± 4.32	12.08 ± 4.41	10.98 ± 5.28	10.12 ± 5.98	8.56 ± 5.10	9.07 ± 5.26
GH (ng/ml)	1.84 ± 1.45	1.79 ± 1.42	2.46 ± 1.38	5.10 ± 1.52	5.82 ± 1.60	5.16 ± 1.57
PRL (ng/ml)	13.71 ± 1.82	12.22 ± 1.67	13.60 ± 2.01	14.25 ± 1.86	12.18 ± 1.54	13.91 ± 1.90
NE (pg/ml)	302.18 ± 56.7	311.26 ± 59.4	219.25 ± 60.4	190.06 ± 79.8	200.14 ± 74.6	198.59 ± 79.8
SBP (mm/Hg)	112.20 ± 08.6	108.60 ± 06.9	101.20 ± 06.2	96.50 ± 08.1	98.20 ± 06.6	100.00 ± 08.6
DBP (mm/HG)	72.00 ± 05.2	72.20 ± 03.8	62.60 ± 14.2	56.40 ± 14.2	60.20 ± 15.3	66.70 ± 16.5

From *Psychoneuroendocrinology*.

hypersympathetic normal subjects.[1] The raised CRT and PRL plasma levels observed during exacerbation periods are compatible with the hypothesis that during exacerbation a stress factor plays some role in the underlying physiopathological mechanisms.

Although raised norepinephrine and CRT plasma levels have also been observed in many depressed subjects[2] the challenge of clonidine provokes in them no CRT or GH changes.[2,24]Siever and Uhde[25] showed that plasma CRT decreases after clonidine challenge in depressed patients; however, the decreases were not large. The i.v. administration of clonidine[25] perhaps represents a more robust stimulus, so that peripheral mechanisms (outside the blood brain barrier) may be triggered. Thus, the abrupt suppression of β-adrenergic activity at the median eminence level induced by i.v. clonidine could result in a sudden reduction of hypothalamic-hypophysal-adrenal corticol activity.[26] The very low incidence of depression in our patients during exacerbation (6.7%) is consistent with the large CRT and GH changes induced in them by clonidine.

The strong positive correlation during the exacerbation periods between norepinephrine and DBP reductions, but not between norepinephrine and SBP reductions, confirms similar findings obtained in studies of hypersympathetic normal subjects.[1] This close correlation suggests that both clonidine-induced changes share, at least partially, a common mechanism. Several arguments support the hypothesis that the dual effects of clonidine are exerted through the drug interference with presynaptic NE activity. The raised basal plasma norepinephrine levels during the exacerbation periods are indicative of a high norepinephrine release by sympathetic nerves, since these nerves are the main origin of norepinephrine circulating during the resting state.[27] In addition, if, as is widely accepted, clonidine exerts its pharmacological effects preferentially through central mechanisms,[28,29] the lowering of blood pressure and plasma norepinephrine would be secondary to the central inhibition of sympathetic nervous discharge evoked by clonidine.[30] It has been demonstrated that peripheral plasma norepinephrine levels correlate positively with central norepinephrine levels, in cerebrospinal fluid.[31] This correlation also has been associated with the activation of a sympathoexcitatory NE circuit in the locus coeruleus-posterior hypothalamus.[32-35]

Further, the close negative correlation between changes in plasma GH and CRT levels also suggests that clonidine-induced effects on both hormones share, at least partially, a common mechanism. The axons of both GH-releasing and GH-inhibiting factor neurons end in the median eminence, a hypothalamic area located outside the BBB. Norepinephrine exerts a dual effect on the regulation of GH secretion. It stimulates GH-inhibiting factor through β-adreno-ceptors[36-38] and α_1-adrenoceptors[39] at the median eminence level (outside the BBB), and it also stimulates GH-releasing factor through α_2-postsynaptic adrenoceptors located in ventromedial and anterior hypothalamic areas (inside the BBB).[13,39] The median eminence is innervated by norepinephrine axons arising from the locus coeruleus (LC),[40] while the anterior hypothalamic area is innervated by NE axons integrated in the ventral NE bundle which arise from the noncoeruleus NE nuclei (A1 cell group, mainly).[41-45]These two NE systems display opposite effects[46-48] and interchange inhibitory axons.[49-51]

If we assume that peripheral hypersympathetic activity is positively correlated with overactivity of the LC-posterior hypothalamic pathway and is negatively correlated with the central sympathoinhibitory noradrenergic nuclei (A1 and others), then peripheral hypersympathetic activity would occur simultaneously with an excess of norepinephrine at the posterior hypothalamic level and a deficit of norepinephrine at the anterior hypothalamic level. In this case, clonidine would act presynaptically to inhibit the LC and, by extension, central and peripheral sympathetic activity (plasma norepinephrine levels, DBP). Furthermore, suppression of norepinephrine release from LC axons at the median eminence level would result in a reduction of GH-inhibiting factor. Suppression of LC activity would also result in a reduction of GH-inhibiting factor. Suppression of LC activity would also result in disinhibition of the medullary NE-A1 cell group, which innervates the anterior hypothalamus and ventromedial arcuate nuclei. The NE released by these A1 axons would stimulate supersensitive α_2-postsynaptic adrenoceptors located in these areas and would be responsible for the secretion of GH-releasing factor. Hence, the sharp increase in plasma GH levels registered in our hypersympathetic patients (exacerbation patients) would result from the sudden suppression of GH-inhibiting factor plus the sudden secretion of GH-releasing factor. The fact that a reduction of norepinephrine in the arcuate, ventromedial, and preoptic nuclei is found during experimentally induced stress[3] is consistent with the above mentioned hypothesis. In addition, the demonstration that clonidine blocks the stress-induced release of ACTH and the increase of plasma corticosteroids in man[52] reinforces our hypothesis.

A clonidine-induced reduction of norepinephrine release at the posterior hypothalamic level, along with a clonidine-induced increase of norepinephrine release at the anterior hypothalamic level, as postulated herein, is consistent with the findings that microinjection of norepinephrine in the posterior hypothalamus increases blood pressure and peripheral sympathetic activity, while norepinephrine microinjection in the anterior hypothalamus decreases blood pressure and peripheral sympathetic activity.[53,54]

Although experimental evidence has demonstrated that clonidine can exert a direct stimulatory action on postsynaptic α_2-adrenoceptors in norepinephrine-depleted animals,[39,55] this effect is obtained only with a much higher dose of the drug, at least three times higher than that producing a presynaptic effect.[39] Moreover, recent studies have demonstrated that the clonidine-induced GH plasma increase disappears after destruction of norepinephrine axons reaching the mediobasal hypothalamus and anterior hypothalamus.[56]

The high plasma NE and CRT levels in severely chronically ill patients during exacerbation periods reflect a central and peripheral hypersympathetic activity, along with an overactivity of the pituitary-adrenocortical axis. Both physiopathological findings are compatible with the hypothesis that a stress factor plays some role in the exacerbation mechanisms. In addition, the strong reduction of DBP, norepinephrine, and CRT, along with the large increase of GH, might be useful indicators in assessing the exacerbation and progression of severe chronic illness.[57]

V. SUMMARY

SBP, DBP, and plasma norepinephrine, CRT, GH, and PRL were studied before and after clonidine (2.5 μg/kg intramuscular) adminsitration in 193 chronic severely ill patients and 193 normal subjects matched by age and sex. During exacerbation periods (positive manifestations of impairment and progressive disease), the patients showed higher norepinephrine, CRT, and DBP than the normals or when they were investigated during nonexacerbation periods (92 of the 913). Clonidine induced sharp, marked reductions of norepinephrine, CRT, and DBP, plus a sudden increase of GH, in all the patients during exacerbation periods. Nonsignificant reductions of norepinephrine, CRT, and DBP were observed in normals and in patients during nonexacerbation periods. On the other hand, the GH increase registered during exacerbation periods was of an order of magnitude higher than that registered in normals and in patients during nonexacerbation periods. Significant reduction of SBP was registered both in normals and in patients (exacerbation and nonexacerbation periods). Some tendency to PRL lowering was observed during exacerbation periods only. A high positive correlation between norepinephrine and DBP (pre- and postclonidine values) was obtained during exacerbation periods in patients, but not in normals during nonexacerbation periods. No significant correlation was found between norepinephrine and SBP in any group of subjects. The clonidine-induced changes in GH and CRT observed in these patients during exacerbation periods were in striking contrast to the absence of these changes in depressed patients. This finding is consistent with the low rate of depression (6.7%) registered among our patients during exacerbation periods.

The high plasma NE and CRT levels registered in chronic severely ill patients during exacerbation periods reflect a central and peripheral sympathetic hyperactivity, accompanied by an overactivity of the pituitary-adrenocortical axis. The strong reduction of DBP, NE, and CRT, along with the sharp and great increase of GH, might be useful as indicators in assessing the exacerbation and progression of severe chronic illnesses.

REFERENCES

1. **Lechin F, van der Dijs B, Jakubowicz D, Camero RE, Villa S, Lechin E, Gómez F.** Effects of clonidine on blood pressure, noradrenaline, cortisol, growth hormone and prolactin plasma levels in high and low intestinal tone subjects. *Neuroendocrinology* 40: 253-261, 1985.
2. **Lechin F, van der Dijs B, Jakubowicz D, Camero RE, Villa S, Arocha L, Lechin AE.** Effects of clonidine on blood pressure, noradrenaline, cortisol, growth hormone and prolactin plasma levels in high and low intestinal tone depressed patients. *Neuroendocrinology* 41: 156-162, 1985.
3. **Kvetnansky R, Mitro A, Palkovits M, Brownstein M.** Catecholamines in individual hypothalamic nuclei in stressed rats, in *Catecholamines and Stress.* Usdin E, Kvetnansky R, Kopin IJ. Eds., Pergamon Press, Elmsford, NY, 1976, 39.
4. **Kobayashi RM, Palkovits M, Kizer JS, Jacobowitz DM, Kopin IJ.** Selective alterations of catecholamines and tyrosine hydroxylase activity in the hypothalamus following acute and chronic stress, in *Catecholamines and Stress.* Usdin E, Kvetnansky R, Kopin IJ. Eds., Pergamon Press, Elmsford, NY, 1976, 29.
5. **Anisman H, Pizzino A, Sklar LS.** Coping with stress, norepinephrine depletion and scape performance. *Brain Res* 191: 583-588, 1980.
6. **Tanaka M, Kohno Y, Nakagawa R, Ida Y, Takeda S, Nagasaki N.** Time-related differences in noradrenaline turnover in rat brain regions by stress. *Pharmacol Biochem Behav* 16: 315-319, 1982.
7. **Roth KA, Mefford IM, Barchas JD.** Epinephrine, norepinephrine, dopamine and serotonin: differential effects of acute and chronic stress on regional brain amines. *Brain Res* 239: 417-424, 1982.
8. **Ramade F, Bayle JD.** Thalamic-hypothalamic interrelationships and stress-induced rebounding adrenocortical response in the pigeon. *Neuroendocrinology* 34: 7-13, 1982.
9. **Glavin GB.** Regional rat brain noradrenaline turnover in response to restraint stress. *Pharmacol Biochem Behav* 19: 287-290, 1983.
10. **Ida Y, Tanaka M, Tsuda A, Kohno Y, Hoaki Y, Nakagawa R, Iimori K, Nagasaki N.** Recovery of stress-induced increases in noradrenaline turnover is delayed in specific brain region of old rats. *Life Sci* 34: 2357-2363, 1984.
11. **Willoughby JO, Terry LC, Brazeau P, Martin JB.** Pulsatile growth hormone, prolactin and thyrotropin secretion in rats with hypothalamic deafferentation. *Brain Res* 127: 137-152, 1977.
12. **Eden S, Boile P, Modigh K.** Monoaminergic control of episodic growth hormone secretion in the rat: effects of reserpine, alpha-methyl-*p*-tyrosine, *p*-chlorophenylalanine and haloperidol. *Endocrinology* 105: 523-529, 1979.
13. **Eriksson E, Eden S, Modigh K.** Up- and down-regulation of central postysynaptic alpha$_2$ receptors reflected in the growth hormone response to clonidine in reserpine-pretreated rats. *Psychopharmacology* 77: 327-331, 1982.
14. **Engberg G, Elam M, Svensson TH.** Clonidine withdrawal: activation of brain noradrenergic neurons with specifically reduced alpha$_2$ receptor sensitivity. *Life Sci* 30: 235-243, 1982.
15. **Abe K, Critchlow V.** Effect of corticosterone, dexamethasone and surgical isolation of the medial basal hypothalamus on rapid feedback control of stress-induced corticotropin secretion in female rats. *Endocrinology* 101: 498-505, 1977.
16. **Stith RD, Person RJ.** Effect of central catecholamine depletion on 3H-dexamethasone binding in the dog. *Neuroendocrinology* 34: 410-414, 1982.
17. **Swenson RM, Vogel WH.** Plasma catecholamine and corticosterone as well as brain catecholamine changes during coping in rats exposed to stressful footshock. *Pharmacol Biochem Behav* 18: 689-693, 1983.
18. **American Psychiatric Association.** Affective Disorders, in *Diagnostic and Statistical Manual of Mental Disorders,* 3rd ed. American Psychiatric Association, Washington, D.C., 1980, 205.
19. **Sole MJ, Hussein MN.** A simple, specific radioenzymatic assay for the measurement of picogram quantities of norepinephrine, epinephrine and dopamine in plasma and tissues. *Biochem Med* 18: 301-307, 1977.
20. **Carey RM, Van Loon GR, Baines AD, Kaiser DL.** Suppression of basal and stimulated noradrenergic activities by the dopamine agonist bromocriptine in man. *J Clin Endocrinol Metab* 56: 595-602, 1983.
21. **Murphy BEP.** Some studies of the protein-binding of steroids and their application to the routine micro and ultramicro measurement of various steroids in body fluids by competitive protein-binding in radioassay. *J Clin Endocrinol Metab* 27: 973-990, 1967.
22. **Wide L.** Radioimmunoassays employing absorbents. *Acta Endocrinol* 63 (Suppl. 142), 207-210, 1982.
23. **Sinha YN, Selby FW, Lewis UJ, Vanderlaan WP.** A homologous radioimmunoassay for human prolactin. *J Clin Endocrinol Metab* 36: 509-512, 1973.

24. **Checkley SA, Slade AP, Shur E.** Growth hormone and other responses to clonidine in patients with endogenous depression. *Br J Psychiatry* 138: 51-55, 1981.

25. **Siever LJ, Uhde TW.** New studies and perspectives on the noradrenergic receptor system in depression: effects of the alpha-2-adrenergic agonist clonidine. *Biol Psychiatry* 19: 131-156, 1984.

26. **Brown MR, Fisher LA, Webb V, Vale WW, Rivier JE.** Corticotropin releasing factor: a physiologic regulator of adrenal epinephrine secretion. *Brain Res* 328: 355-357, 1985.

27. **Lake CR, Ziegler MG, Kopin IJ.** Use of plasma norepinephrine for evaluation of sympathetic neuronal function in man. *Life Sci* 18: 1315-1326, 1976.

28. **Starke K, Montel H.** Involvement of alpha-receptors in clonidine-induced inhibition of transmitter release from central monoamine neurons. *Neuropharmacology* 12: 1073-1080, 1973.

29. **Langer SZ.** Presynaptic receptors in the regulation of transmitter release in the peripheral and central nervous system: physiological and pharmacological significance, in *Catecholamines: Basic and Clinical Frontiers.* Usdin E, Kopin IJ, Barchas J. Eds., Pergamon Press, Elmsford, NY, 1979, 387.

30. **Koss MC, Christensen HD.** Evidence for a central postsynaptic action of clonidine. *Arch Pharmacol* 307: 45-50, 1979.

31. **Ziegler MG, Lake CR, Wood JH.** Relationship between norepinephrine in blood and cerebrospinal fluid in the presence of a blood-cerebrospinal fluid barrier for norepinephrine. *J Neurochem* 28: 677-679, 1977.

32. **Przuntek M, Philippu A.** Reduced pressor responses to stimulation of locus coeruleus after lesion of the posterior hypothalamus. *Arch Pharmacol* 276: 119-122, 1973.

33. **Daiguji M, Mikuni M, Okada F, Yamashita I.** The diurnal variations of dopamine-beta-hydroxylase activity in the hypothalamus and locus coeruleus of the rat. *Brain Res* 155: 409-412, 1978.

34. **Crawley JN, Roth RH, Maas JW.** Locus coeruleus stimulation increases noradrenergic metabolite levels in rat spinal cord. *Brain Res* 166: 180-184, 1979.

35. **Gurtu S, Pant KK, Sinha JN, Bhargava KP.** An investigation into the mechanism of cardiovascular responses elicited by electrical stimulation of locus coeruleus and subcoeruleus in the cat. *Brain Res* 301: 59-64, 1984.

36. **Blackard W, Heidingsfelder S.** Adrenergic receptor control mechanism for growth hormone secretion. *J Clin Invest* 47: 1407-1414, 1968.

37. **Schaub C, Delbarre B, Blue-Pajot MT, Casset-Senon D, Lornet-Videau C, Ferger A.** Effects of beta-adrenergic agonists and antagonists on growth hormone and prolactin secretion in the monkey. *Neuroendocrinol Lett* 2: 45-49, 1980.

38. **Torres I, Guaza C, Fernández-Durango R, Borell J, Charro AI.** Evidence for a modulator role of catecholamine on hypothalamic somatostatin in the rat. *Neuroendocrinology* 35: 159-162, 1982.

39. **Krulich L, Mayfield MA, Steele MK, McMillen BA, McCann SM, Koenig JI.** Differential effects of pharmacological manipulations of central alpha-1 and alpha-2 adrenergic receptors on the secretion of thyrotropin and growth hormone in male rats. *Endocrinology* 110: 796-804, 1982.

40. **Day TA, Willoughby JO.** Noradrenergic afferent to median eminence: inhibitory role in the rythmic growth hormone secretion. *Brain Res* 202: 335-345, 1980.

41. **Kobayashi RM, Palkovits M, Kizer JS, Jacobowitz DM, Kopin IJ.** Biochemical mapping of the noradrenergic projection from the locus coeruleus. A model for studies of brain neuronal pathways. *Neurology* 25: 223-233, 1975.

42. **Palkovits M, Fekete M, Makara GB, Herman JP.** Total and partial hypothalamic deafferentation for topographical identification of catecholaminergic innervation of certain preoptic and hypothalamic nuclei. *Brain Res* 127: 127-136, 1977.

43. **Jones BE, Moore RY.** Ascending projection of the locus coeruleus in the rat. II. Autoradiographic study. *Brain Res* 127: 23-53, 1977.

44. **O'Donohue TL, Crowley WR, Jacobowitz DM.** Biochemical mapping of the noradrenergic ventral bundle projection sites: evidence for a noradrenergic dopaminergic interaction. *Brain Res* 172: 87-100, 1979.

45. **Moore RY, Bloom FE.** Central catecholamine neurone systems: anatomy and physiology of the norepinephrine and epinephrine systems. *Annu Rev Neurosci* 2: 113-168, 1979.

46. **Kostowski W, Jerlicz M, Bidzinski A, Hauptmann M.** Evidence for the existence of two opposite noradrenergic brain systems controlling behavior. *Psychopharmacology* 59: 311-312, 1978.

47. **Kostowski W.** Two noradrenergic systems in the brain and their interactions with other monoaminergic neurons. *Pol J Pharmacol Pharm* 31: 425-436, 1979.

48. **Kostowski W.** Noradrenergic interactions among central neurotransmitters, in *Neurotransmitters, Receptors and Drug Action.* Essman W. Ed., Spectrum, New York, 1980, 47.

49. **Morgane PJ.** Historical and modern concepts of hypothalamic organization and function, in *Handbook of the Hypothalamus.* Morgane PJ, Pannksepp J. Eds., Marcel Dekker, New York, 1979, 1.

50. **Sawchenko PE, Swanson LW.** Central noradrenergic pathways for the integration of hypothalamic neuroendocrine and autonomic responses. *Science* 214: 685-687, 1981.

51. **McKellar S, Lowey AD.** Efferent projections of the A1 catecholamine cell group in the rat: an autoradiographic study. *Brain Res* 241: 11-29, 1982.

52. **Masala A, Salta G, Alagua S, Anania V, Frasetto GA, Rovasio PP, Semiani A.** Effect of clonidine on stress-induced cortisol release in man during surgery. *Pharmacol Res Commun* 17: 293-298, 1985.

53. **Nakamura K, Nakamura K.** Role of brainstem and spinal noradrenergic and adrenergic neurons in the development and maintenance of hypertension in spontaneously hypertensive rats. *Exp Pathol Pharmakol* 305: 127-133, 1978.

54. **Benarroch EE, Balda MS, Finkelman S, Nahmod VE.** Neurogenic hypertension after depletion of norepinephrine in anterior hypothalamus induced by 6-hydroxydopamine administration into the ventral pons: role of serotonin. *Neuropharmacology* 22: 29-34, 1983.

55. **McWilliam JR, Meldrum BS.** Noradrenergic regulation of growth hormone secretion in the baboon. *Endocrinology* 112: 254-259, 1983.

56. **Karteszi M, Fiok J, Makara GB.** Lack of episodic growth hormone secretion in rats with anterolateral deafferentation of the mediobasal hypothalamus. *J Endocrinol* 94: 77-81, 1982.

57. **Lechin F, van der Dijs B, Jakubowicz D, Camero RE, Lechin S, Villa S, Reinfeld B, Lechin ME.** Role of stress in the exacerbation of chronic illness: effects of clonidine administration on blood pressure and plasma norepinephrine, cortisol, growth hormone, and prolactin concentrations. *Psychoneuroendocrinology* 12: 117-129, 1987.

INDEX